Application of Integrable Systems to Phase Transitions

C.B. Wang

Application of Integrable Systems to Phase Transitions

C.B. Wang
Institute of Analysis
Troy, USA

ISBN 978-3-642-44024-3   ISBN 978-3-642-38565-0 (eBook)
DOI 10.1007/978-3-642-38565-0
Springer Heidelberg New York Dordrecht London

Mathematics Subject Classification: 82B26, 82B27, 81V22, 81V05, 81V15, 34M03, 33C45, 34M55, 35Q15, 34E05, 34F10, 91B80

Printed on acid-free paper

Springer is part of Springer Science+Business Media (www.springer.com)

This book is dedicated to my parents, my wife Ling Shen and my lovely children Shela and Henry.

Preface

This book is aimed at providing a unified eigenvalue density formulation of matrix models and discussing the corresponding phase transitions and critical phenomena. The purpose of this book is to systematically classify the transition models based on the potential functions that define the consequent integrable systems, string equations, eigenvalue densities, free energies and critical points considered in the transitions.

The unification of different couplings or interactions is one of the important research topics in quantum chromodynamics (QCD) theory that can help to study the dominance of confinement or asymptotic freedom in low or high energy scale. Hilbert space theory and Fourier analysis are the necessary mathematical tools to work on the physical problems. As a foundation of the Hilbert space theory, orthogonal polynomials can provide a new mathematical background for further investigating the fundamental concepts such as position and momentum described in the Heisenberg uncertainty principle that is tightly related to matrix models which are important in QCD.

In quantum physics the probability amplitude for a particle to travel from one point to another in a given time is characterized by propagator. By the Fourier transform, the propagator becomes a singular function in the momentum space, where the singularity is related to the uncertainty. Correlation function in the matrix models can avoid such singularity in large-N asymptotics that leads the eigenvalue density to represent the momentum so that free energy function can be properly formulated. String equation is a tool in the momentum aspect to construct the eigenvalue densities and give parameter relations in the models for finding the phase transitions. The associated integrable systems imply a unified formula for the different densities in the matrix models by using the Lax pair structures obtained from the corresponding orthogonal polynomials.

Analytic results derived from the integrable systems will reduce the mathematical complexity in discussing the phase transition problems of the matrix models. The different density phases can be generated from the scalings of the string equation and associated discrete differential equations in the Lax pair with proper periodic reductions implemented by using an index folding technique in large-N scaling.

The different scalings can have a common case which is the critical point separating two phases with different conditions. The string equation properly establishes the nonlinear relations between the parameters in the model such that behaviors of the free energy around the critical point can be easily obtained, either analytically or by the expansions according to the parameter relations. The first-, second- and third–order transition models can be created by using the string equation, Toda lattice and corresponding integrable systems. Expansions for the coupling parameters in association with the double scalings present a new strategy to find the divergence transitions or critical phenomena with a fractional power-law.

Phase transition models discussed by using the string equations differ from the transitions in traditional models such as the Ising models, but there is a similarity between the periodic reduction that reorganizes wave functions in the momentum aspect using the index folding and the idea of renormalization in statistical mechanics that reorganizes the particles in the position aspect. Typically, the critical point in the bifurcation transition is when the center or radius parameter is bifurcated, while the critical point in the renormalization method is a fixed point in an iteration process. The power-law at the critical point in the momentum aspect is derived from the algebraic equation reduced from the string equation that is the parameter condition(s) different from the renormalization groups considered in the Ising models. In addition, eigenvalue density on multiple disjoint intervals and corresponding free energy can be referred to study Seiberg-Witten differential and prepotential in the Seiberg-Witten theory which is developed to solve the mass gap problem in quantum Yang-Mills theory.

The organization of this book is as follows. Chapter 1 is about the physical background of the matrix models. The unified model proposed in this book is fundamental for the phase transitions. Chapter 2 is for the reduction of eigenvalue densities from the integrable systems. In Chaps. 3 and 4, various transitions will be discussed for the Hermitian matrix models. In Chaps. 5 and 6, we will talk about the transitions and critical phenomena in the unitary matrix models, including the Gross-Witten third-order phase transition. Chapter 7 deals with the Marcenko-Pastur distribution, McKay's law and their generalizations in association with the Laguerre and Jacobi polynomials.

This book is about how to find the phase transitions, which can be used as a reference book for researchers and students in the fields of phase transitions and critical phenomena in quantum physics. I thank Professor J. Bryce McLeod for his directions and help on both the related works in this book and previous works. I thank Craig A. Tracy for introducing the random matrix theory, and thank Gunduz Caginalp, Xinfu Chen, Palle Jorgensen, Juan Manfredi and William Troy for the useful discussions or suggestions. And I thank Xing-biao Hu, Zaijiu Shang and Lianwen Zhang for the helpful discussions and encouragement.

Troy, USA   Chie Bing Wang
April 2012

Contents

Chapter 1
Introduction

Phase transition problems in the one-matrix models in QCD are discussed in this book by using string equations. The phase transition models are formulated by using eigenvalue density which represents the momentum operator described in quantum mechanics. By using orthogonal polynomials, the eigenvalue densities in various matrix models can be unified. The string equation establishes a connection between the position and momentum aspects described in the Heisenberg uncertainty principle, which is important and usually hard to find in other methods. The recursion formula of the orthogonal polynomials can be applied to introduce an index folding technique to reorganize the wave functions in order to achieve a renormalization in the momentum aspect. It will be discussed that the critical phenomenon associated with the Gross-Witten model can be found by using Toda lattice and double scaling. The hypergeometric-type differential equations improve on some shortages of integrable systems to work on physical problems, such as the fact that a soliton system does not have a differential equation along the spectrum direction, and illustrate a new background to study the singularities of physical quantities, such as mass.

1.1 Unified Model for the Eigenvalue Densities

The well known Wigner semicircle [59, 60] was found in 1955 to represent the energy spectrum distributed on the real line, and now it has been applied as a fundamental eigenvalue density model in many scientific areas. Distribution shapes of the eigenvalues reflect a scientific character of the natural phenomena that have motivated scientists to find further density models to reveal new sciences. And many new discoveries have been made, including the Marcenko-Pastur distribution [36] and McKay's law [37] found in the research of random matrices and random graphs in 1967 and 1981, respectively. These models have been widely applied in quantum physics and complexity research in recent years in order to find the natural rules behind the random phenomena. In this book, we are going to discuss that orthogonal polynomials can be applied to derive these important models and introduce a unified model for these different density models that further confirms the significance

C.B. Wang, *Application of Integrable Systems to Phase Transitions*,
DOI 10.1007/978-3-642-38565-0_1, © Springer-Verlag Berlin Heidelberg 2013

of eigenvalue densities in the quantum models and complexity sciences. The method can be generalized to get various density models, including the phase models in the phase transition problems. Let us first use some samples to see how these models are related to orthogonal polynomials.

Consider the Hermite polynomials $H_n(z)$ satisfying $\int_{-\infty}^{\infty} H_m(z)H_n(z)e^{-V(z)}dz = 2^n n! \sqrt{\pi} \delta_{m,n}$, where $V(z) = z^2$. In some literatures, $V(x)$ is called external potential. In this book, we will simply call it potential or potential function. Let $p_n(z) = H_n(z)/2^n = z^n + \cdots$, and $\Phi_n(z) = e^{-V(z)/2}(p_n(z), p_{n-1}(z))^T$. By the derivative formula of the Hermite polynomials from the textbook, it can be verified that $\Phi_n(z)$ satisfies an equation $\frac{d}{dz}\Phi_n = A_n(z)\Phi_n$, where

$$A_n(z) = \begin{pmatrix} -z & n \\ -2 & z \end{pmatrix}, \tag{1.1}$$

and $\operatorname{tr} A_n(z) = 0$. Let $z = \sqrt{n}\eta$. Then

$$\frac{1}{n\pi}\sqrt{\det A_n(z)}dz = \frac{1}{\pi}\sqrt{2 - \eta^2}d\eta, \tag{1.2}$$

with $|\eta| \leq \sqrt{2}$, which gives the Wigner semicircle density [59, 60] for the rescaled potential $W(\eta) = \eta^2$. It should be noted that the semicircle is introduced here just for the basic density concept. The Wigner semicircle model that has been widely considered in the random researches such as random matrix theory, does not have a phase transition. The phase transitions or critical phenomena discussed in this book are all for the generalized models from the basic models presented in this section.

If the potential is changed to $V(z) = tz^2$, then

$$A_n(z) = \begin{pmatrix} -tz & 2tv_n \\ -2t & tz \end{pmatrix}, \tag{1.3}$$

where v_n satisfies

$$2tv_n = n. \tag{1.4}$$

This trivial algebraic equation is a simple string equation. Complicated string equations will appear when the potential $V(x)$ is changed to a higher degree polynomial. In the following samples, there will be similar string equations if the potential is changed by adding some parameter(s) like the t above. Now, let us just show how orthogonal polynomials can be applied to provide a unified structure to construct different fundamental density models.

The next sample is the Marcenko-Pastur distribution. Consider the Laguerre polynomials $L_n^{(\alpha)}(z)$ satisfying $\int_0^{\infty} L_m^{(\alpha)}(z)L_n^{(\alpha)}(z)z^\alpha e^{-z}dz = \Gamma(\alpha+1)\binom{n+\alpha}{n}\delta_{m,n}$, where $\alpha > -1$, and $\Gamma(\cdot)$ is the Gamma function. For the Laguerre polynomials, choose $\Phi_n(z) = z^{\alpha/2}e^{-z/2}(L_n^{(\alpha)}(z), L_{n-1}^{(\alpha)}(z))^T$. By using the differential equation for the Laguerre polynomials, we have that $\Phi_n(z)$ satisfies an equation $\frac{d}{dz}\Phi_n =$

$A_n(z)\Phi_n$, where

$$A_n(z) = \frac{1}{z}\begin{pmatrix} -\frac{z-\alpha}{2}+n & -n-\alpha \\ n & \frac{z-\alpha}{2}-n \end{pmatrix}, \tag{1.5}$$

and tr $A_n(z) = 0$. Let $z = n\eta$, $q = \frac{n}{n+\alpha}$, $\eta_+ = (1+\frac{1}{\sqrt{q}})^2$ and $\eta_- = (1-\frac{1}{\sqrt{q}})^2$. Then there is [39]

$$\frac{1}{(n+\alpha)\pi}\sqrt{\det A_n(z)}dz = \frac{q}{2\pi\eta}\sqrt{(\eta_+ - \eta)(\eta - \eta_-)}d\eta, \tag{1.6}$$

with $\eta_- \leq \eta \leq \eta_+$, which gives the Marcenko-Pastur distribution [36]. This density function is also discussed in [51], and it has been popularly applied in econophysics to study the distribution of the positive eigenvalues.

Another fundamental model, McKay's law [37], is associated with the Jacobi polynomials $P_n^{(\alpha,\beta)}(z)$ satisfying $\int_{-1}^{1} P_m^{(\alpha,\beta)}(z)P_n^{(\alpha,\beta)}(z)w(z)dz = h_n\delta_{mn}$, defined on the interval $[-1, 1]$ with the weight $w(z) = (1-z)^\alpha(1+z)^\beta$, where $\alpha > -1$ and $\beta > -1$. Let $\Phi_n(z) = w(z)^{1/2}(P_n^{(\alpha,\beta)}(z), P_{n-1}^{(\alpha,\beta)}(z))^T$. If we choose $\beta = \alpha$, then $\Phi_n(z)$ satisfies an equation $\frac{d}{dz}\Phi_n = A_n(z)\Phi_n$, where

$$A_n(z) = \frac{1}{1-z^2}\begin{pmatrix} -(n+\alpha)z & n+\alpha \\ -\frac{n(n+2\alpha)}{n+\alpha} & (n+\alpha)z \end{pmatrix}, \tag{1.7}$$

obtained by using the differential equation of the Jacobi polynomials. Let $z = \eta/c$ and $\alpha/n = c/2 - 1$, then we get

$$\frac{1}{n\pi}\sqrt{\det A_n(z)}dz = \frac{c}{2\pi(c^2-\eta^2)}\sqrt{4(c-1)-\eta^2}d\eta, \tag{1.8}$$

with $|\eta| \leq 2\sqrt{c-1}$ and $c > 1$, where the density function on the right hand side above is called McKay's law, see [37].

Now, let us consider the simple orthogonal polynomials $p_n(z) = z^n$ on the unit circle $z = e^{i\theta}$ in the complex plane, satisfying $\oint p_m(z)\overline{p_n(z)}\frac{dz}{2\pi iz} = \delta_{m,n}$, where the integral is over the unit circle $|z| = 1$, and $\overline{p_n(z)} = z^{-n}$ is the complex conjugate of $p_n(z)$. It can be shown that $\Phi_n(z) = (z^{-n/2}p_n(z), z^{n/2}\overline{p_n(z)})^T$ satisfies an equation $\frac{d}{dz}\Phi_n = A_n(z)\Phi_n$, where

$$A_n(z) = \begin{pmatrix} \frac{n}{2z} & 0 \\ 0 & -\frac{n}{2z} \end{pmatrix}. \tag{1.9}$$

The weight is equal to 1, the potential is $U(z) = 0$, and $p_n(0) = 0$. It is easy to see that

$$\frac{1}{n\pi}\sqrt{\det A_n(z)}dz = \frac{1}{2\pi}d\theta, \quad -\pi \leq \theta \leq \pi. \tag{1.10}$$

The right hand side is a uniform distribution, which is a degenerate case of the Gross-Witten strong coupling density model in the unitary matrix model [22] when the temperature is equal to infinity. The Gross-Witten weak and strong coupling density models can be obtained by changing the potential to $U(z) = s(z + z^{-1})$, and the weight for the orthogonality is then changed to $\exp(s(z + z^{-1}))$ implying the following generalized coefficient matrix [39]

$$A_n(z) = \begin{pmatrix} \frac{s}{2} + \frac{s}{2z^2} + \frac{n - 2sx_n x_{n+1}}{2z} & s(x_{n+1} - \frac{x_n}{z})z^{-1} \\ s(x_n - \frac{x_{n+1}}{z}) & -\frac{s}{2} - \frac{s}{2z^2} - \frac{n - 2sx_n x_{n+1}}{2z} \end{pmatrix}, \tag{1.11}$$

where $x_n = p_n(0) \in [0, 1]$ satisfying another string equation,

$$s(1 - x_n^2)\frac{x_{n+1} + x_{n-1}}{-x_n} = n. \tag{1.12}$$

The phase transitions considered in QCD are about the change from one density model to another one. So the density models in the matrix model theory in QCD are no longer as simple as we have seen above, but multiple reductions from the coefficient matrix A_n. It will be discussed in detail in Chaps. 5 and 6 that if $x_{n+1} \sim -x_n \sim x_{n-1}$, then $T = n/s = 2(1 - x_n^2)(\leq 2)$, and

$$\frac{1}{n\pi}\sqrt{\det A_n(z)}dz \sim \frac{2}{\pi T}\cos\frac{\theta}{2}\sqrt{\frac{T}{2} - \sin^2\frac{\theta}{2}}d\theta, \tag{1.13}$$

where $z = e^{i\theta}$. If $T > 2$, then $x_n \to 0$, and

$$\frac{1}{n\pi}\sqrt{\det A_n(z)}dz \sim \frac{1}{2\pi}\left(1 + \frac{2}{T}\cos\theta\right)d\theta. \tag{1.14}$$

These are the weak and strong coupling eigenvalue densities in the Gross-Witten third-order phase transition model [22], which is the fundamental reference for the transition models discussed in this book.

We have seen that the eigenvalue densities, which will be denoted by ρ in later discussions, can always be obtained from the unified model

$$\frac{1}{n\pi}\sqrt{\det A_n(z)}, \tag{1.15}$$

where the matrix A_n is derived from the orthogonal polynomials. The idea of the above unified density model was first introduced in [39]. We use the terminology "unified model" here as a guidance for the readers to see how the densities will be constructed in this book. By this unified model of the eigenvalue densities, we can study relations between the different states or couplings of the physical model. Phase transition is one of the theories to find the critical point with discontinuous property of the free energy in order to separate different states. This book is planned to study a type of phase transition models based on the unified model and string equations. In Chap. 2, we will use generalized Hermite polynomials to get many density models to extend the Wigner semicircle that will be applied to study transition problems in Chaps. 3 and 4 for the Hermitian matrix models.

The orthogonality is a typical case of consistency considered in integrable systems. Orthogonal polynomials provide an easy method to find integrable system for string equations or Toda lattice, which is then a generalization of orthogonal polynomial system. In an integrable system, the parameters are allowed to vary in a wider space than in orthogonal polynomial system. For example, for the potential $V(x) = tx^2$ discussed above, the parameter t needs to be positive in the orthogonal polynomials. While in integrable system, t can be negative. Such extensions also apply to other potentials. In phase transition models, the parameters can be anywhere in the space with complicated nonlinear relations that are generally very hard to figure out in other methods. The analysis based on string equations naturally follows the orthogonality or consistency structure to get nonlinear relations and avoids solving complicated equations that traditional methods often depend on.

When the eigenvalue density problems are solved, elliptic integrals in the free energy function are the next consideration for the phase transition problems, which are generally not easy. We will use contour integrals, asymptotics, initial value problems, recursive relations, hypergeometric differential equations and Legendre's relation to discuss the elliptic integrals. However, these methods are not enough to always solve behaviors of the free energy function at the critical point. The ε-expansion method for the parameter equations obtained from the string equations and associated large-N double scaling method are necessary techniques to derive the discontinuity or divergence in transition, and it works for all the transition problems to be discussed in this book.

1.2 String Equation and Matrix Models

Now let us generalize the weight function e^{-z^2} of the Hermite polynomials to a new weight function $e^{-V(z)}$ where $V(z) = tz^2 + z^4$, and consider orthogonal polynomials $p_n(z) = z^n + \cdots$ satisfying $\int_{-\infty}^{\infty} p_m(z)p_n(z)e^{-(tz^2+z^4)}dz = h_n\delta_{mn}$. The quantity Z_n,

$$Z_n = n!h_0h_1\cdots h_{n-1}, \tag{1.16}$$

called partition function, is a function of t and n. This formula holds when $V(z)$ is changed to a general potential, and it will be used to define free energy in the transition problem. The orthogonal polynomials can be applied to get the derivative(s) of the logarithm of the partition function, $\ln Z_n$. It will be discussed in Sect. 3.2 that the derivatives of h_n and $v_n = h_n/h_{n-1}$ have simple formulations, and the partition function will be proved to satisfy the following relation,

$$\frac{d^2}{dt^2}\ln Z_n = v_n(v_{n-1} + v_{n+1}), \tag{1.17}$$

for $n \geq 2$. The first-order derivative has a complicated formula than the second order derivative (1.17), that will be explained in Sect. 3.2. These formulas are based on derivative formula of the v_n, called Toda lattice.

The parameters or functions in the orthogonal polynomials need to be rescaled to study the physical problems in the matrix models. Specially, $\ln Z_n$ needs to be rescaled to get the free energy function, and then all the corresponding relations need to be rescaled consequently. The questions include which equations need to be involved and how to implement the rescaling. For the phase transition problems in our consideration, we need a string equation in the Hermitian matrix model,

$$2tv_n + 4(v_{n+1} + v_n + v_{n-1})v_n = n, \qquad (1.18)$$

(called discrete Painlevé I equation in [16]) obtained by Fokas, Its and Kitaev [16] in 1991 by using the orthogonality and recursion formula $zp_n = p_{n+1} + v_n p_{n-1}$ [56], which is a three-term recursion formula by counting how many different indexes in the formula. This discrete equation will be related to the parameter condition in the planar diagram model, which is important for studying the free energy function. The string equation can be reduced to continuum integrable equation in large-N asymptotics to study the 2D quantum gravity problems, for example, see [6, 11, 12, 16, 20]. The phase transition problems discussed in this book will only involve discrete or continuum equations, not the solutions, to solve the nonlinear relations of parameters, that are part of the mathematical difficulties in the matrix models. It is believed that the survey of QCD usually does not start from perturbation analysis or analytical methods, but more likely from lattice gauge theories or matrix models as explained in the following.

In 1978, Brezin, Itzykson, Parisi and Zuber [7] introduced the planar diagram theory to approximate the field theory in order to ultimately provide a mean of performing reliable computation in the large coupling phase of non-Abelian gauge fields in four dimensions. The theory is based on the work of 't Hooft that the only diagrams which survive the large-N limit of an SU(N) gauge theory are planar. The planar ideas work for Yang-Mills field and other more general fields. The method is to introduce a $N \times N$ matrix $M = (M_{ij})$ such that the Lagrangian is represented in terms of M. Hermitian matrix model is the case when M is a Hermitian matrix as studied in the planar diagram theory. Free energy E is led to the following relation,

$$e^{-n^2 E(g)} = \lim_{n \to \infty} \int \exp\left\{ -\frac{1}{2} \operatorname{tr} M^2 - \frac{g}{n} \operatorname{tr} M^4 \right\} dM, \qquad (1.19)$$

where g is the coupling parameter, and

$$dM = \prod_i dM_{ii} \prod_{i<j} d(\operatorname{Re} M_{ij}) d(\operatorname{Im} M_{ij}). \qquad (1.20)$$

Because of the complexity of the formulations as seen above, it is developed to study the partition function

$$Z_n = \int_{-\infty}^{\infty} \cdots \int_{-\infty}^{\infty} e^{-\Sigma_{i=1}^n V(z_i)} \Delta_n^2 dz_1 \cdots dz_n, \qquad (1.21)$$

where $V(z) = t_2 z^2 + t_4 z^4$ and $\Delta_n = \prod_{j<k}(z_j - z_k)$, which is in the scope of non-perturbative theory. The polynomial $p_n(z)$ has a representation [56]

$$p_n(z) = Z_n^{-1} \int_{-\infty}^{\infty} \cdots \int_{-\infty}^{\infty} e^{-\Sigma_{i=1}^{n} V(z_i)} \Delta_n^2 \prod_{j=1}^{n} (z - z_j) dz_1 \cdots dz_n, \qquad (1.22)$$

and the orthogonality tells that $h_n = \int_{-\infty}^{\infty} p_n^2(z) e^{-V(z)} dz = \int_{-\infty}^{\infty} p_n(z) z^n e^{-V(z)} dz$. For Z_{n+1}, we have $n+1$ opportunities to select one of $z_1, \ldots, z_{n+1}$ to do the following, and each one is equivalent to the operation for $z_{n+1} = z$,

$$\frac{Z_{n+1}}{n+1} = \int_{-\infty}^{\infty} \left(\int_{-\infty}^{\infty} \cdots \int_{-\infty}^{\infty} e^{-\Sigma_{i=1}^{n} V(z_i)} \Delta_n^2 \prod_{i=1}^{n} (z - z_i) dz_1 \cdots dz_n (z^n + \cdots) \right) e^{-V(z)} dz$$

$$= \int_{-\infty}^{\infty} Z_n p_n(z)(z^n + \cdots) e^{-V(z)} dz = Z_n h_n,$$

where "$\cdots$" in "$z^n + \cdots$" means lower order terms, and $Z_1 = h_0$. This has shown the equation (1.16) for the partition function, which is a basic property in this research field. The steepest descent method is used in [7] to study the free energy $E = -\lim_{n \to \infty} \frac{1}{n^2} \ln Z_n$ in the following way,

$$E = \lim_{n \to \infty} \frac{1}{n^2} \left\{ \sum_j \left(\frac{1}{2} z_j^2 + \frac{g}{n} z_j^4 \right) - \sum_{j \neq k} \ln |z_j - z_k| \right\}, \qquad (1.23)$$

where $t_2/n = 1/2$ and $t_4 = g$, with the stationary condition $\frac{1}{2} z_j + 2 \frac{g}{n} z_j^3 = \sum_{k \neq j} \frac{1}{z_j - z_k}$, which can be further scaled to a variational equation

$$\frac{1}{2} \eta + 2g \eta^3 = (P) \int_{-2b}^{2b} \frac{\rho(\lambda)}{\eta - \lambda} d\lambda, \qquad (1.24)$$

where $\rho(\eta) = \frac{dx}{d\eta}$ is the density of eigenvalues satisfying $\int_{-2b}^{2b} \rho(\eta) d\eta = 1$, x is introduced in $\eta(x)$ during the scaling $z_j = \sqrt{n} \eta(j/n)$, $[-2b, 2b]$ is the interval where eigenvalues are distributed in, and (P) stands for the principle value of the integral. The free energy then becomes

$$E = \int_{-2b}^{2b} \left(\frac{1}{2} \eta^2 + g \eta^4 \right) \rho(\eta) d\eta - \int_{-2b}^{2b} \int_{-2b}^{2b} \ln |\lambda - \eta| \rho(\eta) \rho(\lambda) d\lambda d\eta. \qquad (1.25)$$

The density $\rho(\eta)$ above is called planar diagram eigenvalue density model with the following explicit formula [7]

$$\rho(\eta) = \frac{1}{\pi} \left(\frac{1}{2} + 4g b^2 + 2g \eta^2 \right) \sqrt{4b^2 - \eta^2}, \qquad (1.26)$$

for $\eta \in [-2b, 2b]$, where

$$b^2 + 12gb^4 = 1. \tag{1.27}$$

The free energy function then has the following explicit formula

$$E(g) = E(0) + \frac{1}{24}(b^2 - 1)(9 - b^2) - \frac{1}{2}\ln b^2. \tag{1.28}$$

It can be calculated that $E(0) = 3/4$. And E has a singular point at $g = g^c$ where $g^c = -1/48$. In next chapter, we will discuss that the eigenvalue density ρ above can be obtained from the unified model $\frac{1}{n\pi}\sqrt{\det A(z)}$ for a generalized matrix $A_n(z)$ from (1.3).

Since the free energy is obtained in large-N scaling from $\ln Z_n$, there should be some relations between the formulas before the scaling and after. If we compare the string equation (1.18) with (1.27), it can be found that these two algebraic equations have the same nonlinear structure except the coefficients t and g, that can be changed to be consistent if we consider the general potential $V(z) = t_2 z^2 + t_4 z^4$, and the parameter n can be removed by scaling. This idea inspires us to find a connection between the conditions of the parameters in the eigenvalue density and the string equations. It is found [39] that the string equation properly matches with the restriction conditions for the parameters to solve the stationary condition or variational equation in the Hermitian matrix models, that will be discussed in detail in next chapter.

The method here using string equations to get parameter conditions for the density of eigenvalues differs from the double scaling method in the matrix models developed in early 90's in last century. The early double scaling mainly deals the limit at a singular point about how a discrete equation can be scaled to a continuum integrable equation. The previous scaling is in some sense to transform to another aspect by the double scaling, whereas the string equation method discussed in this book keeps the problem in the same aspect. Interested readers can refer or compare the discussions, for example, in [6, 12, 20] for the Hermitian matrix models about the double scaling limit by using the orthogonal polynomials on the real line, as well as in [46–48] for the unitary matrix models which will be discussed in the next section. The planar diagram model has been generalized to study the regularization [31] and phase transition [52] problems. Bleher and Eynard [4] studied a third-order transition model in the Hermitian matrix model in association with the orthogonal polynomials. The double scaling limit of eigenvalue correlations at the critical points studied in [4] is related to Painlevé II equation, universal kernel and a nonlinear hierarchy of ordinary differential equations. The Lax pair method [39] for the eigenvalue densities in the Hermitian matrix models is to find the mathematical formulation of the density models and the associated parameter relations by using string equation. The string equation provides an much easier method to formulate the eigenvalue densities than other methods such as loop equations discussed in some literatures. The parameter relations so obtained are the necessary formulas to study the phase transition problems that can be seen in later discussions.

The planar diagram model discussed above is a typical Hermitian one-matrix model, and there are also multiple matrix models, such as two-matrix model [27, 41] by considering the integral over two matrices,

$$Z_n = \int \exp\left\{ -\text{tr}(M_1^2 + M_2^2) - \frac{g}{n} \text{tr}(M_1^4 + M_2^4) + 2c \, \text{tr}(M_1 M_2) \right\} dM_1 dM_2. \quad (1.29)$$

The two-matrix model is believed to describe all of (p, q) conformal minimal models in quantum physics, whereas the one-matrix model is for $(p, 2)$ minimal models. The partition function given above can be simplified to $Z_n = \text{const} \cdot n! \Pi_{j=0}^{n-1} h_j$ by using the orthogonal polynomials $p_j(x) = x^j + \cdots$ defined by the bi-orthogonality,

$$\int_{-\infty}^{\infty} \int_{-\infty}^{\infty} p_j(x) p_k(y) \exp\left\{ -(x^2 + y^2) - \frac{g}{n}(x^4 + y^4) + 2cxy \right\} dx dy = h_j \delta_{jk}. \quad (1.30)$$

The orthogonal polynomials now satisfy a four-term recursion formula $xp_j(x) = p_{j+1} + R_j p_{j-1}(x) + T_j p_{j-3}(x)$, in which there are four different indexes $j + 1$, j, $j - 1$ and $j - 3$. Discussions for the eigenvalue distribution problems for the two-matrix models can be found, for example, in [19, 27, 32, 43], including the distribution with a gap for constructing the coupling models. The four-term recursion formula for the bi-orthogonal polynomials is important in the two-matrix models for creating density model with gap(s) according to the literatures. The differential equations [41] and τ-function theory have been applied to study partition function in two-matrix model. Bertola [2] computed second- and third-order derivatives of the free energy in planar limit for arbitrary genus spectral curves using τ-function theory and Bergman kernel which is a bi-differential form, and the method in principle allows to compute the derivatives of any order. In 2009, Bertola and Marchal [3] proved that the partition function in the Itzykson-Zuber-Eynard-Mehta two-matrix model is an isomonodromic τ-function as a generalization of the Jimbo-Miwa-Ueno's τ-function [28]. Correlation function and large-N expansion are also researched for the two-matrix models in association with the Yang-Baxter equation [14, 15]. Bethe ansatz for the correlation functions discussed in [15] about the cyclic property is interesting. We will explain later that the cyclic or periodic behaviors are important for the transition problems.

Split densities with gap(s) in the one-matrix models discussed in this book are obtained by three-term recursion formula of orthogonal polynomials using the index folding technique. The three-term recursion formula can be applied to itself to derive sufficient terms for constructing the eigenvalue density on multiple disjoint intervals to be explained in Sect. 1.4 with concrete recursion formula. Different recursive relations can be derived depending how to apply the recursion to itself, and then different phases in the transition problems can be constructed, including the density models with gap(s). In this sense, the three-term recursion formula can achieve a similar role as expected on the four-term recursion formula. Three-term recursion has a wide application background. The three-term recursion formula is a discrete version of the Schrödinger equation, which is fundamental in the soliton integrable systems [45], to be discussed in Sect. 4.3.3 by using large-N double scaling method. In statistical physics, Laplace transform of the Heisenberg equations gives a three-term recursion formula [34, 50]. Density models obtained in this book are all based on the three-term recursion formula.

Eigenvalue densities with or without gap(s) in the one Hermitian matrix models are based on the corresponding analytic function with single or multiple disjoint cut(s) on the real line or in the complex plane with a general potential $W(\eta) = \sum_{j=1}^{2m} g_j \eta^j$. The density models here are the generalization of the planar diagram density (1.26) and always take the form as a product of a polynomial and the square root of another polynomial. These two special polynomials can be obtained from the coefficient matrix in the Lax pair for the string equation to be given in Chap. 2 in detail. The square root of the polynomial in the density formula will give positive and negative signs alternatively on the multiple intervals. The polynomial in the outside of the square root then also need to have alternative positive and negative signs along the intervals such that the density function is not negative. This is a property that is often neglected in some literatures. In the Seiberg-Witten theory, this property is described by using the Riemann surface. For the Hermitian matrix models, we can directly experience this property by considering some special models, such as the potential $W(\eta) = g_2\eta^2 + g_4\eta^4$ or $W(\eta) = g_2\eta^2 + g_4\eta^4 + g_6\eta^6$ for the density on two or three disjoint intervals, that will be discussed in detail in the later chapters.

The importance of string equations also includes the ε-expansion for studying the discontinuity of the derivative(s) of free energy. The free energy is generally expressed in terms of elliptic integrals since density is generally on multiple disjoint intervals, and it is hard to reduce the integrals to simpler formulas. Algebraic equations derived from the string equation can be applied to determine the ε-expansions of the functions or parameters by properly choosing the order of the ε in the potential parameter expansion so that all the coefficients in the expansions can be determined. The algebraic equations can be applied to find the criticality in the transition such that discontinuity or power-law divergence including the critical phenomena can be obtained if the parameters, g_j, a_j and b_j, say, are properly expanded in terms of ε. The transition can be in the g_2 or g_4 direction for the potential $W(\eta) = g_2\eta^2 + g_4\eta^4$, for instance, or in the temperature direction T if we choose the potential as $W(\eta) = T^{-1}(\eta^2 + c_0\eta^4)$ with certain constant c_0. Details of the transitions for the Hermitian matrix models will be discussed in the Chaps. 3 and 4 with some technical backgrounds given in Appendices B, C and D.

1.3 Critical Point in Gross-Witten Model

If we generalize the orthogonal polynomials $p_n(z) = z^n$ on the unit circle in the complex plane discussed in Sect. 1.1 by changing the weight 1 to a weight function $w(z) = \exp(s(z + z^{-1}))$, the orthogonal polynomials $p_n(z) = z^n + \cdots$ now satisfy [56] $\oint p_m(z)\bar{p}_n(z)w(z)\frac{dz}{2\pi i z} = h_n\delta_{mn}$. Partition function in the corresponding unitary matrix model [22, 46–48], $Z_n = \int \exp\{s\,\mathrm{Tr}(U + U^\dagger)\}dU$, can be simplified to

$$Z_n = n! h_0 h_1 \cdots h_{n-1},$$

similar to the Hermitian matrix model discussed in last section, where U is a $N \times N$ unitary matrix and dU is similar to the dM discussed in last section, see [22]. Recursion formula for the polynomials now becomes [38] $z(p_n + v_n p_{n-1}) = p_{n+1} + u_n p_n$, derived from the Szegö's equation [56]. By orthogonality of the polynomials, $x_n = p_n(0, s)$ satisfy a different string equation [38, 46, 47, 58]

$$\frac{n}{s} x_n = -(1 - x_n^2)(x_{n+1} + x_{n-1}). \tag{1.31}$$

Similar to the Hermitian matrix model, the partition function now satisfies the following relation

$$\frac{d^2}{ds^2} \ln Z_n = 2(1 - x_n^2)(1 - x_{n-1} x_{n+1}), \tag{1.32}$$

for $n \geq 2$, which can be applied to study continuity or discontinuity of the derivatives of free energy function. In this book, we will use the notation s instead of t to denote the potential parameter in the unitary matrix model(s) to distinguish from the Hermitian matrix models when discussing the partition functions. And in the unitary matrix models, the eigenvalue densities will be defined on complement of the cut(s) in the unit circle to be discussed in detail in Chap. 5, that is an important difference from the Hermitian matrix models in which the eigenvalues are distributed on the cut(s) in the real line.

It has been obtained by Gross and Witten [22] in 1980 by using steepest descent method that the free energy $E = \lim_{n \to \infty} \frac{-1}{n^2} \ln Z_n$ can be reduced to the following integral formula

$$E = -\frac{2}{T} \int \cos\theta \rho(\theta) d\theta - \iint \ln\left|\sin\frac{\theta - \theta'}{2}\right| \rho(\theta)\rho(\theta') d\theta d\theta' - \ln 2, \tag{1.33}$$

where $T = n/s$. The eigenvalue density ρ has different formulas for $T > 2$ and $T < 2$, called strong and weak coupling densities respectively. The strong coupling density is

$$\rho(\eta) = \frac{1}{2\pi}\left(1 + \frac{2}{T}\cos\theta\right) d\theta, \quad \theta \in [-\pi, \pi], \ T \geq 2, \tag{1.34}$$

and the weak coupling density is

$$\rho(\eta) = \frac{2}{\pi T}\cos\frac{\theta}{2}\sqrt{\frac{T}{2} - \sin^2\frac{\theta}{2}} d\theta, \quad |\sin\theta/2| \leq \sqrt{T/2}, T \leq 2. \tag{1.35}$$

Based on these densities, the free energy function can be explicitly solved and the results are $E = -\frac{1}{T^2}$ for $T \geq 2$, and $E = -\frac{2}{T} - \frac{1}{2}\ln\frac{T}{2} + \frac{3}{4}$ for $T \leq 2$. The free energy has continuous first- and second-order derivatives, but the third-order derivative is discontinuous at the critical point $T = 2$. This is a brief introduction of the well known Gross-Witten third-order phase transition model [22].

It is interesting that the transition models based on the densities are associated with the critical phenomena (second-order divergence). It is seen that the formulas (1.17) and (1.32) for the second-order derivative of $\ln Z_n$ do not involve the potential parameters. If we consider the scaling only in the n direction, the second-order derivative of free energy does not have singularity. If the potential parameter is involved in the formula, then singularity will appear. It is shown in [38] based on Toda lattice that the $x_{n-1}x_{n+1}$ term in (1.32) can be expressed in terms of $(dx_n/ds)^2$, $x_{n-1}x_{n+1} = ((nx_n)^2/s^2 - (dx_n/ds)^2)/(4(1 - x_n^2)^2)$. Then (1.32) can be changed to

$$\frac{d^2}{ds^2} \log Z_n = 2(1 - x_n^2) - \frac{(nx_n)^2/s^2 - (dx_n/ds)^2}{2(1 - x_n^2)}. \tag{1.36}$$

This formula is identical to formula (1.32) in the integrable systems. But in large-N asymptotics, these two formulas will go to different ways that will not be consistent. The non-consistency is because of the non-simultaneous reductions in the large-N asymptotics. The discrete integrable system for the string equation is reduced to the integrable system for a continuum integrable equation at the critical point by large-N double scaling, whereas the integrable system for the Toda lattice can not join with the double scaling to become a new integrable system. Then there must be a singularity in the model, that is how we search for critical phenomenon. Briefly, if we take $x_{n-1} \sim -x_n \sim x_{n+1}$, then string equation (1.31) is reduced to $1 - x_n^2 \sim \frac{n}{2s}$ as $s \to \frac{n}{2}+0$, which implies $dx_n/ds \sim \pm\frac{1}{\sqrt{2n}}(s - \frac{n}{2})^{-1/2}$. In this case, (1.36) becomes

$$\frac{d^2}{ds^2} \ln Z_n = O\left(\frac{1}{s - \frac{n}{2}}\right), \tag{1.37}$$

as s approaches to $n/2$ from the right side, that gives a critical phenomenon, a power-law divergence of the second-order derivative of $\ln Z_n$. Note that the critical point $s = n/2$ is remarkably like the critical point $T^c = \lim_{n,s\to\infty} n/s = 2$ in the Gross-Witten third-order phase transition model. The physical background behind this unusual property is still the uncertainty. At the critical point, the original entire integrable system is broken into several pieces. Some are in the position space, and others are in the momentum space. The transition in the s direction is local and in the position space. The temperature variable T is macroscopic and the corresponding density models neglect the local properties. The critical phenomenon for the Hermitian matrix model will be discussed in Sect. 4.2.2 in association with the planar diagram model.

Critical phenomenon is an important subject researched in physics due to the universality represented by the power-law divergence, $|T - T^c|^{-\nu}$, say, that is mostly for the correlation function which is related to the second-order derivative of the free energy [61]. In the matrix models, the second-order derivatives of the free energy are usually continuous based on the derivatives of the partition function obtained in the integrable system as shown above. The third-order discontinuity or the absence of the second-order phase transition implies that quantum chromodynamics is both asymptotically free at short distance (weak coupling) and confining at large distance (strong coupling) [22], that are related to many complicated physical theories. The integrable systems can be applied to introduce some new mathematical properties and connections, such as the "tiny" difference between the s and T variables, for further researches in physics.

In some sense, string equations play a role as the equation of state in statistical mechanics. Mathematically, we can compare the string equations (1.4) and (1.18) with the equation of state $pV = NRT$ for the ideal gas. And the string equation (1.31) can be written as $(1 + 1/(u_n u_{n-1}))v_n = n/s$ which can be further changed to

$$\left(1 + \frac{1}{u_{n-1}u_n}\right)(u_n + x_n x_{n+1}) = \frac{n}{s}, \tag{1.38}$$

with $u_n = -x_{n+1}/x_n$ and $v_n = u_n + x_n x_{n+1}$, where $x_n x_{n+1}$ is negative in our discussions. We can mathematically compare this equation with the formula of the Van der Waals equation [61] $(p + a/V^2)(V - b) = NRT$. The string equation seems to play a role of the Van der Waals equation in the momentum aspect. The index change from $n - 1$ to n is likely to perform a "pressure" role, that in some sense indicates the importance of the index change in the integrable systems. The equations of state are the physical rules in the thermodynamics with the thermodynamic laws that characterize the dynamics of the model in terms of the differentials of energy, entropy, magnetic field, magnetization and volume, and the free energy is studied based on the fundamental laws [61]. Also see [33, 55] for the general backgrounds. In the matrix models, the free energy will be discussed based on the integrable systems for the string equation and Toda lattice that characterize the changes of the wave functions in the directions of n, eigenvalue and potential parameters, and the details will be discussed in the later chapters.

The string equation and the associated orthogonal polynomials on the unit circle as well as the related theories have been studied in many literatures, specially for discussing the unitary matrix models. Interested readers can find the discussions on, for example, the orthogonal polynomials in [53, 54, 56], unitary matrix models in [5, 17, 21, 22, 25, 26, 38, 44, 46–48] and string theory in [9, 11, 57]. There are too many publications in this field to list all of them here. The difference this book is going to discuss is to use the string equations to solve the algebraic relations of the parameters and finally derive the transition models.

1.4   Phase Transitions in the Momentum Aspect

For $\Phi_n(z) = e^{-\frac{1}{2}V(z)}(p_n(z), p_{n-1}(z))^T$ considered in the matrix models where the p_n's are the orthogonal polynomials, there is $(\frac{d}{dz} - A_n)\Phi_n = 0$, which indicates that the quantity $\frac{\hbar}{2\pi}\sqrt{\det A_n(z)}$ stands for the momentum represented by the operator $\frac{\hbar}{2\pi i}\frac{d}{dz}$ as known in quantum mechanics, where $\operatorname{tr} A_n = 0$. See the explanations in

[42] (Appendix A.9). When this idea is applied to the orthogonal polynomials with a general potential, the eigenvalue densities discussed in Sect. 1.1 can be widely generalized. Consequently, the phase transitions based on the eigenvalue densities can be formulated in the momentum aspect, in a different way if we compare with the transition models in statistical mechanics.

To construct the A_n, the recursion formula of the orthogonal polynomials is fundamental. The number of the terms in the recursion tells how many layers the eigenvalues are coupled to form. In some sense, the index range in the recursion is a indication for the space size that the random variables exist in. The construction of split eigenvalue densities is one of the main concerns in the phase transitions in the matrix models, which require multiple layers of couplings. The three-term recursion formula structure has limitation to spread the eigenvalues separately in order to split the distribution since three indexes do not fill a sufficiently large range for the random performance. The method here is to apply the recursion formula to itself to get wider range of indexes. For example, for the recursion formula $z p_n = p_{n+1} + v_n p_{n-1}$, we can change it to the following form,

$$\begin{pmatrix} p_{n+1} \\ p_n \end{pmatrix} = \begin{pmatrix} z & -v_n \\ 1 & 0 \end{pmatrix} \begin{pmatrix} z & -v_{n-1} \\ 1 & 0 \end{pmatrix} \begin{pmatrix} p_{n-1} \\ p_{n-2} \end{pmatrix}, \tag{1.39}$$

which can achieve the formulation of the density model on two disjoint intervals. As $n \to \infty$, if v_n and v_{n-1} are accumulated to different spots by choosing $n = 2n_1$, say, the gap in the density can be generated. On the other hand, if we just use the three-term recursion formula to get one spot accumulation, then the density has no gap. The common situation for these two cases is the critical point, which can be found from the bifurcation of the reduced parameter(s), and the transition is then obtained by studying the corresponding string equation.

For the model with the potential $W(\eta) = g_2 \eta^2 + g_4 \eta^4$, for instance, we will use

$$J^{(1)} = \begin{pmatrix} 0 & 1 \\ -b^2 & \eta \end{pmatrix} \quad \text{and} \quad J^{(2)} = \begin{pmatrix} 0 & 1 \\ -b_1^2 & \eta \end{pmatrix} \begin{pmatrix} 0 & 1 \\ -b_2^2 & \eta \end{pmatrix} \tag{1.40}$$

to write $W(\eta)$ in the following two different forms,

$$g_2 \big(\operatorname{tr} J^{(1)} \big)^2 + g_4 \big(\operatorname{tr} J^{(1)} \big)^4 \quad \text{and} \quad \big(g_2 + 2g_4 \big(b_1^2 + b_2^2 \big) \big) \operatorname{tr} J^{(2)} + g_4 \big(\operatorname{tr} J^{(2)} \big)^2, \tag{1.41}$$

under separate conditions. By changing the representation of the potential function, the eigenvalue density, the parameter conditions and the critical point can be directly derived by using the string equation and the associated Lax pair structure. The details will be discussed in Chaps. 2 and 3. The bifurcation transitions based on $J^{(1)}$ and $J^{(2)}$, for instance, are due to the phase change from

$$\sqrt{\big(\operatorname{tr} J^{(1)} \big)^2 - 4 \det J^{(1)}} \quad \text{to} \quad \sqrt{\big(\operatorname{tr} J^{(2)} \big)^2 - 4 \det J^{(2)}}, \tag{1.42}$$

that will cause a third-order discontinuity. For the unitary matrix models, the eigenvalues are distributed on the arcs of the unit circle, and the corresponding problems can be solved by discussing the orthogonality on the unit circle.

When the v_{n-1} and v_n are scaled to b_1 and b_2, we need to refer the uncertainty principle. For example, for the recursion formula $zp_n = p_{n+1} + u_n p_n + v_n p_{n-1}$ for the generalized Hermite polynomials, u_n satisfies

$$\int_{-\infty}^{\infty} z \left(\frac{p_n(z)}{\sqrt{h_n}} \right)^2 e^{-V(z)} dz = u_n, \quad \text{with} \quad \int_{-\infty}^{\infty} \left(\frac{p_n(z)}{\sqrt{h_n}} \right)^2 e^{-V(z)} dz = 1, \quad (1.43)$$

that means u_n represents the position, while the density represents the momentum as discussed in Appendix A.9 in [42] and above, where the u_n will be roughly scaled as $n^{1/2m} a$ that is certainly not accurate. If the position is accurate, then the momentum is not accurate according the Heisenberg uncertainty inequality $\Delta p \geq \hbar/(2\Delta x)$, which is based on the fundamental property $[\frac{d}{dz}, z] = 1$ where $[\cdot, \cdot]$ is the Lie bracket in mathematics. For the v_n, the discussion will be more complex since generally the position and momentum are studied by using more complicated operators with fractional powers of operators [10] based on $[P, Q] = 1$ and the generalized KdV hierarchies for the multiple matrix models, for instance.

The density models in the multiple matrix models [19, 27, 32, 43] are possibly related to the corresponding orthogonal polynomials, that satisfy a four- or more term recursion formula. The coefficient matrices in the Lax pair for the two-matrix models are 4×4 matrices, for instance, and the discussions will be much complicated. In this book, we focus on the one-matrix models with the three-term recursion formulas. The index folding technique provides a new method to achieve the multiple recursive goal to get the densities. The method based on the string equations can at least solve a class of the density problems in the one-matrix models by using the unified model of the eigenvalue densities.

Besides the bifurcation transitions, other phase transitions can also be generated, for example, by changing the X variable in the phase

$$\sqrt{\eta^4 - g\eta^2 - 2\eta - X}, \quad (1.44)$$

for the potential $W(\eta) = -g\eta^2 + \frac{2}{3}\eta^3$. Like before, this density model is still obtained from the general structure $\sqrt{\det A_n}$ by a new organization of the terms and the large-N asymptotics. The consequent transitions, of first, second or higher order, are called large-N transitions to distinguish from the bifurcation cases, that will be discussed in Chap. 4. The hypergeometric-type differential equation (Sect. D.1)

$$C(X)\frac{d\mathbf{M}}{dX} = C_0 \mathbf{M}, \quad (1.45)$$

will be applied to study the large-N transitions, where $\mathbf{M}$ is a vector of the elliptic integrals defined as the moment quantities of (1.44), and $C(X)$ and C_0 are the coefficient matrices. See [1, 24] for the elliptic integral theories. The zeros of the equation $\det C(X) = 0$, $X = X(a)$, say, which are the singular points (curves) of the hypergeometric-type differential equation, will give different phases for the model (1.44). As the singular curves meet together, the common point(s) of these X curves will be the critical point(s) of the transition(s).

The first-order and third-order discontinuities in the Hermitian matrix models will be based on the cuts of the associated analytic function on the real line, and the second-order transitions will be based on the cuts in the complex plane. In a first-order discontinuity model, the associated analytic function has two cuts that do not come together during the transition but keep a gap between them that is different from the third-order transition case. The smaller cut will shrink until it disappears at the critical point to form another phase with one cut. In the third-order transition models, the two cuts can merge or split. For example, the two cuts are together in the beginning and then separate at the critical point to become a new phase. In a second-order transition case, there are cuts in the complex plane shrinking at the critical point, but these cuts do not significantly affect the eigenvalue density on the real line. The different transition models are then classified by the cuts in the complex plane. The density models on multiple disjoint intervals are also fundamental in the Seiberg-Witten theory [49]. We will discuss the relation between the density models and the Seiberg-Witten differential [8, 18] in Sect. 3.4.

Another type of discontinuity is the power-law divergence at the critical point for the third- or second-order derivative of the free energy. For example, we can get a fractional power-law divergence of the third-order derivative for the planar diagram model based on the algebraic equation $2gv - v^2/4 = 1$ at the critical point $g_2^c = 1/2$ and $g_4^c = -1/48$ [7], where $g = g_2$. The critical phenomenon (second-order divergence) discussed in Sect. 4.2.2 can be obtained similarly as the sample given in last section. The second-order divergence is in the t_2 direction, and the third-order divergence is in the g_2 direction. The discrete integrable system is reduced to the continuum integrable system at the critical point by the large-N double scaling that leads to the third-order divergence. But the corresponding Toda lattice can not join with the double scaling that gives the divergence of the second-order derivative of $\ln Z_n$ as the Toda lattice is involved in the analysis. The difference can be also analyzed by considering the order of the limit process and the derivative for $\ln Z_n$. There are more interesting properties at the critical point of the transition, such as

$$\frac{dg}{d \ln v} = 0, \tag{1.46}$$

based on the algebraic equation and the criticality, which could be related to the β function in the asymptotic freedom theory. According to the β function theory [22], when $\beta = 0$, a second-order transition is expected, that is what we will discuss in the t_2 direction. If we start from this formula to find the criticality, then the string equation and Toda lattice can directly give the second-order discontinuity as a short cut of the double scaling method. The calculations are usually complicated to get a second-order transition model in the renormalization theories. The integrable systems can simplify the discussions for the expectation of the Wilson loop operator and quickly catch the critical phenomena based on the nonlinear relations, and the details will be discussed in Sects. 4.2 and 4.3.

The double scaling method transforms the integrable systems for the string equations to the soliton integrable systems, that is a connection to the position aspect. The string equation systems deal with the momentum properties, and the soliton systems describe the dynamics in the position aspect. If the steepest descent method for the matrix model is considered as a role of an integral transform changing the lattice model with many z_j variables in scaling to the eigenvalue density model which is in the scope of Lax pair system, then the double scaling method performs a role as an inverse transformation to transform the Lax pair system with one z variable for the string equation to the soliton or equivalent system with another eigenvalue variable obtained from the z variable in the double scaling. And there are different double scalings to achieve such process depending on the type of the corresponding density models with cut(s) on the real line or in the complex plane.

The partition function Z_n and the τ function which is a different version of the partition are in some sense the common part of the two aspects—position and momentum, and they are in the scope of the integrable systems with infinitely many freedoms. To reduce to a physical model of finite freedoms using large-N method, the strategy is sensitive because sometimes a method works for this type model but not for another model. We need to pay attention, as least, to the limit process for $\ln Z_n$, the index range in the formulas, and the stationary point z in the scaling. These are often neglected parts. For example, a derivative of $\ln Z_n$ expressed in terms of v_n or v_{n+1} will have different results. The integrable systems provide a good structure to organize the different elements in the system.

The partition function also takes a determinant representation [23, 27] besides the integral representations discussed in the previous sections. The τ-function with such determinant representation has been studied to describe the integrability in the soliton theories by using the infinite dimensional Lie algebra [28]. In recent years, it is found that the soliton systems can be applied to derive the phase transition models. The second-order phase transition can be obtained by using the periodic induction of the parameters in the τ-function in association with the renormalization theory for the Ising models, for example, see [35]. The similarity between the periodic induction in the soliton systems and the periodic reduction in the recursion formulas of the string equation systems indicates that sometimes the reorganizations of the particles considered in the renormalization theory can be achieved alternatively by reorganizing the wave functions in the string equation systems under certain conditions. Some problems are easy in the position aspect, while others are relatively easy in the momentum aspect, that is a role of the Fourier or integral transform theory which connects the position and momentum aspects.

The Fourier transform and Heisenberg uncertainty principle are also fundamental in many other subjects such as the wavelet and signal theories as shown by the uncertainty inequality $\Delta t \cdot \Delta \omega \geq 1/2$ of the signal processing for the time t and frequency ω, for example, see the graduate textbook in mathematics [29] and related literatures. There are many interesting papers in this field, such as [30] for the comparison between the neighboring fields, and [13] for a connection to the dynamical renormalization [40]. According to Palle Jorgensen [29], the visual situation in Heisenberg's theory was much the same as the one which comes out from the much later wavelet constructions. And Heisenberg's aim for the early quantum theory was quite different from the later wavelet theories. For reader's convenience to see how the uncertainty principle works in the different fields, let us compare the partition function (generating function) in the matrix model with the mother wavelet function in the wavelet theory. Both of them play a role to handle the two sides of the uncertainty. The mother wavelet is used to define the wavelet transform acting on a function to change forward and backward like the Fourier transform. The partition function, however, does not act on functions, but accumulates in different ways in large scalings to go to the different aspects. The two sides of the uncertainty in the matrix model are handled by the partition in a nonlinear way that is different from the direct Fourier transform. The partition model is to formulate the basic physical system for analyzing the states. The Heisenberg uncertainty principle is a common guidance for many subjects that would motivate researchers in various fields to make new discoveries.

The integrable systems can be applied to introduce a unified structure to solve the variational equations for the bifurcation transitions, large-N transitions and divergence transitions in the matrix models as explained above, and then the analysis on the free energy functions become easier in studying the discontinuity, criticality, order of transition and divergences. In this book we focus on the application of the string equations and Toda lattices to the discontinuities in the phase transition problems. One should associate the transition models with other physical theories, such as string theory, unified theory, renormalization group, lattice models, 2D quantum gravity and quantum chromodynamics, for instance, for a better understanding of the physical backgrounds.

References

1. Akhiezer, N.I.: Elements of the Theory of Elliptic Functions. Translations of Math. Monographs, vol. 79. AMS, Providence (1990)
2. Bertola, M.: Free energy of the two-matrix model/dToda tau-function. Nucl. Phys. B **669**, 435–461 (2003)
3. Bertola, M., Marchal, O.: The partition of the two-matrix models as an isomonodromic τ function. J. Math. Phys. **50**, 013529 (2009)
4. Bleher, P., Eynard, B.: Double scaling limit in random matrix models and a nonlinear hierarchy of differential equations. J. Phys. A **36**, 3085–3106 (2003)
5. Brézin, E., Hikami, S.: Intersection theory from duality and replica. Commun. Math. Phys. **283**, 507–521 (2008)
6. Brézin, E., Kazakov, V.A.: Exactly solvable field theories of closed strings. Phys. Lett. B **236**, 144–150 (1990)
7. Brézin, E., Itzykson, C., Parisi, G., Zuber, J.B.: Planar diagrams. Commun. Math. Phys. **59**, 35–51 (1978)
8. Chekhov, L., Mironov, A.: Matrix models vs. Seiberg–Witten/Whitham theories. Phys. Lett. B **552**, 293–302 (2003)
9. Dijkgraaf, R., Moore, G.W., Plesser, R.: The partition function of 2-D string theory. Nucl. Phys. B **394**, 356–382 (1993)
10. Douglas, M.R.: Strings in less than one dimension and the generalized KdV hierarchies. Phys. Lett. B **238**, 176–180 (1990)
11. Douglas, M.R., Kazakov, V.A.: Large N phase transition in continuum QCD in two-dimensions. Phys. Lett. B **319**, 219–230 (1993)
12. Douglas, M.R., Shenker, S.H.: Strings in less than one dimension. Nucl. Phys. B **335**, 635–654 (1990)

13. Dutkay, D.E., Jorgensen, P.E.T.: Hilbert spaces built on a similarity and on dynamical renormalization. J. Math. Phys. **47**(5), 053504 (2006)
14. Eynard, B.: Large-N expansion of the 2 matrix model. J. High Energy Phys. **1**, 051 (2003)
15. Eynard, B., Orantin, N.: Mixed correlation functions in the 2-matrix model, and the Bethe ansatz. J. High Energy Phys. **08**, 028 (2005)
16. Fokas, A.S., Its, A.R., Kitaev, A.V.: Discrete Painlevé equations and their appearance in quantum gravity. Commun. Math. Phys. **142**, 313–344 (1991)
17. Fuji, H., Mizoguchi, S.: Remarks on phase transitions in matrix models and $N = 1$ supersymmetric gauge theory. Phys. Lett. B **578**, 432–442 (2004)
18. Gorsky, A., Krichever, I., Marshakov, A., Mironov, A., Morozov, A.: Integrability and Seiberg-Witten exact solution. Phys. Lett. B **355**, 466–477 (1995)
19. Gross, D.J.: Some remarks about induced QCD. Phys. Lett. B **293**, 181–186 (1992)
20. Gross, D.J., Migdal, A.A.: Nonperturbative two-dimensional quantum gravity. Phys. Rev. Lett. **64**, 127–130 (1990)
21. Gross, D.J., Newman, M.J.: Unitary and Hermitian matrices in an external field II: the Kontsevich model and continuum Virasoro constraints. Nucl. Phys. B **380**, 168–180 (1992)
22. Gross, D.J., Witten, E.: Possible third-order phase transition in the large-N lattice gauge theory. Phys. Rev. D **21**, 446–453 (1980)
23. Harish-Chandra: Differential operators on a semisimple Lie algebra. Am. J. Math. **79**, 87–120 (1957)
24. Hille, E.: Analytic Function Theory, vols. 1, 2. Chelsea, New York (1974)
25. Hisakado, M.: Unitary matrix model and the Painlevé III. Mod. Phys. Lett. A **11**, 3001–3010 (1996)
26. Hisakado, M., Wadati, M.: Matrix models of two-dimensional gravity and discrete Toda theory. Mod. Phys. Lett. A **11**, 1797–1806 (1996)
27. Itzykson, C., Zuber, J.B.: The planar approximation II. J. Math. Phys. **21**, 411–421 (1980)
28. Jimbo, M., Miwa, T., Ueno, K.: Monodromy preserving deformation of linear ordinary differential equations with rational coefficients. I. Physica D **2**, 306–352 (1981)
29. Jorgensen, P.E.T.: Analysis and Probability: Wavelets, Signals, Fractals. Graduate Texts in Mathematics, vol. 234. Springer, New York (2006)
30. Jorgensen, P.E.T., Song, M.-S.: Comparison of Discrete and Continuous Wavelet Transforms. Springer Encyclopedia of Complexity and Systems Science. Springer, Berlin (2008)
31. Jurkiewicz, J.: Regularization of one-matrix models. Phys. Lett. B **235**, 178–184 (1990)
32. Kazakov, V.A., Migdal, A.A.: Induced QCD at large N. Nucl. Phys. B **397**, 214–238 (1993)
33. Landau, L.D., Lifshitz, E.M.: Statistical Physics, Part 1. Course of Theoretical Physics, vol. 5, 3rd edn. Butterworth–Heinemann, Oxford (1980)
34. Lee, M.H.: Solutions of the generalized Langevin equation by a method of recurrence relations. Phys. Rev. B **26**, 2547–2551 (1982)
35. Loutsenko, I.M., Spiridonov, V.P.: A critical phenomenon in solitonic Ising chains. SIGMA **3**, 059 (2007)
36. Marcenko, V.A., Pastur, L.A.: Distribution of eigenvalues for some sets of random matrices. Mat. Sb. **72(114)**(4), 507–536 (1967)
37. McKay, B.D.: The expected eigenvalue distribution of a large regular graph. Linear Algebra Appl. **40**, 203–216 (1981)
38. McLeod, J.B., Wang, C.B.: Discrete integrable systems associated with the unitary matrix model. Anal. Appl. **2**, 101–127 (2004)
39. McLeod, J.B., Wang, C.B.: Eigenvalue density in Hermitian matrix models by the Lax pair method. J. Phys. A, Math. Theor. **42**, 205205 (2009)
40. McMullen, C.T.: Complex Dynamics and Renormalization. Annals of Mathematics Studies, vol. 135. Princeton University Press, Princeton (1994)
41. Mehta, M.L.: A method of integration over matrix variables. Commun. Math. Phys. **79**, 327–340 (1981)
42. Mehta, M.L.: Random Matrices, 3rd edn. Academic Press, New York (2004)
43. Migdal, A.A.: Phase transitions in induced QCD. Mod. Phys. Lett. A **8**, 153–166 (1993)

44. Mironov, A., Morozov, A., Semeno, G.W.: Unitary matrix integrals in the framework of generalized Kontsevich model. 1. Brézin-Gross-Witten model. Int. J. Mod. Phys. A **11**, 5031–5080 (1996)
45. Newell, A.C.: Solitons in Mathematics and Physics. SIAM, Philadelphia (1985)
46. Periwal, V., Shevitz, D.: Unitary-matrix models as exactly solvable string theories. Phys. Rev. Lett. **64**, 1326–1329 (1990)
47. Periwal, V., Shevitz, D.: Exactly solvable unitary matrix models: multicritical potentials and correlations. Nucl. Phys. B **344**, 731–746 (1990)
48. Rossi, P., Campostrini, M., Vicari, E.: The large-N expansion of unitary-matrix models. Phys. Rep. **302**, 143–209 (1998)
49. Seiberg, N., Witten, E.: Electric-magnetic duality, monopole condensation, and confinement in $N = 2$ supersymmetric Yang-Mills theory. Nucl. Phys. B **426**, 19–52 (1994). Erratum, ibid. B **430**, 485–486 (1994)
50. Sen, S.: Exact solution of the Heisenberg equation of motion for the surface spin in a semi-infinite $S = \frac{1}{2} XY$ chain at infinite temperatures. Phys. Rev. B **44**, 7444–7450 (1991)
51. Sengupta, A.M., Mitra, P.P.: Distributions of singular values for some random matrices. Phys. Rev. E **60**, 3389–3392 (1991)
52. Shimamune, Y.: On the phase structure of large N matrix models and gauge models. Phys. Lett. B **108**, 407–410 (1982)
53. Simon, B.: Orthogonal Polynomials on the Unit Circle, vol. 1: Classical Theory. AMS Colloquium Series. AMS, Providence (2005)
54. Simon, B.: Orthogonal Polynomials on the Unit Circle, vol. 2: Spectral Theory. AMS Colloquium Series. AMS, Providence (2005)
55. Stanley, H.E.: Introduction to Phase Transitions and Critical Phenomena. Oxford University Press, Oxford (1971)
56. Szegö, G.: Orthogonal Polynomials, 4th edn. American Mathematical Society Colloquium Publications, vol. 23. AMS, Providence (1975)
57. Vafa, C.: Geometry of grand unification arXiv:0911.3008 (2009)
58. Wang, C.B.: Orthonormal polynomials on the unit circle and spatially discrete Painlevé II equation. J. Phys. A **32**, 7207–7217 (1999)
59. Wigner, E.P.: Characteristic vectors of bordered matrices with infinite dimensions. Ann. Math. **62**, 548–564 (1955)
60. Wigner, E.P.: On the distribution of the roots of certain symmetric matrices. Ann. Math. **67**, 325–328 (1958)
61. Yeomans, J.M.: Statistical Mechanics of Phase Transitions. Oxford University Press, London (1994)

Chapter 2
Densities in Hermitian Matrix Models

Orthogonal polynomials are traditionally studied as special functions in mathematical theories such as in the Hilbert space theory, differential equations and asymptotics. In this chapter, a new purpose of the generalized Hermite polynomials will be discussed in detail. The Lax pair obtained from the generalized Hermite polynomials can be applied to formulate the eigenvalue densities in the Hermitian matrix models with a general potential. The Lax pair method then solves the eigenvalue density problems on multiple disjoint intervals, which are associated with scalar Riemann-Hilbert problems for multi-cuts. The string equation can be applied to derive the nonlinear algebraic relations between the parameters in the density models by reformulating the potential function in terms of the trace function of the coefficient matrix obtained from the Lax pair and using the Cayley-Hamilton theorem in linear algebra. The Lax pair method improves the traditional methods for solving the eigenvalue densities by reducing the complexities in finding the nonlinear relations, and the parameters are then well organized for further analyzing the free energy function to discuss the phase transition problems.

2.1 Generalized Hermite Polynomials

Let us consider the Hermitian matrix model with a general potential

$$V(z) = \sum_{j=0}^{2m} t_j z^j, \tag{2.1}$$

where z is a real or complex variable, t_j are real, and $t_{2m} > 0$ such that the partition function

$$Z_n = \int_{-\infty}^{\infty} \cdots \int_{-\infty}^{\infty} e^{-\Sigma_{i=1}^n V(z_i)} \prod_{j<k} (z_j - z_k)^2 dz_1 \cdots dz_n \tag{2.2}$$

is well defined. The free energy function is defined as [1] $E = -\lim_{n\to\infty} \frac{1}{n^2} \ln Z_n$. By the scaling transformation $z = n^{\frac{1}{2m}} \eta$ and $t_j = n^{1-\frac{j}{2m}} g_j$, the potential $V(z)$ be-

C.B. Wang, *Application of Integrable Systems to Phase Transitions*,
DOI 10.1007/978-3-642-38565-0_2, © Springer-Verlag Berlin Heidelberg 2013

comes a new potential $W(\eta) = \sum_{j=0}^{2m} g_j \eta^j$. The eigenvalue density $\rho(\eta)$ on l_1 interval(s) $\Omega = \bigcup_{j=1}^{l_1} [\eta_-^{(j)}, \eta_+^{(j)}]$ is defined to minimize the free energy function

$$E = \int_\Omega W(\eta)\rho(\eta)d\eta - \int_\Omega \int_\Omega \ln|\lambda - \eta|\rho(\lambda)\rho(\eta)d\lambda d\eta. \tag{2.3}$$

The density $\rho(\eta)$ is required to satisfy the following conditions [1, 7]:

(i) ρ is non-negative when $\eta \in \Omega$,

$$\rho(\eta) \geq 0; \tag{2.4}$$

(ii) ρ is normalized,

$$\int_\Omega \rho(\eta)d\eta = 1; \tag{2.5}$$

(iii) ρ satisfies the following variational equation for an inner point η of Ω,

$$(P) \int_\Omega \frac{\rho(\lambda)}{\eta - \lambda} d\lambda = \frac{1}{2} W'(\eta), \tag{2.6}$$

where (P) stands for the principal value of the integral, for example,

$$(P) \int_{\eta_-}^{\eta_+} \frac{\rho(\lambda)}{\eta - \lambda} d\lambda \doteq \lim_{\varepsilon \to 0} \left(\int_{\eta_-}^{\eta-\varepsilon} \frac{\rho(\lambda)}{\eta - \lambda} d\lambda + \int_{\eta+\varepsilon}^{\eta_+} \frac{\rho(\lambda)}{\eta - \lambda} d\lambda \right).$$

The generalized Hermite polynomials will be applied to find the eigenvalue density $\rho(\eta)$. The last two conditions will be satisfied based on the asymptotic of an analytic function $\omega(\eta)$ as $\eta \to \infty$, and the first condition needs separate discussions. The density generally takes a form as the product of a polynomial in η and the square root of another polynomial in η of degree $l \geq l_1$, and the parameters in the density are generally restricted by complicated nonlinear relations that can be obtained by using the string equation associated with the Hermitian matrix models.

Now, let us use the planar diagram model [1] to briefly explain how the above basic concepts are connected each other. For the planar diagram eigenvalue density shown in Sect. 1.2 for $W(\eta) = \frac{1}{2}\eta^2 + g\eta^4$, if we define

$$\omega(\eta) = \left(\frac{1}{2} + 4gb^2 + 2g\eta^2 \right)\sqrt{\eta^2 - 4b^2}, \tag{2.7}$$

for $\eta \in \mathbb{C} \backslash \Omega$ where $\Omega = [-2b, 2b]$ and $\mathbb{C}$ stands for the complex plane, then there is the following asymptotics

$$\omega(\eta) = \frac{1}{2}\left(\eta + 4g\eta^3\right) - \left(b^2 + 12gb^4\right)\frac{1}{\eta} + o\left(\frac{1}{\eta^2}\right), \tag{2.8}$$

as $\eta \to \infty$. Let $\Omega^* = \Omega^- \cup \Omega^+$ be a closed counterclockwise contour, where Ω^- and Ω^+ are the lower and upper edges of Ω respectively. By Cauchy theorem, there is $\int_{\Omega^*} \omega(\eta)d\eta = -(b^2 + 12gb^4) \int_{|\eta|=R} \frac{d\eta}{\eta} = -2\pi i(b^2 + 12gb^4)$. The analytic function $\omega(\eta)$ has opposite signs on Ω^- and Ω^+. If we define $\rho(\eta) = \frac{1}{\pi i}\omega(\eta)|_{\Omega^+}$, then $\rho(\eta)$ satisfies the normalized condition above if the parameters satisfy

$$b^2 + 12gb^4 = 1. \tag{2.9}$$

Further, if Ω^* is changed to Ω_ε^* by just changing the straight lines of the Ω^* at the neighborhood of a inner point η of Ω to the small semicircles of ε radius, then we have

$$\lim_{\varepsilon \to 0} \int_{\Omega_\varepsilon^* \backslash \gamma_\varepsilon} \frac{\omega(\lambda)}{\lambda - \eta} d\lambda = \lim_{\varepsilon \to 0} \int_{\Omega_\varepsilon^*} \frac{\omega(\lambda)}{\lambda - \eta} d\lambda - \lim_{\varepsilon \to 0} \int_{\gamma_\varepsilon} \frac{\omega(\lambda)}{\lambda - \eta} d\lambda, \qquad (2.10)$$

where γ_ε is the circle of ε radius with center η. Since $\omega(\eta)$ has opposite signs on the upper and lower edges, the second limit on the right hand side above is zero. By the asymptotics above, there is

$$\lim_{\varepsilon \to 0} \int_{\Omega^*} \frac{\omega(\lambda)}{\lambda - \eta} d\lambda = \lim_{\varepsilon \to 0} \int_{\Omega^*} \frac{\omega(\lambda) - \frac{1}{2} W'(\lambda)}{\lambda - \eta} d\lambda + \lim_{\varepsilon \to 0} \int_{\Omega_\varepsilon^*} \frac{\frac{1}{2} W'(\lambda)}{\lambda - \eta} d\lambda$$
$$= \pi i W'(\eta). \qquad (2.11)$$

Therefore, $\rho(\eta)$ satisfies the variational equation given above for the eigenvalue density by noting that $\lim_{\varepsilon \to 0} \int_{\Omega_\varepsilon^* \backslash \gamma_\varepsilon} \frac{\omega(\lambda)}{\lambda - \eta} d\lambda = 2\pi i (\text{P}) \int_\Omega \frac{\rho(\eta)}{\eta - \lambda} d\eta$. We have shown the idea to use the $\omega(\eta)$ with the asymptotics to satisfy the conditions for the $\rho(\eta)$. The question then becomes how to get the $\omega(\eta)$ defined in the complex plane except the cut(s) with the corresponding asymptotics. The orthogonal polynomials and string equation can just provide such formulations to construct the $\omega(\eta)$.

In 1991, Fokas, Its and Kitaev [4] obtained that for the potential $V(z) = t_2 z^2 + t_4 z^4$, the coefficients v_n's in the recursion formula $z p_n = p_{n+1} + v_n p_{n-1}$ satisfy the string equation

$$\left(2 t_2 + 4 t_4 (v_{n-1} + v_n + v_{n+1})\right) v_n = n. \qquad (2.12)$$

If all the v_n's are replaced by $n^{1/2} b^2$ and t_j's are replaced by $n^{1-\frac{j}{4}} g_j$, and if we choose $g_2 = 1/2$ and denote $g_4 = g$, then the string equation (2.12) is reduced to the condition (2.9), which is corresponding to (2.5). The relation between (2.12) and (2.9) indicates that the eigenvalue density problem and the string equation have some connections. We will see that these relations are so close not only for their algebraic formulas but also the roles they play in the corresponding models. The condition (2.9) for the eigenvalue density is a property for the exponent of the partition function. The string equation is a property for the exponent n of the orthogonal polynomials $p_n(z) = z^n + \cdots$, that is about the wave function. We will experience more about the connections between these equations in the later discussions.

The main question is how to find the formula of the density ρ. It is discussed in [6] by McLeod and Wang in 2009 that the eigenvalue density can be formulated based on the square root of the determinant of matrix $A_n(z)$, $\sqrt{\det A_n(z)}$, where $A_n(z)$ is the coefficient matrix in the Lax pair

$$\Phi_{n+1} = L_n \Phi_n, \qquad (2.13)$$
$$\frac{\partial}{\partial z} \Phi_n = A_n(z) \Phi_n. \qquad (2.14)$$

Here $\Phi_n(z) = e^{-\frac{1}{2} V(z)} (p_n(z), p_{n-1}(z))^T$, the orthogonal polynomials $p_n = z^n + \cdots$ are defined on the real line with the weight $\exp(-V(z))$: $\langle p_n, p_{n'} \rangle = h_n \delta_{n,n'}$, and

L_n can be obtained from the recursion formula [8] $zp_n(z) = p_{n+1}(z) + u_n p_n(z) + v_n p_{n-1}(z)$. The consistency condition for the Lax pair is the string equation which is a set of two discrete equations for u_n and v_n: $\langle p_n(z), V'(z) p_{n-1}(z) \rangle = nh_{n-1}$ and $\langle p_n(z), V'(z) p_n(z) \rangle = 0$, where $h_n / h_{n-1} = v_n$. These two relations will be used to derive the conditions (2.5) and (2.6). For the planar diagram model, there is no the equation (2.6) since the potential is an even function.

The coefficient matrix $A_n(z)$ above is generally a complicated 2×2 matrix. If we replace all the u_{n-k-1} and v_{n-k} in the Lax pair by x_n and y_n respectively, then $A_n(z)$ can be reduced to

$$\hat{A}_n(z) = D_n \hat{F}_n(z) D_n^{-1} - \frac{1}{2} V'(z) I, \tag{2.15}$$

where the matrix $\hat{F}_n(z)$ is a linear combination of positive powers of a matrix $\hat{J}_n$ derived from L_n,

$$\hat{J}_n = \begin{pmatrix} 0 & 1 \\ -y_n & z - x_n \end{pmatrix}.$$

Here $D_n = \text{diag}(h_n, h_{n-1})$, and I is the identity matrix. By the Cayley-Hamilton theorem for $\hat{J}_n$ as known in the textbooks of linear algebra, we have

$$(z - x_n) I = \hat{J}_n + y_n \hat{J}_n^{-1}. \tag{2.16}$$

Applying this relation to $V'(z) I$ in (2.15), it is found that $D_n^{-1} \hat{A}_n(z) D_n$ can be factorized as a product of a polynomial and a simple matrix

$$D_n^{-1} \hat{A}_n(z) D_n = f_{2m-2}(z) \left(\hat{J}_n(z) - y_n \hat{J}_n^{-1}(z) \right), \tag{2.17}$$

where the polynomial $f_{2m-2}(z)$ will be given in the following sections. There is an important asymptotics

$$\sqrt{-\det \hat{A}_n(z)} = \frac{1}{2} V'(z) - \frac{n}{z} + O\left(\frac{1}{z^2}\right), \tag{2.18}$$

as $z \to \infty$ in the complex plane, obtained by using the string equation. This property will be applied to satisfy the conditions (2.5) and (2.6). Replacing the variable z and the parameters t_j, x_n and y_n by $n^{\frac{1}{2m}} \eta$, $n^{1-\frac{j}{m}} g_j$, $n^{\frac{1}{2m}} a$ and $n^{\frac{1}{m}} b^2$ respectively, the formula of the eigenvalue density $\rho(\eta)$ on the interval $[\eta_-, \eta_+] = [a - 2b, a + 2b]$ can be obtained by

$$\frac{1}{n\pi} \sqrt{\det \hat{A}_n(z)} dz = \rho(\eta) d\eta, \tag{2.19}$$

which follows the unified model discussed in Sect. 1.1. The eigenvalue density problem is then solved when condition (2.4) is satisfied.

The $\hat{A}_n(z)$ can be generalized to a new matrix $\hat{A}_n^{(l)}(z)$ to find the eigenvalue densities on multiple disjoint intervals. It should be noted that in some literatures the Lax pair is applied to study the asymptotics of u_n or v_n or the related functions, while the method here is to derive the density formula by referring the matrix $A_n(z)$ because $\sqrt{-\det A_n(z)}$ itself also has the same asymptotics (2.18) as $z \to \infty$. The details of the formulations and the relations to $A_n(z)$ will be discussed in the following sections.

2.2 Integrable System and String Equation

For the Hermitian matrix model with potential $W(\eta) = \sum_{j=1}^{m} g_j \eta^{2j}$, we have discussed in last section that the eigenvalue density $\rho(\eta)$ needs to satisfy the conditions (2.5) and (2.6). In this section, we discuss how to get an analytic function $\omega(\eta)$ with the asymptotics $\frac{1}{2}W'(\eta) - \frac{1}{\eta}$ as $\eta \to \infty$ in the complex plane, where $' = \partial/\partial\eta$.

Consider the orthogonal polynomials $p_n(z) = z^n + \cdots$ on $(-\infty, \infty)$ defined by

$$\langle p_n, p_{n'} \rangle \equiv \int_{-\infty}^{\infty} p_n(z) p_{n'}(z) e^{-V(z)} dz = h_n \delta_{n,n'}, \tag{2.20}$$

where $V(z) = \sum_{j=0}^{2m} t_j z^j$, $t_{2m} > 0$. We have the following asymptotics $e^{-V(z)/2} \times p_n(z) \sim e^{-\frac{1}{2}V(z)+n\ln z}$ as $z \to \infty$. This asymptotics gives a hint that the differential equation for the orthogonal polynomials may help us to find the $\omega(\eta)$. In the following, we introduce the basic Lax pair theory to construct the coefficient matrix $A_n(z)$.

The orthogonal polynomials satisfy a recursion formula [8],

$$p_{n+1}(z) + u_n p_n(z) + v_n p_{n-1}(z) = z p_n(z). \tag{2.21}$$

By multiplying $p_{n-1}(z)e^{-V(z)}$ on both sides of this recursion formula and taking integral, we get $v_n = h_n/h_{n-1}$. This recursion formula will give the first equation in the Lax pair.

For the second equation in the pair, let us consider the differential equation in the z direction. When $n \geq 2m - 1$, express the derivative of p_n with respect to z as a linear combination of p_j's,

$$\frac{\partial}{\partial z} p_n = a_{n,n-1} p_{n-1} + a_{n,n-2} p_{n-2} + \cdots + a_{n,0} p_0, \tag{2.22}$$

where $a_{n,j}$ are independent of z. By integration by parts, there are

$$a_{n,j} h_j = \int_{-\infty}^{\infty} V'(z) p_j(z) p_n(z) e^{-V(z)} dz, \quad (' = \partial/\partial z)$$

for $j = 0, 1, \ldots, n - 1$, and $a_{n,j} = 0$ when $j < n - 2m + 1$ by the orthogonality. Then, by using the recursion formula, $\frac{\partial}{\partial z} p_n$ can be changed to a linear combination of p_n and p_{n-1}, but the new coefficients are dependent on z.

Denote $\Phi_n(z) = e^{-\frac{1}{2}V(z)}(p_n(z), p_{n-1}(z))^T$. By the discussions above, there are

$$\Phi_{n+1} = L_n \Phi_n, \tag{2.23}$$

where

$$L_n = \begin{pmatrix} z - u_n & -v_n \\ 1 & 0 \end{pmatrix};$$

and

$$\frac{\partial}{\partial z} \Phi_n = A_n(z) \Phi_n, \tag{2.24}$$

for a matrix $A_n(z)$. Equations (2.23) and (2.24) are called the Lax pair for the string equation. This Lax pair structure was given in [4], as well as in [3] (Part 2, Chap. 1).

The Lax pair method for the eigenvalue density starts from the construction of the matrix A_n. For $m \geq 1$ and $n \geq 2m$, consider

$$\frac{\partial}{\partial z} p_n = a_{n,n-1} p_{n-1} + a_{n,n-2} p_{n-2} + \cdots + a_{n,n-2m+1} p_{n-2m+1},$$

$$\frac{\partial}{\partial z} p_{n-1} = a_{n-1,n-2} p_{n-2} + a_{n-1,n-3} p_{n-3} + \cdots + a_{n-1,n-2m} p_{n-2m},$$

where

$$a_{n',n'-k} h_{n'-k} = \int_{-\infty}^{\infty} V'(z) p_{n'-k} p_{n'} e^{-V(z)} dz, \qquad (2.25)$$

for $n' = n$ or $n-1$, and $k = 1, 2, \ldots, 2m-1$, with $V'(z) = \sum_{j=1}^{2m} j t_j z^{j-1}$. It follows that

$$\frac{\partial}{\partial z} \begin{pmatrix} p_n \\ p_{n-1} \end{pmatrix} = \sum_{k=1}^{2m-1} C_{n-k} \begin{pmatrix} p_{n-k} \\ p_{n-k-1} \end{pmatrix}, \qquad (2.26)$$

where

$$C_{n-k} = \begin{pmatrix} a_{n,n-k} h_{n-k} & 0 \\ 0 & a_{n-1,n-k+1} h_{n-k+1} \end{pmatrix}, \qquad (2.27)$$

for $k = 1, \ldots, 2m-1$. And $P_j = p_j / h_j$ satisfy

$$\begin{pmatrix} P_j \\ P_{j-1} \end{pmatrix} = J_{j+1} \begin{pmatrix} P_{j+1} \\ P_j \end{pmatrix}, \qquad J_{j+1} = \begin{pmatrix} 0 & 1 \\ -v_{j+1} & z - u_j \end{pmatrix}, \qquad (2.28)$$

by using (2.23) and $v_{j+1} = h_{j+1} / h_j$. Denote $D_n = \mathrm{diag}(h_n, h_{n-1})$. The above discussion gives

$$\frac{\partial}{\partial z} \begin{pmatrix} p_n \\ p_{n-1} \end{pmatrix} = D_n F_n D_n^{-1} \begin{pmatrix} p_n \\ p_{n-1} \end{pmatrix}, \qquad (2.29)$$

where the matrix F_n is defined by

$$D_n F_n = C_{n-1} J_n + C_{n-2} J_{n-1} J_n + \cdots + C_{n-2m+1} J_{n-2m+2} J_{n-2m+3} \cdots J_n. \quad (2.30)$$

Let I be the 2×2 identity matrix. Then, there is

$$A_n = D_n F_n D_n^{-1} - \frac{1}{2} V'(z) I, \quad n \geq 2m. \qquad (2.31)$$

Let Δ be the operator for the index change acting only on the polynomials $\Delta^k p_n = p_{n+k}$, where k is an integer. This is the basic idea for the index folding technique for constructing the eigenvalue density on one interval. The recursion formula (2.21) becomes $(z - u_n) p_n = (\Delta + v_n \Delta^{-1}) p_n$. In the reduced model, we will consider x_n and y_n, and $(z - x_n)^q p_{n'-k}$ will be associated to

$$(\Delta + y_n \Delta^{-1})^q p_{n'-k} = \sum_{r=0}^{q} \binom{q}{r} y_n^r \Delta^{q-2r} p_{n'-k} = \sum_{r=0}^{q} \binom{q}{r} y_n^r p_{n'-k+q-2r},$$

for $n' = n$ or $n - 1$, $k = 1, 2, \ldots, 2m - 1$, and $q = 0, 1, \ldots, 2m - 1$. Then by orthogonality and

$$V'(z) = \sum_{j=1}^{2m} jt_j (x_n + (z - x_n))^{j-1} = \sum_{j=1}^{2m} jt_j \sum_{q=0}^{j-1} \binom{j-1}{q} x_n^{j-q-1} (z - x_n)^q,$$

we have that $a_{n',n'-k} h_{n'-k}$ is reduced to the following according to (2.25),

$$\sum_{j=1}^{2m} jt_j \sum_{q=0}^{j-1} \binom{j-1}{q} x_n^{j-q-1} \sum_{r=0}^{2[q/2]-\mu_q+1} \binom{q}{r} y_n^r \int_{-\infty}^{\infty} p_{n'-k+q-2r} \, p_{n'} e^{-V(z)} dz$$

$$= \sum_{j=1}^{2m} jt_j \sum_{q=0}^{j-1} \binom{j-1}{q} x_n^{j-q-1} \sum_{r=0}^{[q/2]-\mu_q} \binom{q}{r} y_n^r h_{n'} \delta_{q-k-2r,0}, \qquad (2.32)$$

for $n' = n$ or $n - 1$, where $[\cdot]$ denotes the integer part,

$$\mu_q = \frac{1 + (-1)^q}{2} = \begin{cases} 1, & q \text{ is even}, \\ 0, & q \text{ is odd}, \end{cases} \qquad (2.33)$$

$$q = 2[q/2] - \mu_q + 1, \qquad (2.34)$$

and for $k > 0$,

$$q - k - 2r = 2([q/2] - \mu_q - r) + 1 + \mu_q - k < 0, \quad \text{if } r > [q/2] - \mu_q,$$

which implies that $\delta_{q-k-2r,0} = 0$ when $r > [q/2] - \mu_q$, that is why the upper bound of the index r for the last summation above is changed to $[q/2] - \mu_q$. Consequently, the $D_n F_n$, defined by (2.30) is reduced to

$$\sum_{k=1}^{2m-1} \sum_{j=1}^{2m} jt_j \sum_{q=0}^{j-1} \binom{j-1}{q} x_n^{j-q-1} \sum_{r=0}^{[q/2]-\mu_q} \binom{q}{r} y_n^r \delta_{q-k-2r,0} D_n \hat{J}_n^k$$

$$= \sum_{j=1}^{2m} jt_j \sum_{q=0}^{j-1} \binom{j-1}{q} x_n^{j-q-1} \sum_{r=0}^{[q/2]-\mu_q} \binom{q}{r} y_n^r D_n \hat{J}_n^{q-2r},$$

and then F_n is reduced to

$$\hat{F}_n = \sum_{j=1}^{2m} jt_j \sum_{q=0}^{j-1} \binom{j-1}{q} x_n^{j-q-1} \sum_{r=0}^{[q/2]-\mu_q} \binom{q}{r} y_n^r \hat{J}_n^{q-2r}, \qquad (2.35)$$

where

$$\hat{J}_n = \begin{pmatrix} 0 & 1 \\ -y_n & z - x_n \end{pmatrix}. \qquad (2.36)$$

Let

$$\hat{A}_n(z) = D_n \hat{F}_n D_n^{-1} - \frac{1}{2} V'(z) I, \qquad (2.37)$$

which is the matrix we need for the eigenvalue density on one interval. We use the hat symbol such as $\hat{J}_n$ and $\hat{A}_n$ for the reduced models, to distinguish from the J_n and A_n in the integrable system. Based on the factorization and asymptotics for this matrix, we will discuss how to find the formula for the eigenvalue density in the later sections. In the following, we discuss the condition for the parameters that are reduced from the string equation.

By the orthogonality of the polynomials $p_n(z) = z^n + \cdots$ and integration by parts, we have the following string equation,

$$\langle p_n(z), V'(z) p_{n-1}(z) \rangle = n h_{n-1}, \tag{2.38}$$

$$\langle p_n(z), V'(z) p_n(z) \rangle = 0, \tag{2.39}$$

including two recursion formulas for the u_n's and v_n's. The set of (2.38) and (2.39) is called string equation. The string equation is the consistency condition for the Lax pair (2.23) and (2.24). The consistency can be discussed, for example, by referring the methods in [4, 5]. For the density problems, we only need the equations for restricting the parameters.

If the differential equation is written in the form

$$\frac{\partial}{\partial z} p_n = a_{n,n} p_n + a_{n,n-1} p_{n-1} + \cdots + a_{n,n-2m+1} p_{n-2m+1},$$

where $a_{n,n} = 0$, then the formula (2.25) is still true for $k = 0$. Let us write (2.38) and (2.39) as

$$a_{n,n-1} h_{n-1} = n h_{n-1}, \tag{2.40}$$

$$a_{n,n} h_n = 0. \tag{2.41}$$

Based on the reduction (2.32) with $n' = n$ for $k = 1$ and $k = 0$ respectively, in our method for the eigenvalue density on one interval, we need x_n and y_n to satisfy the following equations

$$\sum_{j=1}^{2m} j t_j \sum_{q=0}^{j-1} \binom{j-1}{q} x_n^{j-q-1} \sum_{r=0}^{[q/2]-\mu_q} \binom{q}{r} y_n^{r+1} \delta_{q,2r+1} = n, \tag{2.42}$$

$$\sum_{j=1}^{2m} j t_j \sum_{q=0}^{j-1} \binom{j-1}{q} x_n^{j-q-1} \sum_{r=0}^{[q/2]-\mu_q} \binom{q}{r} y_n^r \delta_{q,2r} = 0. \tag{2.43}$$

Note that $\delta_{q,2r+1} = 0$ when q is even, and $\delta_{q,2r} = 0$ when q is odd. Take $q = 2p+1$, $r = p$ in (2.42), and $q = 2p$, $r = p$ in (2.43), we get

$$\sum_{j=2}^{2m} j t_j \sum_{p=0}^{[\frac{j}{2}]-1} \binom{j-1}{2p+1} \binom{2p+1}{p} x_n^{j-2p-2} y_n^{p+1} = n, \tag{2.44}$$

$$\sum_{j=1}^{2m} j t_j \sum_{p=0}^{[\frac{j-1}{2}]} \binom{j-1}{2p} \binom{2p}{p} x_n^{j-2p-1} y_n^p = 0. \tag{2.45}$$

These two equations will be rescaled to satisfy (2.5) and (2.6).

Specially, when $V(z)$ is even, $V(-z) = V(z)$, or $t_1 = t_3 = \cdots = t_{2m-1} = 0$, there is $p_n(-z) = p_n(z)$, which implies $u_n = 0$, and it follows that $x_n = 0$. Then (2.45) becomes $0 = 0$, because the terms on the left hand side of (2.45) has either a factor x_n or t_j with odd j. And (2.44) becomes

$$\sum_{j=1}^{m} 2jt_{2j} \binom{2j-1}{j} y_n^j = n, \tag{2.46}$$

by replacing j by $2j$ and taking $p = j - 1$ on the left hand side of (2.44).

For the density on multiple disjoint intervals, consider

$$\hat{j}_n^{(l)} = \begin{pmatrix} 0 & 1 \\ -y_n^{(1)} & z - x_n^{(1)} \end{pmatrix} \cdots \begin{pmatrix} 0 & 1 \\ -y_n^{(l)} & z - x_n^{(l)} \end{pmatrix}. \tag{2.47}$$

According to the Cayley-Hamilton theorem for $\hat{j}_n^{(l)}$, there is the following trace formula

$$(\text{tr}\, \hat{j}_n^{(l)}) I = \hat{j}_n^{(l)} + (\det \hat{j}_n^{(l)}) \hat{j}_n^{(l)-1}. \tag{2.48}$$

We can transform t_j ($j = 1, \ldots, 2m$) into a new set of parameters t'_j ($j = 1, \ldots, 2m$) by a linear transformation, such that

$$V'(z) = \sum_{s=0}^{l-1} z^s \sum_{q=0}^{m_s} t'_{lq+s+1} (\text{tr}\, \hat{j}_n^{(l)})^q, \tag{2.49}$$

where each m_s ($s = 0, \ldots, l-1$) is the largest integer such that $s + lm_s \leq 2m - 1$. In fact, by expanding the above expression in terms of z and comparing the coefficients with $V'(z) = \sum_{j=1}^{2m} jt_j z^{j-1}$, we can get a upper triangle matrix T_{2m} so that $T_{2m} \mathbf{t}' = \mathbf{t}$ where $\mathbf{t} = (t_1, 2t_2, \ldots, 2mt_{2m})^T$ and $\mathbf{t}' = (t'_1, t'_2, \ldots, t'_{2m})^T$. The derivative $\partial p_n / \partial z$ is now expanded as

$$\frac{\partial p_n}{\partial z} = \sum_{s=0}^{l-1} \sum_{q'=1}^{N_0} a_{n,n-lq'+s}^{(l)} z^s p_{n-lq'}(z) + \sum_{k=lN_0+1}^{n} a_{n,n-k}^{(l)} p_{n-k}(z), \tag{2.50}$$

where $n - lN_0 < l$ and the choice of N_0 is dependent on the value of m. This is the idea of the index folding technique for constructing the eigenvalue density on multiple disjoint intervals, so called because of the folding term lq' in the index above, which is one of the folding techniques in this subject area. As discussed in Sect. 1.3, the index change is referred as a role of the pressure. The index folding or periodic reduction reflects an even pressure property or multiple even pressure layers in the system that is often assumed in studying the application problems.

By the index change operator Δ, the coefficient $a_{n,n-lq'+s}^{(l)}$ is reduced to

$$\sum_{q=1}^{m_s} t'_{lq+s} \int_{-\infty}^{\infty} p_{n-lq'+s} z^s \left(\Delta^l + (\det \hat{j}_n^{(l)}) \Delta^{-l} \right)^q p_n e^{-V(z)} dz$$

$$= \sum_{q=1}^{m_s} t'_{lq+s} \sum_{r=0}^{[q/2]-\mu_q} \binom{q}{r} (\det \hat{j}_n^{(l)})^{q-r} \delta_{q-q'-2r,0}, \quad q' \leq m_s. \tag{2.51}$$

Also, $a_{n,n-lq'+s}^{(l)}$ for $q' > m_s$ and $a_{n,n-k}^{(l)}$ for $k = lN_0 + 1, \ldots, n$ are reduced to 0 according to the term $\delta_{q-q'-2r,0}$ above. Then we get another reduced matrix

$$\hat{A}_n^{(l)}(z) = D_n \hat{F}_n^{(l)} D_n^{-1} - \frac{1}{2} V'(z) I, \tag{2.52}$$

where

$$\hat{F}_n^{(l)} = \sum_{s=0}^{l-1} z^s \sum_{q=1}^{m_s} t'_{lq+s+1} \sum_{r=0}^{[q/2]-\mu_q} \binom{q}{r} (\det \hat{J}_n^{(l)})^r (\hat{J}_n^{(l)})^{q-2r}, \tag{2.53}$$

by referring that $(p_{n-lq'}, p_{n-lq'-1})^T$ is connected to

$$D_n (\det \hat{J}_n^{(l)})^{-q'} (\hat{J}_n^{(l)})^{q'} D_n^{-1} (p_n, p_{n-1})^T.$$

The matrix $\hat{A}_n^{(l)}(z)$ will be applied to derive the formula of the density on multiple disjoint intervals to be discussed in Sect. 2.4. The restriction conditions for the parameters are similar to the one-interval case and will be given in Sect. 2.4.

The matrix $\hat{A}_n^{(l)}(z)$ is obtained from A_n by replacing the $u_{n-lq+s-1}$ and v_{n-lq+s} by $x_n^{(s)}$ and $y_n^{(s)}$ respectively. One may ask whether the u_N and v_N functions must have such periodic behaviors. The explanation is that the string equations are applied to reorganize the wave functions, not the particles. In the momentum aspect, the parameters and the corresponding functions such as u_N and v_N control the wave functions of the random variables, so that the asymptotics of these functions are not directly related to the behaviors of the particles. If there is an asymptotic relation, it should be a "relative" asymptotics. These functions are closely connected to the moments of the eigenvalues. Each reduction from the integrable system is not a necessity, but a case of the probability. And the occurrence of each possible case is not based on certainty principle, but the uncertainty principle. The accumulations of the sequences and the distribution of the possibilities would give better explanations for the reduction method.

2.3 Factorization and Asymptotics

We have obtained in last section that $D_n^{-1} \hat{A}_n D_n = \hat{F}_n - \frac{1}{2} V'(z) I$, where $D_n = \mathrm{diag}(h_n, h_{n-1})$. In this section, we are going to show that the matrix $\hat{F}_n - \frac{1}{2} V'(z) I$ can be factorized as a product of a polynomial in z of degree $2m - 2$ and the matrix $\hat{J}_n - y_n \hat{J}_n^{-1}$. The equation $\det(\hat{J}_n - y_n \hat{J}_n^{-1}) = 0$ has only two simply zeros which will be used as the bounds of the eigenvalue domain. In the following, we will denote $(\hat{J}_n^{-1})^k$ by $\hat{J}_n^{-k}$ where k is an integer.

Lemma 2.1 *If x_n, y_n and t_j ($j = 1, \ldots, 2m$) satisfy (2.45), then for $\hat{A}_n(z)$ defined by (2.37) and $\mu_q = (1 + (-1)^q)/2$, there is*

$$D_n^{-1}\hat{A}_n D_n = \frac{1}{2}\sum_{j=1}^{2m} jt_j \sum_{q=0}^{j-1}\binom{j-1}{q}x_n^{j-q-1}$$

$$\times \sum_{r=0}^{[q/2]-\mu_q}\binom{q}{r}y_n^r\big(\hat{J}_n^{q-2r}-(y_n\hat{J}_n^{-1})^{q-2r}\big). \qquad (2.54)$$

Proof Recall the matrix $\hat{J}_n$ defined by (2.36),

$$\hat{J}_n = \begin{pmatrix} 0 & 1 \\ -y_n & z-x_n \end{pmatrix}.$$

Applying the Cayley-Hamilton theorem for $\hat{J}_n$, there is $\hat{J}_n^2 - (z-x_n)\hat{J}_n + y_n I = 0$, which implies

$$(z-x_n)I = \hat{J}_n + y_n\hat{J}_n^{-1}. \qquad (2.55)$$

Then by binomial expansion and $q = 2[q/2]-\mu_q+1$, we have

$$(z-x_n)^q I$$

$$= \left(\sum_{r=0}^{[q/2]-\mu_q} +\mu_q\sum_{r=[q/2]}^{[q/2]} + \sum_{r=[q/2]+1}^{2[q/2]-\mu_q+1}\right)\binom{q}{r}y_n^r\hat{J}_n^{q-2r}$$

$$= \sum_{r=0}^{[q/2]-\mu_q}\binom{q}{r}y_n^r\hat{J}_n^{q-2r} + \mu_q\binom{q}{[q/2]}y_n^{[q/2]}\hat{J}_n^{q-2[q/2]}$$

$$+ \sum_{s=0}^{[q/2]-\mu_q}\binom{q}{s}y_n^{q-s}\hat{J}_n^{-q+2s}$$

$$= \sum_{r=0}^{[q/2]-\mu_q}\binom{q}{r}y_n^r\big(\hat{J}_n^{q-2r}+(y_n\hat{J}_n^{-1})^{q-2r}\big) + \mu_q\binom{q}{[q/2]}y_n^{[q/2]}\hat{J}_n^{q-2[q/2]},$$

where s comes out by the substitution $s = q-r$, and then replaced by r in the last step. Since

$$V'(z) = \sum_{j=1}^{2m} jt_j \sum_{q=0}^{j-1}\binom{j-1}{q}x_n^{j-q-1}(z-x_n)^q,$$

$V'(z)I$ can be expressed as a linear combination of the positive and negative powers of $\hat{J}_n$.

By $D_n^{-1}\hat{A}_n D_n = \hat{F}_n - \frac{1}{2}V'(z)I$ given by (2.37) and (2.35), we then have

$$D_n^{-1}\hat{A}_n D_n = \frac{1}{2}\sum_{j=1}^{2m} jt_j \sum_{q=0}^{j-1}\binom{j-1}{q}x_n^{j-q-1}$$

$$\times \sum_{r=0}^{[q/2]-\mu_q} \binom{q}{r} y_n^r \left(\hat{J}_n^{q-2r} - (y_n \hat{J}_n^{-1})^{q-2r} \right)$$

$$= -\frac{1}{2} \sum_{j=1}^{2m} jt_j \sum_{q=0}^{j-1} \binom{j-1}{q} x_n^{j-q-1} \mu_q \binom{q}{[q/2]} y_n^{[q/2]}.$$

Since $\mu_q = 1$ when q is even, and $\mu_q = 0$ when q is odd, the last part in the above vanishes by taking $q = 2p$ and applying (2.45). So we get the result in this lemma. □

Let

$$\alpha_n = \frac{z - x_n + \sqrt{(z - x_n)^2 - 4y_n}}{2}, \tag{2.56}$$

which satisfies

$$\alpha_n + y_n \alpha_n^{-1} = z - x_n, \tag{2.57}$$

$$\alpha_n - y_n \alpha_n^{-1} = \sqrt{(z - x_n)^2 - 4y_n}. \tag{2.58}$$

And it is easy to check that

$$y_n \hat{J}_n^{-1} = \begin{pmatrix} z - x_n & -1 \\ y_n & 0 \end{pmatrix}.$$

We will need

$$\hat{J}_n - y_n \hat{J}_n^{-1} = \begin{pmatrix} -z + x_n & 2 \\ -2y_n & z - x_n \end{pmatrix},$$

which satisfies

$$\sqrt{-\det\left(\hat{J}_n - y_n \hat{J}_n^{-1}\right)} = \sqrt{(z - x_n)^2 - 4y_n} = \alpha_n - y_n \alpha_n^{-1}. \tag{2.59}$$

Lemma 2.2 *For the $\hat{J}_n$ defined by (2.36) and $k = 1, 2, \ldots$, there are*

$$\hat{J}_n^k + y_n^k \hat{J}_n^{-k} = \left(\alpha_n^k + y_n^k \alpha_n^{-k} \right) I, \tag{2.60}$$

and

$$\hat{J}_n^k - y_n^k \hat{J}_n^{-k} = \frac{\alpha_n^k - y_n^k \alpha_n^{-k}}{\alpha_n - y_n \alpha_n^{-1}} \left(\hat{J}_n - y_n \hat{J}_n^{-1} \right). \tag{2.61}$$

Proof By the relations $\hat{J}_n + y_n \hat{J}_n^{-1} = (z - x_n)I$, and $\alpha_n + y_n \alpha_n^{-1} = z - x_n$, we have

$$\hat{J}_n + y_n \hat{J}_n^{-1} = \left(\alpha_n + y_n \alpha_n^{-1} \right) I, \tag{2.62}$$

which is (2.60) for $k = 1$. Taking square on both sides of (2.62), we get

$$\hat{J}_n^2 + y_n^2 \hat{J}_n^{-2} = \left(\alpha_n^2 + y_n^2 \alpha_n^{-2} \right) I,$$

which is (2.60) for $k = 2$.

Now, by mathematical induction, suppose (2.60) is true for $k - 1$ and k, let us show it is also true for $k + 1$. Multiplying (2.60) with (2.62), we get

$$\hat{J}_n^{k+1} + y_n^{k+1} \hat{J}_n^{-k-1} + y_n\left(\hat{J}_n^{k-1} + y_n^{k-1} \hat{J}_n^{-k+1}\right)$$
$$= \left(\alpha_n^{k+1} + y_n^{k+1} \alpha_n^{-k-1}\right)I + y_n\left(\alpha_n^{k-1} + y_n^{k-1} \alpha_n^{-k+1}\right)I.$$

By the assumption, we see that (2.60) is true for $k + 1$.

Equation (2.61) can also be proved by using mathematical induction. It is easy to check that

$$\hat{J}_n^2 - y_n^2 \hat{J}_n^{-2} = \left(\hat{J}_n + y_n \hat{J}_n^{-1}\right)\left(\hat{J}_n - y_n \hat{J}_n^{-1}\right) = (z - x_n)\left(\hat{J}_n - y_n \hat{J}_n^{-1}\right),$$

and then (2.61) is true for $k = 1$ and 2. Suppose (2.61) is true for $k - 1$ and k. We show it is true for $k + 1$. Multiplying (2.61) with (2.62), we have

$$\hat{J}_n^{k+1} - y_n^{k+1} \hat{J}_n^{-k-1} + y_n\left(\hat{J}_n^{k-1} - y_n^{k-1} \hat{J}_n^{-k+1}\right)$$
$$= \frac{\alpha_n^{k+1} - y_n^{k+1} \alpha_n^{-k-1}}{\alpha_n - y_n \alpha_n^{-1}}\left(\hat{J}_n - y_n \hat{J}_n^{-1}\right) + y_n \frac{\alpha_n^{k-1} - y_n^{k-1} \alpha_n^{-k+1}}{\alpha_n - y_n \alpha_n^{-1}}\left(\hat{J}_n - y_n \hat{J}_n^{-1}\right).$$

By the assumption, we have that (2.61) is true for $k + 1$. $\qquad\square$

Lemma 2.3 *For the α_n defined by (2.56) and $k = 1, 2, \ldots$, there are*

$$\alpha_n^k + y_n^k \alpha_n^{-k} = \frac{1}{2^{k-1}} \sum_{s=0}^{[\frac{k}{2}]} \binom{k}{2s} (z - x_n)^{k-2s}\left((z - x_n)^2 - 4y_n\right)^s, \qquad (2.63)$$

and

$$\frac{\alpha_n^k - y_n^k \alpha_n^{-k}}{\alpha_n - y_n \alpha_n^{-1}} = \frac{1}{2^{k-1}} \sum_{s=0}^{[\frac{k-1}{2}]} \binom{k}{2s + 1} (z - x_n)^{k-2s-1}\left((z - x_n)^2 - 4y_n\right)^s. \quad (2.64)$$

Proof By (2.57) and (2.58), we have

$$\alpha_n = \frac{1}{2}\left(z - x_n + \left((z - x_n)^2 - 4y_n\right)^{1/2}\right),$$
$$y_n \alpha_n^{-1} = \frac{1}{2}\left(z - x_n - \left((z - x_n)^2 - 4y_n\right)^{1/2}\right).$$

Then the binomial formula implies

$$\alpha_n^k + y_n^k \alpha_n^{-k}$$
$$= \frac{1}{2^k} \sum_{j=0}^{k} \binom{k}{j} (z - x_n)^{k-j}\left(\left((z - x_n)^2 - 4y_n\right)^{\frac{j}{2}} + (-1)^j\left((z - x_n)^2 - 4y_n\right)^{\frac{j}{2}}\right)$$
$$= \frac{1}{2^{k-1}} \sum_{s=0}^{[\frac{k}{2}]} \binom{k}{2s} (z - x_n)^{k-2s}\left((z - x_n)^2 - 4y_n\right)^s,$$

where the terms with odd j are canceled, and the terms with even j are combined by taking $j = 2s$. So (2.63) is obtained.

Similarly, there is

$$\alpha_n^k - y_n^k \alpha_n^{-k}$$

$$= \frac{1}{2^k} \sum_{j=0}^{k} \binom{k}{j} (z - x_n)^{k-j} \left(\left((z - x_n)^2 - 4y_n \right)^{\frac{j}{2}} - (-1)^j \left((z - x_n)^2 - 4y_n \right)^{\frac{j}{2}} \right)$$

$$= \frac{1}{2^{k-1}} \sum_{s=0}^{[\frac{k-1}{2}]} \binom{k}{2s+1} (z - x_n)^{k-2s-1} \left((z - x_n)^2 - 4y_n \right)^{s+\frac{1}{2}},$$

where the terms with even j are canceled, and the terms with odd j are combined by taking $j = 2s + 1$. So the lemma is proved.   □

Now, let

$$f_{2m-2}(z) = \frac{1}{2} \sum_{j=1}^{2m} j t_j \sum_{q=0}^{j-1} \binom{j-1}{q} x_n^{j-q-1} \sum_{r=0}^{[q/2]-\mu_q} \binom{q}{r} \frac{y_n^r}{2^{q-2r-1}} f^{(q,r)}(z),$$

(2.65)

where

$$f^{(q,r)}(z) = \sum_{s=0}^{[\frac{q-2r-1}{2}]} \binom{q-2r}{2s+1} (z - x_n)^{q-2r-2s-1} \left((z - x_n)^2 - 4y_n \right)^s. \quad (2.66)$$

Theorem 2.1 *If the x_n, y_n and t_j ($j = 1, \ldots, 2m$) satisfy (2.45), then for any $z \in \mathbb{C}$, there is*

$$D_n^{-1} \hat{A}_n(z) D_n = f_{2m-2}(z) \left(\hat{J}_n(z) - y_n \hat{J}_n^{-1}(z) \right), \quad (2.67)$$

where $\hat{A}_n(z)$ is defined by (2.37), $f_{2m-2}(z)$ is a polynomial of degree $2m - 2$ given by (2.65), and $\hat{J}_n(z)$ is given by (2.36).

Proof By Lemma 2.1 and (2.61) in Lemma 2.2, $D_n^{-1} \hat{A}_n D_n$ is equal to

$$\frac{1}{2} \sum_{j=1}^{2m} j t_j \sum_{q=0}^{j-1} \binom{j-1}{q} x_n^{j-q-1}$$

$$\times \sum_{r=0}^{[q/2]-\mu_q} \binom{q}{r} y_n^r \frac{\alpha_n^{q-2r} - (y_n \alpha_n^{-1})^{q-2r}}{\alpha_n - y_n \alpha_n^{-1}} \left(\hat{J}_n - y_n \hat{J}_n^{-1} \right).$$

Applying (2.54) for $k = q - 2r$ in Lemma 2.3 to the above, we then have the result.   □

The next goal is to study the asymptotics of $(-\det(\hat{A}_n))^{1/2}$ as $z \to \infty$ in the complex plane. The asymptotics comes out based on (2.44) and (2.45) which are reduced from the string equation.

Theorem 2.2 *If the x_n, y_n and t_j ($j = 1, \ldots, 2m$) satisfy (2.44) and (2.45), then as $z \to \infty$ in the complex plane, there is the asymptotics*

$$\sqrt{-\det \hat{A}_n(z)} = \frac{1}{2} V'(z) - \frac{n}{z} + O\left(\frac{1}{z^2}\right), \qquad (2.68)$$

where $V(z) = \sum_{j=0}^{2m} t_j z^j$, $t_{2m} > 0$ and $' = \partial/\partial z$.

Proof By (2.35), (2.36) and (2.37), there is $D_n^{-1} \hat{A}_n(z) D_n \sim m t_{2m} \operatorname{diag}(-z^{2m-1}, z^{2m-1})$ as $z \to \infty$. Since $t_{2m} > 0$, the branch of the square root is determined by $(-\det \hat{A}_n(z))^{1/2} \sim m t_{2m} z^{2m-1}$ as $z \to +\infty$ on the real line.

If we take $k = q - 2r$ in (2.61), then

$$\hat{J}_n^{q-2r} - (y_n \hat{J}_n^{-1})^{q-2r} = \frac{\alpha_n^{q-2r} - (y_n \alpha_n^{-1})^{q-2r}}{\alpha_n - y_n \alpha_n^{-1}} (\hat{J}_n - y_n \hat{J}_n^{-1}).$$

Since $\sqrt{-\det(\hat{J}_n - y_n \hat{J}_n^{-1})} = \alpha_n - y_n \alpha_n^{-1}$, the formula (2.61) implies

$$\sqrt{-\det(\hat{J}_n^{q-2r} - (y_n \hat{J}_n^{-1})^{q-2r})} = \alpha_n^{q-2r} - (y_n \alpha_n^{-1})^{q-2r}.$$

By (2.54), it follows that

$$\sqrt{-\det \hat{A}_n} = \frac{1}{2} \sum_{j=1}^{2m} j t_j \sum_{q=0}^{j-1} \binom{j-1}{q} x_n^{j-q-1}$$

$$\times \sum_{r=0}^{[q/2]-\mu_q} \binom{q}{r} y_n^r (\alpha_n^{q-2r} - (y_n \alpha_n^{-1})^{q-2r}). \qquad (2.69)$$

Here, when $q = 0$, we denote $\sum_{r=0}^{-1} \cdot = 0$ for convenience in the discussions. Let $s = q - r = ([q/2] - \mu_q - r) + [q/2] + 1$ for the terms $(y_n \alpha_n^{-1})^{q-2r}$ above. We arrive

$$\sqrt{-\det \hat{A}_n} = \frac{1}{2} \sum_{j=1}^{2m} j t_j \sum_{q=0}^{j-1} \binom{j-1}{q} x_n^{j-q-1} \left[\sum_{r=0}^{[q/2]-\mu_q} \binom{q}{r} \alpha_n^{q-r} (y_n \alpha_n^{-1})^r \right.$$

$$\left. - \sum_{s=[q/2]+1}^{q} \binom{q}{s} \alpha_n^{q-s} (y_n \alpha_n^{-1})^s \right].$$

Furthermore, by the binomial formula we have

$$\sqrt{-\det \hat{A}_n} = \frac{1}{2} \sum_{j=1}^{2m} j t_j \sum_{q=0}^{j-1} \binom{j-1}{q} x_n^{j-q-1} \left[(\alpha_n + y_n \alpha_n^{-1})^q \right.$$

$$- \mu_q \binom{q}{[q/2]} \alpha_n^{q-[q/2]} (y_n \alpha_n^{-1})^{[q/2]}$$

$$\left. - 2 \sum_{s=[q/2]+1}^{q} \binom{q}{s} y_n^s \alpha_n^{-(2s-q)} \right].$$

For the first part in the bracket, since $\alpha_n + y_n\alpha_n^{-1} = z - x_n$, it is easy to check that

$$\frac{1}{2}\sum_{j=1}^{2m} jt_j \sum_{q=0}^{j-1} \binom{j-1}{q} x_n^{j-q-1}(\alpha_n + y_n\alpha_n^{-1})^q = \frac{1}{2}\sum_{j=1}^{2m} jt_j z^{j-1} = \frac{1}{2}V'(z).$$

The second part in the bracket can be dropped off by considering the outside summations and using (2.45). For $s = [q/2] + 1$ in the third part in the bracket, we have the following by separating the odd q and even q terms and noticing that q starts from $q = 1$,

$$\sum_{j=1}^{2m} jt_j \sum_{q=1}^{j-1} \binom{j-1}{q} x_n^{j-q-1} \binom{q}{[q/2]+1} y_n^{[q/2]+1}\alpha_n^{q-2[q/2]-2}$$

$$= \alpha_n^{-1}\sum_{j=1}^{2m} jt_j \sum_{p=0}^{[\frac{j}{2}]-1} \binom{j-1}{2p+1}\binom{2p+1}{p} x_n^{j-2p-2} y_n^{p+1}$$

$$+ \alpha_n^{-2}\sum_{j=1}^{2m} jt_j \sum_{p=1}^{[\frac{j-1}{2}]} \binom{j-1}{2p}\binom{2p}{p+1} x_n^{j-2p-1} y_n^{p+1},$$

where $q = 2p + 1$ when q is odd, and $q = 2p$ when q is even. As $z \to \infty$, by (2.57) and (2.58), we have

$$\alpha_n^{-1} = \frac{z - x_n - (z - x_n)(1 - \frac{4y_n}{(z-x_n)^2})^{1/2}}{2y_n} = \frac{1}{z - x_n} + O\left(\frac{1}{(z - x_n)^2}\right).$$

Then, combining the discussions above, we obtain

$$\sqrt{-\det(\hat{A}_n)} = \frac{1}{2}\sum_{j=1}^{2m} jt_j z^{j-1} - \frac{n}{z} + O\left(\frac{1}{z^2}\right),$$

by using (2.44), and the theorem is proved.   □

In the following, we show that $(-\det A_n(z))^{1/2}$ has similar asymptotics as discussed for $(-\det \hat{A}_n(z))^{1/2}$ as $z \to \infty$ [6]. Since the restriction conditions for A_n and $\hat{A}_n$ are different in the asymptotics, separate proofs are needed. The Cauchy kernel discussed in [4] is applied in the following proof.

Theorem 2.3 *For A_n defined by (2.31) with $n \geq 2m$, as $z \to \infty$, there is*

$$\sqrt{-\det A_n(z)} = \frac{1}{2}V'(z) - \frac{n}{z} + O\left(\frac{1}{z^2}\right), \qquad (2.70)$$

when the parameters satisfy (2.40).

Proof Denote

$$\hat{p}_n(z) = \int_{-\infty}^{\infty} p_n(\zeta)\frac{e^{-V(\zeta)}}{\zeta - z}d\zeta, \quad \text{and} \quad \Psi_n = \begin{pmatrix} p_n & \hat{p}_n \\ p_{n-1} & \hat{p}_{n-1} \end{pmatrix} e^{-\frac{1}{2}\sigma_3 V(z)}, \quad (2.71)$$

where $\sigma_3 = \text{diag}(1, -1)$. It is not hard to see that $V'(z)$ and $F_n(z)$ are both of degree $2m - 1$ in z. Since $n \geq 2m$, by the orthogonality there is

$$\int_{-\infty}^{\infty} \left[D_n \big(F(\zeta) - F_n(z) \big) D_n^{-1} - \big(V'(\zeta) - V'(z) \big) \right] \begin{pmatrix} p_n(\zeta) \\ p_{n-1}(\zeta) \end{pmatrix} \frac{e^{-V(\zeta)}}{\zeta - z} d\zeta = 0.$$

Then it can be verified that

$$\frac{\partial}{\partial z} \Psi_n = D_n F_n D_n^{-1} \Psi_n - \frac{1}{2} V'(z) \Psi_n, \tag{2.72}$$

that means Ψ_n is a matrix solution for $\frac{\partial}{\partial z} \Psi_n = A_n \Psi_n$ when $n \geq 2m$. The orthogonality of the polynomials also implies

$$\det \Psi_n = \int_{-\infty}^{\infty} \big(p_n(z) p_{n-1}(\zeta) - p_n(\zeta) p_{n-1}(z) \big) \frac{e^{-V(\zeta)}}{\zeta - z} d\zeta = -h_{n-1}.$$

Then there is $\text{tr}(\Psi_n' \Psi_n^{-1}) = (\ln \det \Psi_n)' = 0$ by using the Liouville's formula [2] and $\det \Psi_n = -h_{n-1}$, where $' = \partial/\partial z$. Multiplying Ψ_n^{-1} on both sides of the equation (2.72) and taking trace, we get the following,

$$\text{tr}\, F_n(z) = V'(z), \tag{2.73}$$

that implies $-\det A_n(z) = \frac{1}{4}(V'(z))^2 - \det F_n(z)$.

According to (2.30), there is

$$D_n F_n = \big[C_{n-1} J_{n-1}^{-1} \cdots J_{n-m+1}^{-1} + \cdots$$
$$+ C_{n-2m-1} J_{n-2m+2} \cdots J_{n-m} \big] J_{n-m+1} \cdots J_{n-1} J_n.$$

Considering the leading terms as $z \to \infty$, we have

$$D_n F_n = \big[\det(J_{n-1} \cdots J_{n-m+1})^{-1} z^{m-1} \text{diag}(a_{n,n-1} h_{n-1}, 0) + \cdots$$
$$+ z^{m-1} \text{diag}(0, a_{n-1,n-2m} h_{n-2m}) \big] J_{n-m+1} \cdots J_{n-1} J_n.$$

It can be calculated by using (2.25) that $a_{n-1,n-2m} h_{n-2m} = 2mt_{2m} h_{n-1}$. Since $\det D_n = h_n h_{n-1}$ and $v_n = h_n / h_{n-1}$, there is $\det F_n = 2mt_{2m} a_{n,n-1} z^{2m-2}(1 + O(z^{-1}))$. By (2.40), we have

$$\det F_n(z) = 2mnt_{2m} z^{2m-2} \big(1 + O(z^{-1}) \big). \tag{2.74}$$

Then (2.70) is proved. $\square$

2.4 Density Models

For the density on one interval, denote $z/n^{\frac{1}{2m}}$, $t_j/n^{1-\frac{j}{2m}}$, $x_n/n^{\frac{1}{2m}}$, and $y_n/n^{\frac{1}{m}}$ by η, g_j, a, and b^2 respectively, where $b > 0$. The a and b will be called center and radius parameters respectively in the later discussions. Let $\alpha_n = n^{\frac{1}{2m}} \alpha$, and then $y_n \alpha_n^{-1} = n^{\frac{1}{2m}}(b^2 \alpha^{-1})$, where $\alpha = (\eta - a + \sqrt{(\eta - a)^2 - 4b^2})/2$, and

$b^2\alpha^{-1} = (\eta - a - \sqrt{(\eta - a)^2 - 4b^2})/2$. By Theorem 2.1, it follows that for $z \in \mathbb{C}\backslash[x_n - 2\sqrt{y_n}, x_n + 2\sqrt{y_n}]$,

$$\sqrt{-\det \hat{A}_n(z)} = n^{1-\frac{1}{2m}} k_{2m-2}(\eta)\sqrt{(\eta - a)^2 - 4b^2}, \quad \eta \in \mathbb{C}\backslash[a - 2b, a + 2b], \tag{2.75}$$

where

$$k_{2m-2}(\eta) = \sum_{j=1}^{2m} jg_j \sum_{q=0}^{j-1} \binom{j-1}{q} a^{j-q-1} \sum_{r=0}^{[\frac{q}{2}]-\mu_q} \binom{q}{r} \frac{b^{2r}}{2^{q-2r}} k^{(q,r)}(\eta), \tag{2.76}$$

and

$$k^{(q,r)}(\eta) = \sum_{s=0}^{[\frac{q-2r-1}{2}]} \binom{q-2r}{2s+1} (\eta - a)^{q-2r-2s-1}((\eta - a)^2 - 4b^2)^s, \tag{2.77}$$

where $\binom{m}{n} = m!/(n!(m-n)!)$, $\mu_q = (1 + (-1)^q)/2$, and $[\cdot]$ stands for the integer part.

Define an analytic function [6]

$$\omega(\eta) = k_{2m-2}(\eta)\sqrt{(\eta - a)^2 - 4b^2}, \quad \eta \in \mathbb{C}\backslash[a - 2b, a + 2b]. \tag{2.78}$$

The parameters a, b and $g_j(j = 1, \ldots, 2m)$ are required to satisfy the conditions:

(i) When $\eta \in [\eta_-, \eta_+]$,

$$k_{2m-2}(\eta) \geq 0; \tag{2.79}$$

(ii)
$$\sum_{j=2}^{2m} jg_j \sum_{p=0}^{[\frac{j}{2}]-1} \binom{j-1}{2p+1}\binom{2p+1}{p} a^{j-2p-2}b^{2p+2} = 1; \tag{2.80}$$

(iii)
$$\sum_{j=1}^{2m} jg_j \sum_{p=0}^{[\frac{j-1}{2}]} \binom{j-1}{2p}\binom{2p}{p} a^{j-2p-1}b^{2p} = 0. \tag{2.81}$$

By Theorem 2.2, if a, b, and $g_j(j = 1, \ldots, 2m)$ satisfy (2.80) and (2.81), then for $\eta \in \mathbb{C}\backslash[a - 2b, a + 2b]$ there is

$$\omega(\eta) = \frac{1}{2}\sum_{j=1}^{2m} jg_j \sum_{q=0}^{j-1} \binom{j-1}{q} a^{j-q-1} \sum_{r=0}^{[q/2]-\mu_q} \binom{q}{r} b^{2r}(\alpha^{q-2r} - (b^2\alpha^{-1})^{q-2r}). \tag{2.82}$$

As $\eta \to \infty$, there is

$$\omega(\eta) = \frac{1}{2}W'(\eta) - \frac{1}{\eta} + O\left(\frac{1}{\eta^2}\right). \tag{2.83}$$

In (2.82), the index j actually starts from $j = 2$, and index q starts from 1. We keep this form just for convenience in the later discussion for free energy when we use (2.81) where j is from $j = 1$ and p is from $p = 0$. Let

$$\rho(\eta) = \frac{1}{\pi} k_{2m-2}(\eta) \sqrt{(\eta_+ - \eta)(\eta - \eta_-)}, \quad \eta \in [\eta_-, \eta_+], \tag{2.84}$$

where $\eta_- = a - 2b$, $\eta_+ = a + 2b$, $b > 0$, and $k_{2m-2}(\eta)$ is given by (2.76). By (2.78) and (2.84), there is

$$\omega(\eta)|_{[\eta_-, \eta_+]^\pm} = \pm\pi i \rho(\eta)|_{[\eta_-, \eta_+]}, \tag{2.85}$$

where $[\eta_-, \eta_+]^+$ and $[\eta_-, \eta_+]^-$ stand for the upper and lower edges of the interval $[\eta_-, \eta_+]$ respectively. Since $\rho(\eta)$ is non-negative, we also need

$$k_{2m-2}(\eta) \geq 0, \tag{2.86}$$

for $\eta \in [\eta_-, \eta_+]$.

For the density on multiple disjoint intervals, consider

$$J^{(l)} = \begin{pmatrix} 0 & 1 \\ -b_1^2 & \eta - a_1 \end{pmatrix} \cdots \begin{pmatrix} 0 & 1 \\ -b_l^2 & \eta - a_l \end{pmatrix}, \tag{2.87}$$

where $l \geq 1$. According to the Cayley-Hamilton theorem for $J^{(l)}$, choose $\alpha^{(l)} = (\Lambda + \sqrt{\Lambda^2 - 4b^{(l)2}})/2$, where $\Lambda = \Lambda(\eta) = \operatorname{tr} J^{(l)}$, $b^{(l)2} = \det J^{(l)}$ and $b^{(l)} > 0$. We can transform g_j $(j = 1, \ldots, 2m)$ into a new set of parameters g'_j $(j = 1, \ldots, 2m)$ by a linear transformation so that $W'(\eta) = \sum_{s=0}^{l-1} \eta^s \sum_{q=0}^{m_s} g'_{lq+s+1} \Lambda^q$, where each m_s $(s = 0, \ldots, l - 1)$ is the largest integer such that $s + lm_s \leq 2m - 1$.

Define another analytic function [6]

$$\omega_l(\eta) = \frac{1}{2} \sum_{s=0}^{l-1} \eta^s \sum_{q=1}^{m_s} g'_{lq+s+1} \sum_{r=0}^{[q/2]-\mu_q} \binom{q}{r} b^{(l)2r} \left(\alpha^{(l)q-2r} - (b^{(l)2} \alpha^{(l)-1})^{q-2r} \right), \tag{2.88}$$

for η in the outside of the cuts to be discussed in the following. Then there is $\omega_l(\eta) = \frac{1}{2} W'(\eta) + y(\eta)$, where $-y(\eta)$ is equal to

$$\sum_{s=0}^{l-1} \eta^s \sum_{q=0}^{m_s} g'_{lq+s+1} \left[\frac{\mu_q}{2} \binom{q}{[q/2]} b^{(l)2[q/2]} \alpha^{(l)q-2[q/2]} \right.$$

$$\left. + \sum_{r=[q/2]+1}^{q} \binom{q}{r} b^{(l)2r} \alpha^{(l)q-2r} \right]. \tag{2.89}$$

It is the same argument as discussed for $\omega(\eta)$ that if the parameters satisfy the conditions

$$\sum_{p=0}^{[\frac{m_l-1-1}{2}]} g'_{2lp+2l} \binom{2p+1}{p} b^{(l)2p+2} = 1, \tag{2.90}$$

Fig. 2.1 Multiple cuts and their upper and lower edges

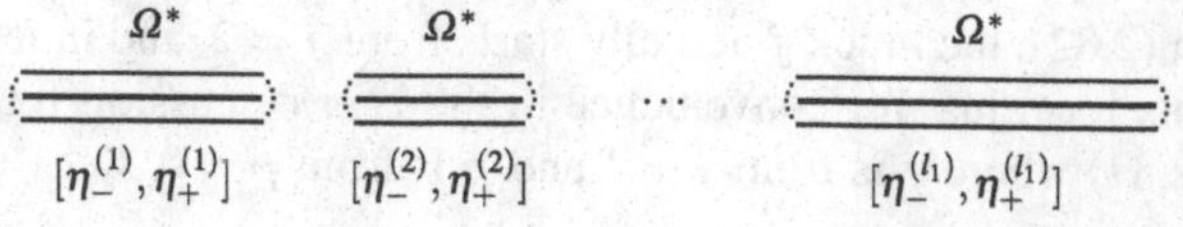

$$\sum_{p=0}^{[\frac{m_s}{2}]} g'_{2lp+s+1} \binom{2p}{p} b^{(l)2p} = 0, \tag{2.91}$$

for $s = 0, 1, \ldots, l-1$, then

$$\omega_l(\eta) = \frac{1}{2} W'(\eta) - \frac{1}{\eta} + O\left(\frac{1}{\eta^2}\right), \tag{2.92}$$

as $\eta \to \infty$.

Now, consider the cuts for $\omega_l(\eta)$, determined by $\alpha^{(l)} - b^{(l)2}\alpha^{(l)-1}$ which is equal to $\sqrt{\Lambda^2 - 4b^{(l)2}}$. Equation $\Lambda^2 - 4b^{(l)2} = 0$ has $2l$ roots, real or complex. If there is a complex root, its complex conjugate is also a root. If there is repeated root, the factor can be moved out from the inside of the square root in the expression of $\omega_l(\eta)$. Therefore, without loss of generality, we consider the equation $\Lambda^2 - 4b^{(l)2} = 0$ has $2l_1$ simple real roots $\eta_-^{(s)}, \eta_+^{(s)}, s = 1, \ldots, l_1$, and $2l_2$ simple complex roots $\eta_s, \bar{\eta}_s$, $s = 1, \ldots, l_2$, where $\bar{\eta}_s$ is the complex conjugate of η_s, $\mathrm{Im}\,\eta_s > 0$, and $l = l_1 + l_2$. Suppose the real roots are so ordered that $[\eta_-^{(s)}, \eta_+^{(s)}], s = 1, \ldots, l_1$, form a set of disjoint intervals, $\Omega = \bigcup_{s=1}^{l_1}[\eta_-^{(s)}, \eta_+^{(s)}]$, see Fig. 2.1. Define

$$\rho_l(\eta) = \mathrm{Re}\,\frac{1}{\pi i}\omega_l(\eta)\Big|_{\Omega+}, \tag{2.93}$$

for $\eta \in \Omega$ as the general eigenvalue density on multiple disjoint intervals in the Hermitian matrix models. It can be seen that when $l = l_1 = 1$, $\omega_1 = \omega$, $\rho_1(\eta) = \rho(\eta)$, and the conditions (2.90) and (2.91) become (2.80) and (2.81) respectively.

Choose l_2 points $\eta_s^{(0)}$ on the real line outside Ω, such that the straight lines Γ_s's, each one connecting η_s and $\eta_s^{(0)}$ for $s = 1, \ldots, l_2$, do not intersect each other. Now, $\omega_l(\eta)$ is well defined and analytic in the outside of $\Omega \cup \bigcup_{s=1}^{l_2}(\Gamma_s \cup \bar{\Gamma}_s)$, where $\bar{\Gamma}_s$ is the straight line connecting $\bar{\eta}_s$ and $\eta_s^{(0)}$. Let Γ_s^* be the closed counterclockwise contour along the edges of $\Gamma_s \cup \bar{\Gamma}_s$, and define

$$I_s = \int_{\Gamma_s^*} \omega_l(\eta)d\eta, \quad \text{and} \quad \hat{I}_s(\eta) = \int_{\Gamma_s^*} \frac{\omega_l(\lambda)}{\lambda - \eta}d\lambda, \quad \eta \in \Omega,$$

for $s = 1, \ldots, l_2$. According to the definition of Γ_s^*, I_s and $\hat{I}_s(\eta)$ are real.

Theorem 2.4 *If the parameters $a_s, b_s (s = 1, \ldots, l)$, and $g_j (j = 1, \ldots, 2m)$ satisfy the conditions (2.90) and (2.91), then $\rho_l(\eta)$ defined by (2.93) on Ω satisfies (2.5) and (2.6).*

Proof Let Γ be a large counterclockwise circle of radius R, and Ω^* be the union of closed counterclockwise contours around the upper and lower edges of all the intervals in Ω. Then by Cauchy theorem and (2.92),

$$\int_{\Omega^*}\left(\omega_l(\eta) - \frac{1}{2}W'(\eta)\right)d\eta + \sum_{s=1}^{l_2} I_s = \int_\Gamma\left(\omega_l(\eta) - \frac{1}{2}W'(\eta)\right)d\eta \to -2\pi i,$$

as $R \to \infty$, which implies $\int_\Omega \rho(\eta)d\eta = 1$ by (2.85), $\int_{\Omega^*} W'(\eta)d\eta = 0$, and I_s are real. So $\rho(\eta)$ satisfies the condition (2.5).

Change the Ω^- and Ω^+ discussed above just at $\eta \in \Omega$ as semicircles of ε radius. By (2.92) and $\int_{\Gamma_s^*} \frac{W'(\lambda)}{\lambda-\eta}d\lambda = 0$, there is

$$\frac{1}{2\pi i}\int_{\Omega^*}\frac{\omega_l(\lambda) - \frac{1}{2}W'(\lambda)}{\lambda - \eta}d\lambda + \frac{1}{2\pi i}\sum_{s=1}^{l_2}\hat{I}_s = \frac{1}{2\pi i}\int_\Gamma\frac{\omega_l(\lambda) - \frac{1}{2}W'(\lambda)}{\lambda - \eta}d\lambda \to 0,$$

as $R \to \infty$. Then taking the real parts of both sides, we get

$$\frac{1}{2}W'(\eta) = \frac{1}{2\pi}\int_{\Omega^*}\frac{\mathrm{Re}\,\frac{1}{i}\omega_l(\lambda)}{\lambda - \eta}d\lambda \to (\mathrm{P})\int_\Omega\frac{\rho(\lambda)}{\eta - \lambda}d\lambda,$$

as $\varepsilon \to 0$ by using (2.93). $\qquad\square$

By the discussions above, it can be seen that when $l_2 = 0$, a_s, b_s $(s = 1, \ldots, l)$, and g_j $(j = 1, \ldots, 2m)$ satisfy the relations (2.90) and (2.91), then $y(\eta) = \omega_l(\eta) - \frac{1}{2}W'(\eta)$ satisfies the following relations: $y(\eta)$ is analytic when $\eta \in \mathbb{C}\backslash\Omega$; $y(\eta)|_{\Omega^+} + y(\eta)|_{\Omega^-} = -W'(\eta)$; $y(\eta) \to 0$, as $\eta \to \infty$. These relations are important in complex analysis, called scalar Riemann-Hilbert problem. If the parameters a_s and b_s can be chosen such that

$$\left(\mathrm{tr}\,J^{(l)}\right)^2 - 4\det J^{(l)} = \prod_{j=1}^l\left(\eta - \eta_-^{(j)}\right)\left(\eta - \eta_+^{(j)}\right), \tag{2.94}$$

then the density models and the corresponding scalar Riemann-Hilbert problems can be well solved. Note that the left hand side of (2.94) is also equal to $-\det(J^{(l)} - (\det J^{(l)})J^{(l)-1})$ by considering $(J^{(l)} - \sqrt{\det J^{(l)}})(J^{(l)} + \sqrt{\det J^{(l)}})$ and calculating the determinants.

By Theorem 2.3, when $n \geq 2m$ and the parameters satisfy (2.40), the $\sigma_n(z)$ defined by

$$\sigma_n(z) = \frac{1}{\pi}\mathrm{Re}\sqrt{\det A_n(z)}, \quad -\infty < z < \infty, \tag{2.95}$$

satisfies $\int_{-\infty}^\infty \sigma_n(z)dz = n$ and $(\mathrm{P})\int_{-\infty}^\infty \frac{\sigma_n(z')}{z-z'}dz' = \frac{1}{2}V'(z)$, as the level density [7], that is consistent with the unified model discussed in Sect. 1.1. When the density involves the parameter n, the string equation and the initial conditions when n is less than $2m$ need to be considered to calculate the functions u_n and v_n.

2.5 Special Densities

When $m = 1$ and $W(\eta) = \eta^2$, there is

$$\rho(\eta) = \frac{1}{\pi}\sqrt{2 - \eta^2},\tag{2.96}$$

for $\eta \in [-\sqrt{2}, \sqrt{2}]$, which is the well known Wigner semicircle.

When $m = 2$ and $W(\eta) = g_2\eta^2 + g_4\eta^4$, by the discussions before we have

$$\rho(\eta) = \frac{1}{\pi}\left(g_2 + 2g_4(\eta^2 + 2b^2)\right)\sqrt{4b^2 - \eta^2},\tag{2.97}$$

for $\eta \in [-2b, 2b]$, with the restriction conditions

$$g_2 + 2g_4(\eta^2 + 2b^2) \geq 0, \quad \eta \in [-2b, 2b],\tag{2.98}$$

$$2g_2b^2 + 12g_4b^4 = 1.\tag{2.99}$$

The results are consistent with the case $W(\eta) = \frac{1}{2}\eta^2 + g\eta^4$ obtained by Brezin, Itzykson, Parisi and Zuber [1] that

$$\rho(\eta) = \frac{1}{\pi}\left(\frac{1}{2} + 4gb^2 + 2g\eta^2\right)\sqrt{4b^2 - \eta^2},\tag{2.100}$$

for $\eta \in [-2b, 2b]$, where

$$b^2 + 12gb^4 = 1.\tag{2.101}$$

When $W(\eta) = g_0 + g_1\eta + g_2\eta^2 + g_3\eta^3 + g_4\eta^4$, by Theorem 2.4, we have the following general density formula

$$\rho(\eta) = \frac{1}{2\pi}\left(2g_2 + 3g_3(\eta + a) + 4g_4(\eta^2 + a\eta + a^2 + 2b^2)\right)\sqrt{4b^2 - (\eta - a)^2},\tag{2.102}$$

where the parameters satisfy the following conditions

$$2g_2 + 3g_3(\eta + a) + 4g_4(\eta^2 + a\eta + a^2 + 2b^2) \geq 0, \quad \eta \in [\eta_-, \eta_+],\tag{2.103}$$

$$2g_2b^2 + 6g_3ab^2 + 12g_4(a^2 + b^2)b^2 = 1,\tag{2.104}$$

$$g_1 + 2g_2a + 3g_3(a^2 + 2b^2) + 4g_4a(a^2 + 6b^2) = 0,\tag{2.105}$$

obtained from (2.4), (2.5) and (2.6), where $\eta_- = a - 2b$, $\eta_+ = a + 2b$. The density formula and the conditions coincide with the results (45) and (46) in [1] for the case $W(\eta) = \frac{1}{2}\eta^2 + g_3\eta^3$.

When $g_1 = g_2 = 0$, i.e., $W(\eta) = g_0 + g_3\eta^3 + g_4\eta^4$, the conditions become

$$3g_3(\eta + a) + 4g_4(\eta^2 + a\eta + a^2 + 2b^2) \geq 0, \quad \eta \in [\eta_-, \eta_+],\tag{2.106}$$

$$g_3 = -\frac{8a(a^2 + 6b^2)}{3b^2(5a^4 + 3(a^2 - 4b^2)^2)},\tag{2.107}$$

$$g_4 = \frac{2(a^2 + 2b^2)}{b^2(5a^4 + 3(a^2 - 4b^2)^2)}.\tag{2.108}$$

The first condition (2.106) is satisfied if and only if $\tau = \frac{4b^2}{a^2}$ is restricted in the interval $0 < \tau \le \tau_-$ or $\tau_+ \le \tau$, where $\tau_+ = 1 + \sqrt{5}$, and τ_- is uniquely determined by the conditions: $0 < \tau_- < 1/2$ and $1 - 2\tau_-^{1/2} + \frac{3}{4}\tau_-^2 = 0$. Approximately we have $\tau_- \approx 0.28$ and $\tau_+ \approx 3.24$. The density function in this case can be further rescaled into the following forms. Let $\eta = ax$ and $\tau = c^2$ ($c > 0$). Then

$$\rho(\eta)d\eta = \frac{16}{\pi} \frac{(\frac{x}{c} - \frac{c}{2})^2 + \frac{x^2-1}{2}}{5 + 3(1 - c^2)^2} \sqrt{c^2 - (x-1)^2}dx, \qquad (2.109)$$

for $x \in [1 - c, 1 + c]$, where $c \in (0, c_-] \cup [c_+, \infty)$, $c_- = \sqrt{\tau_-}$ and $c_+ = \sqrt{\tau_+}$. On the other hand, if $\eta = -ax$ and $\tau = c^2$ ($c > 0$), then

$$\rho(\eta)d\eta = \frac{16}{\pi} \frac{(\frac{x}{c} + \frac{c}{2})^2 + \frac{x^2-1}{2}}{5 + 3(1 - c^2)^2} \sqrt{c^2 - (x+1)^2}dx, \qquad (2.110)$$

for $x \in [-1 - c, -1 + c]$, where $c \in (0, c_-] \cup [c_+, \infty)$, $c_- = \sqrt{\tau_-}$ and $c_+ = \sqrt{\tau_+}$.

The density on two disjoint intervals can also be obtained by using the method discussed before. Briefly, we have

$$\rho(\eta) = \frac{1}{2\pi}(3g_3 + 4g_4(a_1 + a_2 + \eta)) \operatorname{Re}\sqrt{4b_1^2b_2^2 - ((\eta - a_1)(\eta - a_2) - b_1^2 - b_2^2)^2}, \qquad (2.111)$$

where $-\infty < \eta < \infty$, and

$$4g_4b_1^2b_2^2 = 1, \qquad (2.112)$$

$$2g_2 + (3g_3 + 4g_4(a_1 + a_2))(a_1 + a_2) - 4g_4(a_1a_2 - b_1^2 - b_2^2) = 0, \quad (2.113)$$

$$g_1 - (3g_3 + 4g_4(a_1 + a_2))(a_1a_2 - b_1^2 - b_2^2) = 0. \qquad (2.114)$$

It can be checked that if we take $a_1 = a_2 = a$ and $b_1 = b_2 = b$ in the above, then a and b satisfy (2.104) and (2.105). In addition, the non-negative condition for this $\rho(\eta)$ will be discussed in next chapter.

When $m = 3$, by Theorem 2.4, there is

$$\rho(\eta) = \frac{1}{\pi}(g_2 + 2g_4(\eta^2 + 2b^2) + 3g_6(\eta^4 + 2b^2\eta^2 + 6b^4))\sqrt{4b^2 - \eta^2}, \quad (2.115)$$

for $\eta \in [-2b, 2b]$, and (2.103) and (2.104) become

$$g_2 + 2g_4(\eta^2 + 2b^2) + 3g_6(\eta^4 + 2b^2\eta^2 + 6b^4) \ge 0, \quad \eta \in [-2b, 2b], \quad (2.116)$$

$$2g_2b^2 + 12g_4b^4 + 60g_6b^6 = 1. \qquad (2.117)$$

Generally, for the symmetric density with the potential $W(\eta) = \sum_{j=1}^{m} g_j \eta^{2j}$, we have the following by Theorem 2.4

$$\rho(\eta) = \frac{1}{\pi}k_{2m-2}(\eta)\sqrt{4b^2 - \eta^2}, \quad \eta \in [-2b, 2b], \qquad (2.118)$$

where

$$k_{2m-2}(\eta) = \sum_{j=1}^{m} jg_{2j} \sum_{p=1}^{j} \binom{2j-1}{j-p} \frac{b^{2(j-p)}}{4^{p-1}} \sum_{s=0}^{p-1} \binom{2p-1}{2s+1} \eta^{2(p-s-1)}(\eta^2 - 4b^2)^s, \qquad (2.119)$$

and

$$k_{2m-2}(\eta) \geq 0, \quad \eta \in [-2b, 2b], \tag{2.120}$$

$$\sum_{j=1}^{m} 2j g_{2j} \binom{2j-1}{j} b^{2j} = 1. \tag{2.121}$$

Here, the formula (2.119) is obtained from (2.76) and (2.77) by choosing $g_1 = g_3 = \cdots = g_{2m-1} = 0$ and $a = 0$ first, then replacing j by $2j$ and taking the substitutions $q = 2j - 1$ and $r = j - p$. By the asymptotics (2.83), we also have that for large $R > 0$, there is

$$k_{2m-2}(\eta) = \frac{1}{2\pi i} \oint_{|\lambda|=R} \frac{\omega(\lambda)}{\sqrt{\lambda^2 - 4b^2}} \frac{d\lambda}{\lambda - \eta} = \frac{1}{2\pi i} \oint_{|\lambda|=R} \frac{\frac{1}{2}W'(\lambda)}{\sqrt{\lambda^2 - 4b^2}} \frac{d\lambda}{\lambda - \eta}. \tag{2.122}$$

References

1. Brézin, E., Itzykson, C., Parisi, G., Zuber, J.B.: Planar diagrams. Commun. Math. Phys. **59**, 35–51 (1978)
2. Coddington, E.A., Levinson, N.: Theory of Ordinary Differential Equations. McGraw-Hill, New York (1955)
3. Faddeev, L., Takhtajan, L.: Hamiltonian Methods in the Theory of Solitons. Springer, Berlin (1986)
4. Fokas, A.S., Its, A.R., Kitaev, A.V.: Discrete Painlevé equations and their appearance in quantum gravity. Commun. Math. Phys. **142**, 313–344 (1991)
5. Jimbo, M., Miwa, T.: Monodromy preserving deformation of linear ordinary differential equations with rational coefficients. II. Physica D **2**, 407–448 (1981)
6. McLeod, J.B., Wang, C.B.: Eigenvalue density in Hermitian matrix models by the Lax pair method. J. Phys. A, Math. Theor. **42**, 205205 (2009)
7. Mehta, M.L.: Random Matrices, 3rd edn. Academic Press, New York (2004)
8. Szegö, G.: Orthogonal Polynomials, 4th edn. American Mathematical Society Colloquium Publications, vol. 23. AMS, Providence (1975)

Chapter 3
Bifurcation Transitions and Expansions

It is believed in matrix model theory that when the eigenvalue density on one interval is split to a new density on two disjoint intervals, a phase transition occurs. The complexity for the mathematical details of this physical phenomenon comes not only from the elliptic integral calculations, but also from the organization of the parameters in the model. Generally, the elliptic integrals do not have simple analytic formulations for discussing the transition. The string equations can be applied to find the critical point for the transition from the parameter bifurcation, and the bifurcation clearly separates the different phases for analyzing the free energy. Based on the expansion method for elliptic integrals, the third-order bifurcation transition for the Hermitian matrix model with a general quartic potential is discussed in this chapter by applying the nonlinear relations obtained from the string equations. The density on multiple disjoint intervals for higher degree potential and the corresponding free energy are discussed in association with the Seiberg-Witten differential. In the symmetric cases for the quartic potential, the third-order phase transitions are explained with explicit formulations of the free energy function.

3.1 Free Energy for the One-Interval Case

In this section, we discuss the free energy [16]

$$E = \int_{\Omega} W(\eta)\rho(\eta)d\eta - \int_{\Omega} \int_{\Omega} \ln|\lambda - \eta|\rho(\lambda)\rho(\eta)d\lambda d\eta, \qquad (3.1)$$

for the density

$$\rho(\eta) = \frac{1}{\pi} k_{2m-2}(\eta)\sqrt{(\eta_+ - \eta)(\eta - \eta_-)}, \quad \eta \in \Omega = [\eta_-, \eta_+], \qquad (3.2)$$

C.B. Wang, *Application of Integrable Systems to Phase Transitions*,
DOI 10.1007/978-3-642-38565-0_3, © Springer-Verlag Berlin Heidelberg 2013

obtained in Sect. 2.4 with the parameter conditions

$$\sum_{j=2}^{2m} jg_j \sum_{p=0}^{[\frac{j}{2}]-1} \binom{j-1}{2p+1}\binom{2p+1}{p} a^{j-2p-2}b^{2p+2} = 1, \tag{3.3}$$

$$\sum_{j=1}^{2m} jg_j \sum_{p=0}^{[\frac{j-1}{2}]} \binom{j-1}{2p}\binom{2p}{p} a^{j-2p-1}b^{2p} = 0. \tag{3.4}$$

The following discussions are based on the results in [15]. As always, we assume $\rho(\eta)$ is non-negative in the discussions.

Lemma 3.1 *For $\rho(\eta)$ defined by (3.2) on $[\eta_-, \eta_+]$ with the parameters a, b and g_j $(j = 1, \ldots, 2m)$ satisfying the conditions (3.3) and (3.4), there is*

$$\int_{\eta_-}^{\eta_+} \eta^k \rho(\eta)d\eta = \sum_{j=2}^{2m} jg_j \sum_{q=1}^{j-1} \binom{j-1}{q} a^{j-q-1}b^{q+1}$$

$$\times \sum_{r=0}^{[q/2]-\mu_q} \binom{q}{[q/2]+r+1} R_{2r+\mu_q+1,k} \tag{3.5}$$

where

$$R_{l,k} = \frac{i}{\pi}\int_{-\pi}^{\pi} (a+2b\cos\theta)^k e^{-il\theta}\sin\theta d\theta, \tag{3.6}$$

with $l = 2r + \mu_q + 1$ and $\mu_q = (1+(-1)^q)/2$.

Proof Let Ω^* be the closed counterclockwise contour around lower and upper edges of $[\eta_-, \eta_+]$, and Γ be a large counterclockwise circle. Since Ω^* is counterclockwise, by the relation $\omega(\eta)|_{[\eta_-,\eta_+]^\pm} = \pm\pi i\rho(\eta)|_{[\eta_-,\eta_+]}$ and the Cauchy theorem we have

$$\int_{\eta_-}^{\eta_+} \eta^k \rho(\eta)d\eta = -\frac{1}{2\pi i}\int_{\Omega^*} \eta^k \omega(\eta)d\eta = -\frac{1}{2\pi i}\int_{\Gamma} \eta^k \omega(\eta)d\eta.$$

So the problem becomes the calculation of the integral $\int_\Gamma \eta^k \omega(\eta)d\eta$.

By using the binomial formula and $\int_\Gamma \eta^k(\alpha + b^2\alpha^{-1})^q d\eta = \int_\Gamma \eta^k(\eta - a)^q d\eta = 0$, we can obtain

$$\int_{\eta_-}^{\eta_+} \eta^k \rho(\eta)d\eta = \frac{1}{2\pi i}\sum_{j=2}^{2m} jg_j \sum_{q=1}^{j-1} \binom{j-1}{q} a^{j-q-1}$$

$$\times \sum_{s=[q/2]+1}^{q} \binom{q}{s} b^{2s}\int_\Gamma \eta^k \alpha^{-(2s-q)}d\eta. \tag{3.7}$$

Notice that the index q is changed to start from 1, and j is changed to start from 2.

On Ω^*, there is $\eta = a + 2b\cos\theta$, $-\pi \leq \theta \leq \pi$, where $a = (\eta_+ + \eta_-)/2$ and $2b = (\eta_+ - \eta_-)/2 > 0$. Then $\alpha^{-1} = b^{-1}e^{-i\theta}$, where the square root takes positive and negative imaginary value on upper and lower edge of $[\eta_-, \eta_+]$ respectively. By Cauchy theorem, the integral along Γ can be changed to along Ω^*, that implies

$$\int_\Gamma \eta^k \alpha^{-(2s-q)}d\eta = -2b^{q-2s+1}\int_{-\pi}^{\pi}(a+2b\cos\theta)^k e^{-i(2s-q)\theta}\sin\theta d\theta.$$

Let $r = s - [q/2] - 1$ in (3.7). Because the range of s is from $[q/2]+1$ to q, and $q = 2[q/2] - \mu_q + 1$, the range of r is from 0 to $[q/2] - \mu_q$. Since $2s - q = 2r + \mu_q + 1$, this lemma is proved. □

Lemma 3.2 *For $\rho(\eta)$ defined by (3.2) on $[\eta_-, \eta_+]$ with the parameters a, b and g_j $(j = 1, \ldots, 2m)$ satisfying the conditions (3.3) and (3.4), there is*

$$\int_{\eta_-}^{\eta_+}\ln|\eta - a|\rho(\eta)d\eta$$

$$= \ln(2b) - \sum_{j=2}^{2m}jg_j\sum_{q=1}^{j-1}\binom{j-1}{q}a^{j-q-1}b^{q+1}$$

$$\times \sum_{r=0}^{[q/2]-\mu_q}\binom{q}{[q/2]+r+1}\Theta_{2r+\mu_q+1} \tag{3.8}$$

where

$$\Theta_l = \mathrm{Re}\,\frac{i}{\pi}\int_0^\pi \theta e^{i\theta}[(e^{i\theta} + \sqrt{e^{2i\theta}-1})^l - (e^{i\theta} - \sqrt{e^{2i\theta}-1})^l]d\theta, \tag{3.9}$$

with $l = 2r + \mu_q + 1$ and $\mu_q = (1 + (-1)^q)/2$.

Proof Let $\gamma = \gamma_1 \cup \gamma_2 \cup \gamma_3$ be a closed counterclockwise contour, where γ_1 is the upper edges of $[\eta_-, a]$, γ_2 is the upper edges of $[a, \eta_+]$, and γ_3 is the semi-circle of radius $2b$ with center a. Applying Cauchy theorem for $\ln(\eta - a)\omega(\eta)$, we have

$$\int_{\gamma_1}(\ln|\eta - a| + \pi i)\omega(\eta)d\eta + \int_{\gamma_2}\ln|\eta - a|\omega(\eta)d\eta + \int_{\gamma_3}\ln(2be^{i\theta})\omega(\eta)d\eta = 0.$$

When $\eta \in \gamma_1 \cup \gamma_2$, $\omega(\eta) = \pi i\rho(\eta)$. Then taking imaginary part for the above equation, we get

$$\int_{\eta_-}^{\eta_+}\ln|\eta - a|\rho(\eta)d\eta - \ln(2b) + \frac{1}{\pi}\mathrm{Re}\int_{\gamma_3}\theta\omega(\eta)d\eta = 0, \tag{3.10}$$

where we have used $\int_{\gamma_3}\omega(\eta)d\eta = -\int_{\gamma_1\cup\gamma_2}\omega(\eta)d\eta = -\pi i\int_{\gamma_1\cup\gamma_2}\rho(\eta)d\eta = -\pi i$. So the problem becomes the calculation of the integral $\int_{\gamma_3}\theta\omega(\eta)d\eta$.

Rewrite the formula of $\omega(\eta)$ given in Sect. 2.4 for the one-interval case as

$$\omega(\eta) = \frac{1}{2}\sum_{j=2}^{2m} jg_j \sum_{q=1}^{j-1}\binom{j-1}{q}a^{j-q-1}\sum_{s=0}^{[q/2]-\mu_q}\binom{q}{s}b^{2s}\left(\alpha^{q-2s}-(b^2\alpha^{-1})^{q-2s}\right).$$

Let $r = [q/2] - \mu_q - s$. The range of r is from 0 to $[q/2] - \mu_q$. Since $q = 2[q/2] - \mu_q + 1$, and $q - 2s = 2([q/2] - \mu_q - s) + \mu_q + 1$, we have the following,

$$\omega(\eta) = \frac{1}{2}\sum_{j=2}^{2m} jg_j \sum_{q=1}^{j-1}\binom{j-1}{q}a^{j-q-1}$$

$$\times \sum_{r=0}^{[q/2]-\mu_q}\binom{q}{[q/2]+r+1}b^{2([q/2]-\mu_q-r)}\left(\alpha^{2r+\mu_q+1}-(b^2\alpha^{-1})^{2r+\mu_q+1}\right).$$

On γ_3, we have $\eta - a = 2be^{i\theta}$, which implies $\alpha = b(e^{i\theta} + \sqrt{e^{2i\theta}-1})$, $b^2\alpha^{-1} = b(e^{i\theta} - \sqrt{e^{2i\theta}-1})$. It follows that

$$\int_{\gamma_3}\theta\left(\alpha^{2r+\mu_q+1}-(b^2\alpha^{-1})^{2r+\mu_q+1}\right)d\eta$$

$$= 2ib^{2r+\mu_q+2}\int_0^\pi \theta e^{i\theta}\left[\left(e^{i\theta}+\sqrt{e^{2i\theta}-1}\right)^{2r+\mu_q+1}\right.$$

$$\left. - \left(e^{i\theta}-\sqrt{e^{2i\theta}-1}\right)^{2r+\mu_q+1}\right]d\theta.$$

We finally have

$$\frac{1}{\pi}\,\mathrm{Re}\int_{\gamma_3}\theta\omega(\eta)d\eta$$

$$= \sum_{j=2}^{2m} jg_j \sum_{q=1}^{j-1}\binom{j-1}{q}a^{j-q-1}b^{q+1}\sum_{r=0}^{[q/2]-\mu_q}\binom{q}{[q/2]+r+1}\Theta_{2r+\mu_q+1}.$$

Then by (3.10), the lemma is proved.   □

The Θ_l in the above lemma can be further simplified by the recursion for some elementary integrals as described in the following.

Lemma 3.3 *For $k = 0, 1, 2, \ldots$, there are*

$$\int_0^\pi \theta e^{2i\theta}\left(1-e^{2i\theta}\right)^{k+\frac{1}{2}}d\theta = \frac{\pi}{(2k+3)i},\tag{3.11}$$

$$\int_0^\pi \theta e^{i\theta}\left(1-e^{2i\theta}\right)^{k+\frac{1}{2}}d\theta = -2\int_0^1\int_0^1\left(1-x^2y^2\right)^{k+\frac{1}{2}}dxdy + \frac{\pi i}{2}B\left(\frac{1}{2},k+\frac{3}{2}\right),\tag{3.12}$$

where $B(\cdot,\cdot)$ is the Euler beta function.

Proof The first equation in this lemma can be easily verified by using integration by parts,

$$\int_0^\pi \theta e^{2i\theta}\left(1-e^{2i\theta}\right)^{k+\frac{1}{2}}d\theta = \frac{1}{(2k+3)i}\int_0^\pi\left(1-e^{2i\theta}\right)^{k+\frac{3}{2}}d\theta = \frac{\pi}{(2k+3)i}.$$

To prove the second equation, consider the initial value problems for $J(\gamma)$ and $I(\gamma)$ defined by

$$J(\gamma) = \int_0^\pi e^{i\theta}\left(1-\gamma e^{2i\theta}\right)^{k+\frac{1}{2}}d\theta, \qquad I(\gamma) = \int_0^\pi \theta e^{i\theta}\left(1-\gamma e^{2i\theta}\right)^{k+\frac{1}{2}}d\theta,$$

for $0 \le \gamma \le 1$. It can be calculated that $(\gamma^{\frac{1}{2}}J(\gamma))' = i\gamma^{-\frac{1}{2}}(1-\gamma)^{k+\frac{1}{2}}$, where $' = d/d\gamma$. Then $\gamma^{\frac{1}{2}}J(\gamma) = i\int_0^\gamma t^{-\frac{1}{2}}(1-t)^{k+\frac{1}{2}}dt$, which implies

$$J(\gamma) = 2i\int_0^1\left(1-\gamma x^2\right)^{k+\frac{1}{2}}dx, \tag{3.13}$$

by taking $t = \gamma x^2$.

It can be calculated that $(\gamma^{\frac{1}{2}}I(\gamma))' = \frac{\pi i}{2}\gamma^{-\frac{1}{2}}(1-\gamma)^{k+\frac{1}{2}} - \frac{1}{2i}\gamma^{-\frac{1}{2}}J(\gamma)$. Then by (3.13) and taking integral from 0 to 1, we have

$$I(1) = \frac{\pi i}{2}\int_0^1\gamma^{-\frac{1}{2}}(1-\gamma)^{k+\frac{1}{2}}d\gamma - \int_0^1\gamma^{-\frac{1}{2}}\left(\int_0^1\left(1-\gamma x^2\right)^{k+\frac{1}{2}}dx\right)d\gamma,$$

which gives the second equation in this lemma by taking $\gamma = y^2$. $\qquad\square$

To further calculate the real part of the right hand side of (3.12), consider the following line and double integrals for $k = 0, 1, 2, \ldots$,

$$l_k = \int_0^1\left(1-x^2\right)^{k+\frac{1}{2}}dx, \qquad d_k = \int_0^1\int_0^1\left(1-x^2y^2\right)^{k+\frac{1}{2}}dxdy.$$

First, $l_0 = \frac{\pi}{4}$, and

$$d_0 = \frac{1}{2}\int_0^1\left(\sqrt{1-y^2}+\frac{1}{y}\sin^{-1}y\right)dy = \frac{\pi}{8}+\frac{\pi}{4}\ln 2. \tag{3.14}$$

When $k \ge 1$, by integration by parts, we can verify that l_k satisfy a recursive relation $l_k = l_{k-1} - \frac{1}{2k+1}l_k$, which gives $l_k = \frac{(2k+1)!!}{(2k+2)!!}\frac{\pi}{2}$. Also by integration by parts, we have $d_k = d_{k-1} + \frac{1}{2k+1}(l_k - d_k)$, which implies

$$d_k = \frac{2k+1}{2k+2}d_{k-1} + \frac{(2k+1)!!}{(2k+2)!!}\frac{\pi}{4(k+1)}. \tag{3.15}$$

Specially

$$d_1 = \frac{9\pi}{64} + \frac{3\pi}{16} \ln 2,$$ (3.16)

which will be used in the non-symmetric density discussed later. By combining the above discussions, we have the following result for the free energy.

Theorem 3.1 *For $\rho(\eta)$ defined by (3.2) on $[\eta_-, \eta_+]$ with the parameters a, b and g_j ($j = 1, \ldots, 2m$) satisfying the conditions (3.3) and (3.4), there is the following formula for the free energy:*

$$E = \frac{1}{2} W(a) - \ln(2b) + \sum_{j=2}^{2m} j g_j \sum_{q=1}^{j-1} \binom{j-1}{q} a^{j-1-q} b^{q+1}$$

$$\times \sum_{r=0}^{[q/2]-\mu_q} \binom{q}{[q/2] + r + 1} Y_{2r+\mu_q+1},$$ (3.17)

where

$$Y_l = \frac{1}{2} \sum_{k=0}^{2m} g_k R_{l,k} + \Theta_l,$$ (3.18)

with $l = 2r + \mu_q + 1$ and $\mu_q = (1 + (-1)^q)/2$.

Proof By taking the integral on the variational equation (P) $\int_\Omega \frac{\rho(\lambda)}{\eta - \lambda} d\lambda = \frac{1}{2} W'(\eta)$ from a to η for the variable η, we have $\int_{\eta_-}^{\eta_+} \ln|\lambda - \eta| \rho(\lambda) d\lambda = \frac{1}{2} W(\eta) - \frac{1}{2} W(a) + \int_{\eta_-}^{\eta_+} \ln|\lambda - a| \rho(\lambda) d\lambda$. Multiplying $\rho(\eta)$ and taking $\int_{\eta_-}^{\eta_+} d\eta$ on both sides of this equation, we get the following by using $\int_{\eta_-}^{\eta_+} \rho(\eta) d\eta = 1$,

$$\int_{\eta_-}^{\eta_+} \int_{\eta_-}^{\eta_+} \ln|\lambda - \eta| \rho(\lambda) \rho(\eta) d\lambda d\eta$$

$$= \frac{1}{2} \int_{\eta_-}^{\eta_+} W(\eta) \rho(\eta) d\eta - \frac{1}{2} W(a) + \int_{\eta_-}^{\eta_+} \ln|\lambda - a| \rho(\lambda) d\lambda.$$

According to the definition of the free energy, there is

$$E = \frac{1}{2} W(a) + \frac{1}{2} \sum_{k=0}^{2m} g_k \int_{\eta_-}^{\eta_+} \eta^k \rho(\eta) d\eta - \int_{\eta_-}^{\eta_+} \ln|\eta - a| \rho(\eta) d\eta.$$

By Lemma 3.1 and Lemma 3.2, the integrals above can be expressed in terms of $R_{l,k}$ and Θ_l. After simplifications, the result is proved.   □

For the potential $W(\eta) = g_0 + g_1\eta + g_2\eta^2 + g_3\eta^3 + g_4\eta^4$, based on the above results and the restriction conditions

$$2g_2 + 6g_3a + 12g_4(a^2 + b^2) - b^{-2} = 0, \tag{3.19}$$

$$g_1 + 2g_2a + 3g_3(a^2 + 2b^2) + 4g_4a(a^2 + 6b^2) = 0. \tag{3.20}$$

simplified from (3.3) and (3.4), there are

$$Y_1 = \frac{1}{2}\left(W(a) + \frac{1}{2} - 4g_4b^4\right) + \frac{1}{2} + \ln 2,$$

$$Y_2 = -2(g_3 + 4g_4a)b^3,$$

$$Y_3 = \frac{1}{2}\left(\frac{1}{2} - 3g_4b^4\right) - \frac{3}{4}.$$

The integrals in the free energy function are discussed in Appendix A. Therefore, by (3.17) and the parameter conditions (3.19) and (3.20), the free energy function in this case becomes

$$E = \frac{1}{2}W(a) - \ln(2b) + Y_1 + 3(g_3 + 4g_4a)b^3Y_2 + 4g_4b^4Y_3,$$

which can be further simplified to

$$E = W(a) + \frac{3}{4} - \ln b - 4g_4b^4 - 6(g_3 + 4g_4a)^2b^6 - 6g_4^2b^8. \tag{3.21}$$

If we choose g_2 as a variable, g_0, g_1, g_3 and g_4 as constants, and a and b as functions of g_2, then we have

$$\frac{\partial^2}{\partial g_2^2}E = -2b^2(2a^2 + b^2), \tag{3.22}$$

where the derivatives of a and b have fractional formulas, but there are some common factors in the calculations that can be canceled, and finally we get the above result. It will be seen that this formula is consistent with the formula of $\partial^2/\partial t_2^2 \ln Z_n$ to be discussed in next section. The domain of the above free energy function is determined by the condition $k_2(\eta) \geq 0$ when $\eta \in [\eta_-, \eta_+]$, which is generally not trivial. Let us consider a simple case in the following.

When $W(\eta) = g_0 + g_3\eta^3 + g_4\eta^4$, the free energy function is

$$E = g_0 + \frac{3}{8} - \ln b - \frac{8}{3\tau\tau_1} - \frac{15\tau + 32}{3\tau_1} - \frac{140\tau - 40}{3\tau_1^2}, \tag{3.23}$$

where

$$\tau = \frac{4b^2}{a^2}, \quad \tau_1 = 5 + 3(1 - \tau)^2.$$

As discussed in Sect. 2.5, the parameter τ is restricted in the intervals $(0, \tau_-]$ and $[\tau_+, \infty)$, where $\tau_- \approx 0.28$ and $\tau_+ \approx 3.24$.

When $W(\eta) = g_2\eta^2 + g_4\eta^4$, the free energy becomes

$$E = g_0 + \frac{3}{4} - \ln b + \frac{1}{24}(2g_2b^2 - 1)(9 - 2g_2b^2), \qquad (3.24)$$

which agrees with the result

$$E(g) = E(0) + \frac{1}{24}(b^2 - 1)(9 - b^2) - \frac{1}{2}\ln b^2. \qquad (3.25)$$

obtained in [3] for $W(\eta) = \frac{1}{2}\eta^2 + g\eta^4$. If we think E as a function of $2g_2b^2$, it can be seen that E has an extreme minimum point at $2g_2b^2 = 2$, or at $g_4 = g_4^c$, where $g_4^c = -\frac{g_2^2}{12}$, which is always not positive. For the non-symmetric density discussed above, g_4 is positive at such point.

If $W(\eta) = g_3\eta^3 + g_4\eta^4$ is degenerated to $W(\eta) = g_4\eta^4$ by taking $a \to 0$, the free energy becomes $E = 3/8 - \ln b$. It is the same result as $W(\eta) = g_2\eta^2 + g_4\eta^4$ is degenerated to $W(\eta) = g_4\eta^4$. So the results obtained above are consistent. In this chapter, we talk about the third-order phase transitions for the potential

$$W(\eta) = g_1\eta + g_2\eta^2 + g_3\eta^3 + g_4\eta^4, \qquad (3.26)$$

caused by the bifurcation of the parameter(s) in the density models, or when the density on one-interval is split to a two-interval case, that will be discussed in the following sections.

3.2 Partition Function and Toda Lattice

We have seen in last section that the free energy in the one-interval case can be explicitly derived. For the multiple-interval cases, the free energy is expressed in terms of elliptic integrals that are generally hard to be simplified. However, the derivatives of $\ln Z_n$, which is how the free energy is defined, have some general properties that are useful to discuss the multiple-interval cases. In this section, we are going to discuss the derivatives of $\ln Z_n$ by using the equations obtained from the orthogonal polynomials for the potential

$$V(z) = \sum_{j=0}^{4} t_j z^j, \quad t_4 > 0. \qquad (3.27)$$

Let us first consider the derivatives $\partial u_k/\partial t_1$ for $k = 0, 1, \ldots, n-1$, where $u_0h_0 = \int_{-\infty}^{\infty} e^{-V(z)}dz$, and $\partial v_k/\partial t_1$ for $k = 1, 2, \ldots, n-1$. Since $h_0 = \int_{-\infty}^{\infty} e^{-V(z)}dz$, there is $\frac{\partial h_0}{\partial t_1} = -\int_{-\infty}^{\infty} z e^{-V(z)} dz = -u_0 h_0$, which implies $\frac{\partial}{\partial t_1} \ln h_0 = -u_0$. When $k \geq 1$, $h_k = \int_{-\infty}^{\infty} p_k^2 e^{-V(z)} dz$, and

$$\frac{\partial h_k}{\partial t_1} = -\int_{-\infty}^{\infty} z p_k^2 e^{-V(z)} dz = -u_k h_k.$$

Therefore, we have

$$\frac{\partial}{\partial t_1} \ln h_k = -u_k, \tag{3.28}$$

for $k \geq 0$. Since $p_0 = 1$, $z p_0 = p_1 + u_0 p_0$, $p_1 = z - u_0$, and $p_{1,t_1} = v_1 p_0$ obtained from the orthogonality $\int_{-\infty}^{\infty} p_{1,t_1} p_0 e^{-V(z)} dz = \int_{-\infty}^{\infty} p_{1,t_1} p_0 V_{t_1}(z) e^{-V(z)} dz = v_1 h_0$, we have

$$\frac{\partial u_0}{\partial t_1} = -v_1. \tag{3.29}$$

When $k \geq 1$, taking $\partial/\partial t_1$ on both sides of the recursion formula $z p_k = p_{k+1} + u_k p_k + v_k p_{k-1}$ and applying the relation $p_{k,t_1} = v_k p_{k-1}$, there is

$$\frac{\partial u_k}{\partial t_1} = v_k - v_{k+1}. \tag{3.30}$$

In addition, since $v_k = h_k/h_{k-1}$ for $k \geq 1$, there is

$$\frac{\partial v_k}{\partial t_1} = v_k \left(\frac{\partial \ln h_k}{\partial t_1} - \frac{\partial \ln h_{k-1}}{\partial t_1} \right) = v_k (u_{k-1} - u_k). \tag{3.31}$$

The differential equations above and in the following for the u_k and v_k are generally called Toda lattice in the Hermitian matrix models. In the t_1 direction, these equations can be changed to the original Toda lattice for a chain of particles with nearest neighbor interaction by using the Flaschka's variables. Since the partition function

$$Z_n = \int_{-\infty}^{\infty} \cdots \int_{-\infty}^{\infty} e^{-\Sigma_{i=1}^n V(z_i)} \prod_{j<k} (z_j - z_k)^2 dz_1 \cdots dz_n$$

can be expressed as

$$Z_n = n! h_0 h_1 \cdots h_{n-1}, \tag{3.32}$$

we then get

$$\frac{\partial}{\partial t_1} \ln Z_n = -(u_0 + \cdots + u_{n-1}), \tag{3.33}$$

and

$$\frac{\partial^2}{\partial t_1^2} \ln Z_n = v_n. \tag{3.34}$$

Next, let us consider the derivatives with respect to t_2. According to (3.32), we first need $\partial h_k/\partial t_2$ for $k = 0, 1, \ldots, n-1$. Since $h_0 = -\int_{-\infty}^{\infty} e^{-V(z)} dz$, there is

$$\frac{\partial h_0}{\partial t_2} = -\int z^2 e^{-V(z)} dz = -\int (p_1 - u_0)^2 e^{-V(z)} dz = -(u_0^2 + v_1) h_0,$$

which implies

$$\frac{\partial}{\partial t_2} \ln h_0 = -(u_0^2 + v_1). \tag{3.35}$$

When $k \geq 1$, there is

$$\frac{\partial}{\partial t_2} \ln h_k = -(u_k^2 + v_k + v_{k+1}), \tag{3.36}$$

by applying the recursion formula twice to $\partial h_k/\partial t_2 = -\int z^2 p_k^2 e^{-V(z)} dz$. Hence, we get

$$\frac{\partial}{\partial t_2} \ln Z_n = -(u_0^2 + v_1) - \sum_{k=1}^{n-1} (u_k^2 + v_k + v_{k+1}). \tag{3.37}$$

To get the second-order derivative of $\ln Z_n$, we need to consider the derivatives of u_k and v_k. Since $u_0 h_0 = \int z e^{-V(z)} dz$, we can get $\partial u_0/\partial t_2 = -(u_0 + u_1) v_1$, and similarly

$$\frac{\partial v_1}{\partial t_2} = v_1 \left(\frac{\partial \ln h_1}{\partial t_2} - \frac{\partial \ln h_0}{\partial t_2} \right) = -v_1 (u_1^2 - u_0^2 + v_2).$$

Then, there is

$$\frac{\partial}{\partial t_2} (u_0^2 + v_1) = -v_1 ((u_0 + u_1)^2 + v_2). \tag{3.38}$$

When $k \geq 1$, by $u_k h_k = \int z p_k^2 e^{-V(z)} dz$ we can get

$$\frac{\partial u_k}{\partial t_2} = -(u_{k+1} + u_k) v_{k+1} + (u_k + u_{k-1}) v_k. \tag{3.39}$$

Similar to $\partial v_1/\partial t_2$, there is

$$\frac{\partial v_k}{\partial t_2} = v_k \left(\frac{\partial \ln h_k}{\partial t_2} - \frac{\partial \ln h_{k-1}}{\partial t_2} \right) = -v_k (u_k^2 - u_{k-1}^2 + v_{k+1} - v_{k-1}). \tag{3.40}$$

When all the u_k's vanish in the even potential case, this equation becomes the equation discussed by Fokas, Its and Kitaev in 1991. The following result is true for $k \geq 1$,

$$\frac{\partial}{\partial t_2} (u_k^2 + v_k + v_{k+1}) = -v_{k+2} v_{k+1} + v_k v_{k-1} - v_{k+1} (u_{k+1} + u_k)^2 + v_k (u_k + u_{k-1})^2. \tag{3.41}$$

By taking derivative with respect to t_2 on both sides of (3.37), and applying the formulas (3.38) and (3.41) above, we get the following identity after canceling the like terms,

$$\frac{\partial^2}{\partial t_2^2} \ln Z_n = v_n \big((u_n + u_{n-1})^2 + v_{n-1} + v_{n+1}\big). \tag{3.42}$$

We can also study the second-order derivatives in the t_3 or t_4 direction. It can be found that the second-order derivatives have simple formulations in terms of u_n's and v_n's. As we have experienced and to be discussed later that the u_n and v_n formulation for the second-order derivatives of $\ln Z_n$ always leads to the continuity of the second-order derivatives of the free energy. If the second-order derivative of the free energy has a singularity, then the center or radius parameter reduced from u_n or v_n also has singularity, and vice versa. These observations partially explain the cause of the third-order transitions we will discuss next by using the string equation and Toda lattice.

3.3 Merged and Split Densities

In this section, we discuss the properties of the densities on one or two intervals to get ready for the analysis for the phase transition problems. In some literatures, the densities are classified as weak or strong densities. For a convenience of the geometrical imagination, the densities discussed here are just called merged or split density depending the parameters in the density is in the merged or split state. Also, we will explain that it is generally hard to have a explicit or simple formula for the free energy function or the derivative the free energy to analyze the discontinuity at the critical point since for the split density the free energy is generally expressed in terms of the elliptic integrals, even in some special cases, like the symmetric case to be discussed in Sect. 3.5, the free energy has a simple formula as the one-interval case given in Sect. 3.1. Therefore, the ε-expansion method discussed in Sect. 3.4 would be generally applicable.

Let us consider the potential $W(\eta) = \sum_{j=1}^{4} g_j \eta^j$ and the densities discussed in Chap. 1 on $\Omega_1 = [\eta_-, \eta_+]$ and $\Omega_2 = [\eta_-^{(1)}, \eta_+^{(1)}] \cup [\eta_-^{(2)}, \eta_+^{(2)}]$ on the real line. For the purpose in this section, let us write the density on Ω_1 in the following form,

$$\rho_1(\eta) = \frac{1}{2\pi} \big(2g_2 + 3g_3(\eta + a) + 4g_4(\eta^2 + a\eta + a^2 + 2b^2)\big)\sqrt{4b^2 - (\eta - a)^2}, \tag{3.43}$$

where $\eta \in \Omega_1 = [a - 2b, a + 2b]$ and the parameters are restricted to satisfy the following relations

$$2g_2 + 6g_3 w_1 + 12g_4(w_1^2 + w_2) - w_2^{-1} = 0, \tag{3.44}$$

$$g_1 + 2g_2 w_1 + 3g_3(w_1^2 + 2w_2) + 4g_4 w_1(w_1^2 + 6w_2) = 0, \tag{3.45}$$

and

$$w_1 = a, \qquad w_2 = b^2. \tag{3.46}$$

For the density on Ω_2, let us first write it in the following form,

$$\rho_2(\eta) = \frac{1}{2\pi}\left(3g_3 + 4g_4(\eta + a_1 + a_2)\right)$$

$$\times \operatorname{Re}\sqrt{e^{-\pi i}\left[((\eta - a_1)(\eta - a_2) - b_1^2 - b_2^2)^2 - 4b_1^2 b_2^2\right]}, \tag{3.47}$$

subject to the following conditions

$$4g_4 w_2^2 - 1 = 0, \tag{3.48}$$

$$2g_2 + (3g_3 + 4g_4 w_1)w_1 - 4g_4 w_3 = 0, \tag{3.49}$$

$$g_1 - (3g_3 + 4g_4 w_1)w_3 = 0, \tag{3.50}$$

where

$$w_1 = (a_1 + a_2)/2, \qquad w_2 = b_1 b_2, \qquad w_3 = a_1 a_2 - b_1^2 - b_2^2. \tag{3.51}$$

The w_j parameters are introduced in an attempt to have a unified density formula for the merged and split phases. The parameters of the polynomial in the square root in (3.47) can be written in terms of the w_j's according to the following factorization,

$$((\eta - a_1)(\eta - a_2) - b_1^2 - b_2^2)^2 - 4b_1^2 b_2^2$$

$$= \left[\left(\eta - \frac{a_1 + a_2}{2}\right)^2 - \frac{1}{4}(a_1 - a_2)^2 - (b_1 - b_2)^2\right]$$

$$\times \left[\left(\eta - \frac{a_1 + a_2}{2}\right)^2 - \frac{1}{4}(a_1 - a_2)^2 - (b_1 + b_2)^2\right]$$

$$= \left[(\eta - w_1)^2 - w_1^2 + 2w_2 + w_3\right]\left[(\eta - w_1)^2 - w_1^2 - 2w_2 + w_3\right]$$

$$= \left(\eta - \eta_-^{(1)}\right)\left(\eta - \eta_+^{(1)}\right)\left(\eta - \eta_-^{(2)}\right)\left(\eta - \eta_+^{(2)}\right), \quad \eta_-^{(1)} < \eta_+^{(1)} < \eta_-^{(2)} < \eta_+^{(2)}. \tag{3.52}$$

The real part of the square root in (3.47) takes a negative sign when $\eta \in [\eta_-^{(1)}, \eta_+^{(1)}]$, and it takes a positive sign when $\eta \in [\eta_-^{(2)}, \eta_+^{(2)}]$. In fact, if we denote $Q(\eta) = (\eta - \eta_-^{(1)})(\eta - \eta_+^{(1)})(\eta - \eta_-^{(2)})(\eta - \eta_+^{(2)})$, then

$$\sqrt{e^{-\pi i}Q(\eta)} = \begin{cases} |\sqrt{Q(\eta)}|e^{\pi i}, & \eta_-^{(1)} < \eta < \eta_+^{(1)}, \\ |\sqrt{Q(\eta)}|, & \eta_-^{(2)} < \eta < \eta_+^{(2)}, \end{cases} \tag{3.53}$$

obtained according to the following argument values that if $\eta \in (\eta_-^{(1)}, \eta_+^{(1)})$, then $\arg(\eta - \eta_-^{(1)}) = 0$, and $\arg(\eta - \eta') = \pi$ for $\eta' = \eta_+^{(1)}, \eta_-^{(2)}, \eta_+^{(2)}$; and if $\eta \in (\eta_-^{(2)}, \eta_+^{(2)})$, there are $\arg(\eta - \eta') = 0$ for $\eta' = \eta_-^{(1)}, \eta_+^{(1)}, \eta_-^{(2)}$, and $\arg(\eta - \eta_+^{(2)}) = \pi$. Therefore, we also need $3g_3 + 4g_4(\eta + a_1 + a_2) \le 0$ when $\eta \in [\eta_-^{(1)}, \eta_+^{(1)}]$, and $3g_3 + 4g_4(\eta + a_1 + a_2) \ge 0$ when $\eta \in [\eta_-^{(2)}, \eta_+^{(2)}]$ such that ρ_2 keeps non-negative on Ω_2.

When $a_1 = a_2 = a = a_c$ and $b_1 = b_2 = b = b_c$ for the fixed constants a_c and $b_c > 0$, both of the density functions above are degenerated to

$$\rho_c(\eta) = \frac{1}{\pi} 2g_4^c(\eta - a_c)^2 \sqrt{4b_c^2 - (\eta - a_c)^2}, \quad \eta \in [a_c - 2b_c, a_c + 2b_c], \quad (3.54)$$

at the critical point, and the parameters should take the values satisfying the following relations

$$4g_4^c b_c^4 = 1, \tag{3.55}$$

$$g_3^c + 4g_4^c a_c = 0, \tag{3.56}$$

$$g_2^c + 3g_3^c a_c + 2g_4^c(3a_c^2 + 2b_c^2) = 0, \tag{3.57}$$

$$g_1^c - (3g_3^c + 8g_4^c a_c)(a_c^2 - 2b_c^2) = 0, \tag{3.58}$$

or equivalently

$$g_1^c = a_c(2b_c^2 - a_c^2)b_c^{-4}, \qquad 2g_2^c = (3a_c^2 - 2b_c^2)b_c^{-4},$$
$$3g_3^c = -3a_c b_c^{-4}, \qquad 4g_4^c = b_c^{-4}. \tag{3.59}$$

Here, the relation (3.56) is obtained based on the non-negative requirement for ρ_1 and ρ_2 as discussed above.

One can image that it is hard to get the explicit formulas of the free energy function for the density ρ_2. If we consider the difference of the derivatives of the free energy for ρ_1 and ρ_2 at the critical point, the computations may be easier since the densities have a unified structure that many common terms could be canceled to simplify the calculations. In the following, let us discuss whether this strategy works.

Recall that the density ρ_1 on $\Omega_1 = [\eta_-, \eta_+] = [a - 2b, a + 2b]$ can be written as $\rho_1 = \frac{1}{\pi}\sqrt{\det A^{(1)}}$ as shown in Chap. 1, where

$$A^{(1)} = c_1 J^{(1)} + c_2(J^{(1)})^2 + c_3(J^{(1)})^3 - \frac{1}{2}W'(\eta)I, \tag{3.60}$$

with $c_1 = 2g_2 + 6g_3 a + 12g_4(a^2 + b^2)$, $c_2 = 3g_3 + 12g_4 a$, $c_3 = 4g_4$, I is the identity matrix, and

$$J^{(1)} = \begin{pmatrix} 0 & 1 \\ -b^2 & \eta - a \end{pmatrix}. \tag{3.61}$$

The density ρ_2 can be changed to $\rho_2 = \frac{1}{\pi}\sqrt{\det A^{(2)}}$, where

$$A^{(2)} = (3g_3 + 4g_4(\eta + a_1 + a_2))J^{(2)} - \frac{1}{2}W'(\eta)I, \tag{3.62}$$

with

$$J^{(2)} = \begin{pmatrix} 0 & 1 \\ -b_1^2 & \eta - a_1 \end{pmatrix} \begin{pmatrix} 0 & 1 \\ -b_2^2 & \eta - a_2 \end{pmatrix}. \tag{3.63}$$

Further, we have

$$A^{(1)} = B^{(1)}(J^{(1)})^2 - \frac{1}{2}W'(\eta)I, \quad \text{and} \quad A^{(2)} = B^{(2)}J^{(2)} - \frac{1}{2}W'(\eta)I, \tag{3.64}$$

where

$$B^{(1)} = (3g_3 + 4g_4(\eta + 2a))I + (b^{-4} - 4g_4)\begin{pmatrix} \eta - a & -1 \\ b^2 & 0 \end{pmatrix},$$

and

$$B^{(2)} = (3g_3 + 4g_4(\eta + a_1 + a_2))I.$$

If we define

$$Q(\eta) = \frac{1}{4}(W')^2 - w_2^2(3g_3 + 4g_4(\eta + 2w_1))^2$$

$$- (3g_3 + 4g_4(\eta + 2w_1))(1 - 4g_4w_2^2)(\eta - w_1) - w_2^{-1}(1 - 4g_4w_2^2)^2, \tag{3.65}$$

for $\eta \in \mathbb{C}$, then

$$-\det A^{(1)} = Q|_{\rho_1} \quad \text{and} \quad -\det A^{(2)} = Q|_{\rho_2}, \tag{3.66}$$

since $1 - 4g_4w_2^2 = 0$ for ρ_2, where $Q|_{\rho_j}$ means the parameters in Q are restricted by the parameter conditions for ρ_j, $j = 1, 2$.

Let us consider the continuity or discontinuity of the derivatives of the free energy function as the density is changed from ρ_1 to ρ_2. First, there is

$$\left.\frac{\partial\sqrt{Q}}{\partial g}\right|_{\rho_j} = \left.\frac{\langle Q_{\mathbf{w}}, \mathbf{w}'\rangle + \langle Q_{\mathbf{g}}, \mathbf{g}'\rangle}{2\sqrt{Q}}\right|_{\rho_j}, \quad j = 1, 2,$$

where $\mathbf{w} = (w_1, w_2, w_3)$, $\mathbf{g} = (g_1, g_2, g_3, g_4)$, $Q_{\mathbf{w}}$ and $Q_{\mathbf{g}}$ are the gradients, $' = \partial/\partial g$ and g is one of the g_j's. It follows that

$$\left.\frac{\partial\sqrt{Q}}{\partial g}\right|_{\rho_1, g = g^c} - \left.\frac{\partial\sqrt{Q}}{\partial g}\right|_{\rho_2, g = g^c} = \left.\frac{\langle Q_{\mathbf{w}}, \mathbf{w}'\rangle}{2\sqrt{Q}}\right|_{\rho_1, g = g^c} - \left.\frac{\langle Q_{\mathbf{w}}, \mathbf{w}'\rangle}{2\sqrt{Q}}\right|_{\rho_2, g = g^c},$$

that implies the first-order derivative of the free energy is continuous at $g = g^c$ by the direct calculations from the free energy formula. For the second-order derivatives, we can get

$$\left.\frac{\partial^2}{\partial g^2}E(\rho_1)\right|_{g = g^c} - \left.\frac{\partial^2}{\partial g^2}E(\rho_2)\right|_{g = g^c} = \frac{-1}{2\pi i}\oint \frac{\partial W(\eta)}{\partial g}\frac{\langle Q_{\mathbf{w}}^c, \mathbf{w}^{(1)} - \mathbf{w}^{(2)}\rangle}{2\sqrt{Q^c}}d\eta,$$

where $Q^c = Q|_{g=g^c}$, and $\mathbf{w}^{(1)}$ and $\mathbf{w}^{(2)}$ are the corresponding vectors following the discussion above. It can be obtained that the second-order derivative is also continuous at the critical point.

However, for the higher order derivatives this method will lead to complicated calculations. So the analysis for the phase transition will not be easy if one hopes to use the explicit formulas in such a way to get the discontinuity. Similar complexities have been experienced in other researches, for example, see [1, 12, 20]. In the next section, we discuss a different method to simplify the calculations for analyzing the nonlinear properties in the transitions.

3.4   Third-Order Phase Transition by the ε-Expansion

After we have experienced the different strategies or properties in the previous sections, let us talk about the ε-expansion method. We will see that the ε-expansions for the parameters based on the algebraic equations reduced from the string equation can quickly give the phase transition properties including the critical phenomena to be discussed in the later chapters.

Consider the Hermitian matrix model with the potential $W(\eta) = \sum_{j=1}^{4} g_j \eta^j$. The phase transition problem is to discuss whether a derivative of the free energy function

$$E = \int_{\Omega} W(\eta)\rho(\eta)d\eta - \int_{\Omega}\int_{\Omega} \ln|\lambda - \eta|\rho(\lambda)\rho(\eta)d\lambda d\eta \qquad (3.67)$$

becomes discontinuous as the density ρ is changed from one to another, for example, from ρ_1 to ρ_2. In the following, we will discuss that the discontinuity is mainly caused by the change of the parameter conditions.

Rewrite (3.47) as

$$\rho_2(\eta) = \frac{1}{2\pi}(3g_3 + 4g_4(\eta + 2u)) \operatorname{Re}\sqrt{e^{-\pi i}[(\eta - u)^2 - x_1^2][(\eta - u)^2 - x_2^2]},$$
$$\qquad (3.68)$$

for $\eta \in \Omega_2 = [u - x_2, u - x_1] \cup [u + x_1, u + x_2]$, where

$$x_1^2 = u^2 - w - 2v = \frac{1}{4}(a_1 - a_2)^2 + (b_1 - b_2)^2, \qquad (3.69)$$

$$x_2^2 = u^2 - w + 2v = \frac{1}{4}(a_1 - a_2)^2 + (b_1 + b_2)^2 \qquad (3.70)$$

and

$$u = (a_1 + a_2)/2, \qquad v = b_1 b_2, \qquad w = a_1 a_2 - b_1^2 - b_2^2. \qquad (3.71)$$

Denote

$$\omega_2(\eta) = \frac{1}{2}(3g_3 + 4g_4(\eta + 2u))\sqrt{[(\eta - u)^2 - x_1^2][(\eta - u)^2 - x_2^2]}, \qquad (3.72)$$

for η in the complex plane outside Ω_2.

Fig. 3.1 Contours for the second formula in (3.73)

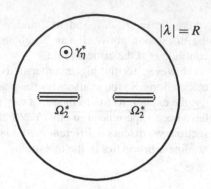

For the models considered in this book, we can discuss the first-order derivative as follows. Consider the following formulas for the density ρ_2,

$$
\begin{cases}
(\mathrm{P}) \int_{\Omega_2} \frac{\rho_2(\lambda)}{\eta-\lambda} d\lambda = \frac{1}{2} W'(\eta), & \eta \in (\eta_-^{(1)}, \eta_+^{(1)}) \cup (\eta_-^{(2)}, \eta_+^{(2)}), \\
\int_{\Omega_2} \frac{\rho_2(\lambda)}{\eta-\lambda} d\lambda = \frac{1}{2} W'(\eta) - \omega_2(\eta), & \eta \in (\eta_+^{(1)}, \eta_-^{(2)}).
\end{cases}
\tag{3.73}
$$

The first formula above is the variational equation for the eigenvalue density. The second formula is, in fact, true for any η outside the cuts. For the discussion of the free energy, we specially pay attention to $\eta \in (\eta_+^{(1)}, \eta_-^{(2)})$. To prove the second formula, consider the counterclockwise contours Ω_2^* around the cuts Ω_2 and the counterclockwise circle γ_η^* around a point η in the complex plane, see Fig. 3.1.

By Cauchy theorem, there is

$$
\frac{1}{2\pi i} \int_{\Omega_2^* \cup \gamma_\eta^*} \frac{\omega_2(\lambda)}{\lambda-\eta} d\lambda = \frac{1}{2\pi i} \int_{|\lambda|=R} \frac{\omega_2(\lambda)}{\lambda-\eta} d\lambda,
$$

where R is a large number, and the right hand side is equal to $\frac{1}{2} W'(\eta)$ as we have discussed before. For the left hand side, there is

$$
\frac{1}{2\pi i} \int_{\Omega_2^*} \frac{\omega_2(\lambda)}{\lambda-\eta} d\lambda + \frac{1}{2\pi i} \int_{\gamma_\eta^*} \frac{\omega_2(\lambda)}{\lambda-\eta} d\lambda = \int_{\Omega_2} \frac{\rho_2(\lambda)}{\eta-\lambda} d\lambda + \omega_2(\eta),
$$

which shows the second formula above. Note that the point η can be large in the complex plane. The formula $\omega_2(\eta) = \frac{1}{2} W'(\eta) + \int_{\Omega_2} \frac{\rho_2(\lambda)}{\lambda-\eta} d\lambda$ can be connected to the asymptotics $\omega_2(\eta) = \frac{1}{2} W'(\eta) - \eta^{-1} + O(\eta^{-2})$ as $\eta \to \infty$. In fact, when $\eta \to \infty$, $\int_{\Omega_2} \frac{\rho_2(\lambda)}{\lambda-\eta} d\lambda$ collects all the rest terms in the expansion of ω_2 after the leading term $\frac{1}{2} W'(\eta)$.

Now, by the free energy formula (3.67) there is

$$
\frac{\partial}{\partial g} E(\rho_2) = \int_{\Omega_2} \frac{\partial W(\eta)}{\partial g} \rho_2(\eta) d\eta
$$

$$
- 2 \left(\int_{\eta_-^{(1)}}^{\eta_+^{(1)}} + \int_{\eta_-^{(2)}}^{\eta_+^{(2)}} \right) \left(\int_{\Omega_2} \ln|\eta - \lambda| \rho_2(\lambda) d\lambda - \frac{1}{2} W(\eta) \right) \frac{\partial \rho_2(\eta)}{\partial g} d\eta.
\tag{3.74}
$$

where g represents any one of the g_j's. We can get that for $\eta \in (\eta_-^{(1)}, \eta_+^{(1)})$, there is

$$\int_{\Omega_2} \ln|\eta - \lambda|\rho_2(\lambda)d\lambda - \frac{1}{2}W(\eta) = \int_{\Omega_2} \ln|\eta_-^{(1)} - \lambda|\rho_2(\lambda)d\lambda - \frac{1}{2}W(\eta_-^{(1)}), \quad (3.75)$$

and for $\eta \in (\eta_-^{(2)}, \eta_+^{(2)})$, there is

$$\int_{\Omega_2} \ln|\eta - \lambda|\rho_2(\lambda)d\lambda - \frac{1}{2}W(\eta)$$

$$= \int_{\Omega_2} \ln|\eta_-^{(1)} - \lambda|\rho_2(\lambda)d\lambda - \frac{1}{2}W(\eta_-^{(1)}) - \int_{\eta_+^{(1)}}^{\eta_-^{(2)}} \omega_2(\eta)d\eta. \quad (3.76)$$

Consequently, by using $\int_{\Omega_2} \frac{\partial}{\partial g}\rho_2(\eta)d\eta = \frac{\partial}{\partial g}\int_{\Omega_2} \rho_2(\eta)d\eta = 0$ there is the following result since ρ_2 is equal to 0 at the end points of Ω_2.

Theorem 3.2 *For the free energy function* (3.67), *we have*

$$\frac{\partial}{\partial g}E(\rho) = \begin{cases} \int_{\Omega_1} \frac{\partial W(\eta)}{\partial g}\rho_1(\eta)d\eta, & \rho = \rho_1, \\ \int_{\Omega_2} \frac{\partial W(\eta)}{\partial g}\rho_2(\eta)d\eta + 2\int_{\eta_+^{(1)}}^{\eta_-^{(2)}} \omega_2(\eta)d\eta \frac{\partial}{\partial g}\int_{\eta_-^{(2)}}^{\eta_+^{(2)}} \rho_2(\eta)d\eta, & \rho = \rho_2, \end{cases}$$
$$(3.77)$$

where g is one of the g_j's.

However, the higher order derivatives of the free energy will involve the elliptic integral calculations. It is experienced [1, 9, 12, 20] that the elliptic integral calculations are complicated. We have also discussed in last section that the algebraic method does not help too much to get a simple result for the free energy on the entire domain of the parameters. For the transition problems, it would be easier to just work on the behaviors around the critical point. The algebraic equations for the parameters obtained from the string equation can give the behaviors on both sides of the critical point by the ε-expansions. Also, when the degree of the potential is higher, the elliptic integrals will become more complicated, but the expansion method still works for the models of higher degree potentials to get at least the behaviors at the critical point(s) for analyzing the continuity or discontinuity.

Let us first consider the parameters in ρ_2. Applying the notations $u = (a_1 + a_2)/2$, $v = b_1 b_2$ and $w = a_1 a_2 - b_1^2 - b_2^2$ into (3.48), (3.49) and (3.50). we see that (3.49) and (3.50) become

$$2g_2 + 6g_3 u + 12g_4(4u^2 - w) = 0, \quad (3.78)$$

$$g_1 - (3g_3 + 8g_4 u)w = 0. \quad (3.79)$$

Since the above equations do not involve v, let us expand u and w in terms of $\varepsilon_1 = g_2 - g_2^c$,

$$u = u_c(1 + \alpha_1\varepsilon_1 + \alpha_2\varepsilon_1^2 + \alpha_3\varepsilon_1^3 + \cdots), \quad (3.80)$$

$$w = w_c\left(1 + \gamma_1\varepsilon_1 + \gamma_2\varepsilon_1^2 + \gamma_3\varepsilon_1^3 + \cdots\right), \tag{3.81}$$

where $u_c = a_c$, $w_c = a_c^2 - 2b_c^2$, $g_j = g_j^c$ for $j = 1, 3, 4$, and

$$g_1^c = a_c\left(2b_c^2 - a_c^2\right)b_c^{-4}, \qquad g_2^c = \frac{1}{2}\left(3a_c^2 - 2b_c^2\right)b_c^{-4},$$

$$g_3^c = -a_cb_c^{-4}, \qquad g_4^c = \frac{1}{4}b_c^{-4}, \tag{3.82}$$

for fixed constants a_c and $b_c > 0$. The ε is denoted as ε_1 when it is negative, and we will explain next why it is negative in this case. The coefficients α_j and γ_j can be determined by using the equations above,

$$\alpha_1 = -\frac{b_c^2}{2}, \qquad \gamma_1 = -b_c^2, \qquad \alpha_2 = -\frac{b_c^4}{2}, \qquad \gamma_2 = 0. \tag{3.83}$$

Note that $u^2 - w - 2v$, denoted as x_1^2 in the later discussion, is never negative because it is equal to $(a_1 - a_2)^2/4 + (b_1 - b_2)^2$. The expansions must be consistent with this property. The expansion results obtained above imply $u^2 - w - 2v = -2b_c^4\varepsilon_1 + O(\varepsilon_1^2)$ that further imply $\varepsilon_1 < 0$ and consequently $g_2 < g_2^c$.

If we consider the g_1 direction by choosing $g_1 = g_1^c + \varepsilon_1$, then we get $\alpha_1 = -\frac{b_c^2}{4a_c}$, $\gamma_1 = -\frac{a_cb_c^2}{2(a_c^2 - 2b_c^2)}$, $\alpha_2 = 0$ and $\gamma_2 = \frac{b_c^3}{4(a_c^2 - 2b_c^2)}$. In this case, it can be verified that $u^2 - w - 2v = -\frac{3}{16}b_c^4\varepsilon_1^2 + O(\varepsilon_1^3)$ which implies $u^2 - w - 2v$ can be negative when ε_1 is small. This is a contradiction since $u^2 - w - 2v \geq 0$. So there is no transition in the g_1 direction in this case. In the following, we keep working on the g_2 direction and consider the derivatives of the free energy in this direction, and g_2 is related to the mass quantity in physics [9].

Now, let us consider the asymptotic expansion of the important term

$$I_0 \equiv 2\int_{\eta_+^{(1)}}^{\eta_-^{(2)}} \omega_2(\eta)d\eta\frac{\partial}{\partial g}\int_{\eta_-^{(2)}}^{\eta_+^{(2)}} \rho_2(\eta)d\eta, \tag{3.84}$$

in the formula of the first-order derivative of free energy (3.77). Basically, as g_2 approaches to the critical value g_2^c, a_1 and a_2 approach to a_c, and b_1 and b_2 approach to b_c, that imply $\int_{\eta_+^{(1)}}^{\eta_-^{(2)}} \omega_2(\eta)d\eta$ is small, and $\frac{\partial}{\partial g}\int_{\eta_-^{(2)}}^{\eta_+^{(2)}} \rho_2(\eta)d\eta$ is $O(1)$. To get the higher order terms in the expansions, we use the contour integral technique in [15] (Sect. 6). In Sect. B.1, there is the following result,

$$\int_{x_1}^{x_2} \sqrt{\left(x_2^2 - x^2\right)\left(x^2 - x_1^2\right)}dx = \frac{x_2^3}{3} + O\left(x_1^2 \ln x_1\right), \tag{3.85}$$

as $x_1 \to 0$ with $0 < x_1 < x_2$, based on which we can show the following lemma.

Lemma 3.4 *For ρ_2 defined by (3.47), there is*

$$\int_{\eta_-^{(2)}}^{\eta_+^{(2)}} \rho_2(\eta)d\eta = \frac{1}{2} - \frac{2}{\pi}a_c b_c \varepsilon_1 + O(\varepsilon_1^2 \ln|\varepsilon_1|), \qquad (3.86)$$

as $\varepsilon_1 \to 0$, where $g_2 = g_2^c + \varepsilon_1$ and $\varepsilon_1 < 0$.

Proof First, we have

$$\int_{\eta_-^{(2)}}^{\eta_+^{(2)}} \rho_2(\eta)d\eta = \frac{1}{2\pi}\int_{\eta_-^{(2)}}^{\eta_+^{(2)}} \left(3g_3 + 4g_4(\eta + a_1 + a_2)\right)\sqrt{(2v - \Lambda)(\Lambda + 2v)}d\eta,$$

where $v = b_1 b_2$ and $\Lambda = (\eta - a_1)(\eta - a_2) - b_1^2 - b_2^2$. Then, $\int_{\eta_-^{(2)}}^{\eta_+^{(2)}} \rho_2(\eta)d\eta$ is equal to

$$\frac{g_4}{\pi}\int_{x_1^2}^{x_2^2} \sqrt{(x_2^2 - \zeta)(\zeta - x_1^2)}d\zeta + \frac{1}{2\pi}(3g_3 + 12g_4 u)\int_{x_1}^{x_2} \sqrt{(x_2^2 - x^2)(x^2 - x_1^2)}dx,$$

where $\zeta = (\eta - u)^2$, $x = \eta - u$ and $u = (a_1 + a_2)/2$ given above. We then obtain the following since $\int_{x_1^2}^{x_2^2} \sqrt{(x_2^2 - \zeta)(\zeta - x_1^2)}d\zeta = 2\pi v^2$,

$$\int_{\eta_-^{(2)}}^{\eta_+^{(2)}} \rho_2(\eta)d\eta = 2g_4 v^2 + \frac{1}{2\pi}(3g_3 + 12g_4 u)\int_{x_1}^{x_2} \sqrt{(x_2^2 - x^2)(x^2 - x_1^2)}dx.$$

By the formula (3.85), we further have

$$\int_{\eta_-^{(2)}}^{\eta_+^{(2)}} \rho_2(\eta)d\eta = 2g_4 b_c^4 + \frac{1}{2\pi}(3g_3 + 12g_4 u)\frac{x_2^3}{3} + O\left((3g_3 + 12g_4 u)x_1^2 \ln x_1\right).$$

Since $4g_4 b_c^4 = 1$, $3g_3 + 12g_4 u = 12g_4 a_c \alpha_1 \varepsilon_1 + O(\varepsilon_1^2)$, $x_1^2 = O(\varepsilon_1)$ and $x_2 = 2b_c + O(\varepsilon_1)$, the above expansion becomes

$$\int_{\eta_-^{(2)}}^{\eta_+^{(2)}} \rho_2(\eta)d\eta = \frac{1}{2} - \frac{2}{\pi}a_c b_c \varepsilon_1 + O(\varepsilon_1^2 \ln|\varepsilon_1|).$$

Then the lemma is proved. $\qquad\square$

Lemma 3.5

$$\int_{\eta_+^{(1)}}^{\eta_-^{(2)}} \omega_2(\eta)d\eta = \frac{3\pi}{2}a_c b_c^3 \varepsilon_1^2 + O(\varepsilon_1^3), \qquad (3.87)$$

as $\varepsilon_1 \to 0$, where $\varepsilon_1 = g_2 - g_2^c < 0$.

Proof The formula

$$\int_{\eta_+^{(1)}}^{\eta_-^{(2)}} \omega_2(\eta)d\eta = \frac{1}{2}\int_{\eta_+^{(1)}}^{\eta_-^{(2)}} (3g_3 + 4g_4(\eta + a_1 + a_2))\sqrt{(\Lambda - 2v)(\Lambda + 2v)}d\eta,$$

can be changed to

$$\int_{\eta_+^{(1)}}^{\eta_-^{(2)}} \omega_2(\eta)d\eta = 2g_4 \int_{u-x_1}^{u+x_1} (\eta - u)\sqrt{(x_2^2 - (\eta - u)^2)(x_1^2 - (\eta - u)^2)}d\eta$$

$$+ \frac{1}{2}(3g_3 + 12g_4 u) \int_{-x_1}^{x_1} \sqrt{(x_2^2 - x^2)(x^2 - x_1^2)}dx,$$

where $x_1^2 = u^2 - w - 2v$, $x_2^2 = u^2 - w + 2v$, and $x = \eta - u$. It is easy to see that $\int_{-x_1}^{x_1} t\sqrt{(x_2^2 - t^2)(x_1^2 - t^2)}dt = 0$ by the symmetry, that implies the first integral on the right hand side is equal to 0. Note that the integral here is from $-x_1$ to x_1, that is different from the integral in the normalization of the density. For the second integral, we first have the expansion $3g_3 + 12g_4 u = 12g_4 a_c \alpha_2 \varepsilon_1 + O(\varepsilon_1^2) = -\frac{3}{2}a_c b_c^{-2}\varepsilon_1 + O(\varepsilon_1^2)$. Then it follows that

$$\int_{\eta_+^{(1)}}^{\eta_-^{(2)}} \omega_2(\eta)d\eta = -\frac{3}{4}a_c b_c^{-2}\varepsilon_1 \int_{-x_1}^{x_1} \sqrt{(x_2^2 - x^2)(x^2 - x_1^2)}dx + O(\varepsilon_1^3).$$

Since $x_2 = 2b_c + O(\varepsilon_1)$ and $x_1^2 = -2b_c^4\varepsilon_1 + O(\varepsilon_1^2)$, where $\varepsilon_1 < 0$, we have

$$\int_{-x_1}^{x_1} \sqrt{(x_2^2 - x^2)(x^2 - x_1^2)}dx = \frac{\pi}{2}x_2 x_1^2 + O(\varepsilon_1^2) = -2b_c^5\varepsilon_1 + O(\varepsilon_1^2).$$

Therefore, we finally get

$$\int_{\eta_+^{(1)}}^{\eta_-^{(2)}} \omega_2(\eta)d\eta = \frac{3\pi}{2}a_c b_c^3 \varepsilon_1^2 + O(\varepsilon_1^3),$$

and the lemma is proved.   □

By the lemmas above and (3.84), we have the following.

Lemma 3.6

$$I_0 = -6a_c^2 b_c^4 \varepsilon_1^2 + O(\varepsilon_1^3 \ln|\varepsilon_1|), \tag{3.88}$$

as $\varepsilon_1 \to 0$, where $\varepsilon_1 = g_2 - g_2^c < 0$.

The remaining work to find the expansion for

$$\frac{\partial E(\rho_2)}{\partial g_2} = \int_{\Omega_2} \eta^2 \rho_2 d\eta + I_0$$

is the ε-expansion for $\int_{\Omega_2} \eta^2 \rho_2 d\eta$. By the asymptotic expansion of $\omega_2 = \frac{1}{2}(3g_3 + 4g_4(\eta + 2u))(\Lambda^2 - 4v^2)^{1/2}$ as $\eta \to \infty$, we can get that the coefficient of the η^{-3} is $-[6g_3u + 4g_4(8u^2 - w)]v^2$, which implies

$$\int_{\Omega_2} \eta^2 \rho_2 d\eta = \frac{-1}{2\pi i} \int_{|\eta|=R} \eta^2 \omega_2 d\eta = [6g_3u + 4g_4(8u^2 - w)]v^2.$$

Since we discuss the changes in the g_2 direction by keeping other g_j's to their critical values and $4g_4v^2 = 1$, we have that v is constant when $g_2 < g_2^c$ for the ρ_2 phase. Then, the ε_1 expansions for u and w obtained above imply

$$\int_{\Omega_2} \eta^2 \rho_2 d\eta = a_c^2 + 2b_c^2 - (4a_c^2 + 2b_c^2)b_c^2 \varepsilon_1 - 3a_c^2 b_c^4 \varepsilon_1^2 + O(\varepsilon_1^3 \ln |\varepsilon_1|). \quad (3.89)$$

Combining the discussions above, we finally get the following result for the free energy for the ρ_2 as g_2 approaches to g_2^c from the left.

Theorem 3.3 *For the ρ_2 defined by (3.47), there is*

$$\frac{\partial E(\rho_2)}{\partial g_2} = a_c^2 + 2b_c^2 - (4a_c^2 + 2b_c^2)b_c^2 \varepsilon_1 - 9a_c^2 b_c^4 \varepsilon_1^2 + O(\varepsilon_1^3 \ln |\varepsilon_1|), \quad (3.90)$$

as $\varepsilon_1 \to 0$, where $\varepsilon_1 = g_2 - g_2^c < 0$.

Now, let us consider the free energy for the density ρ_1. By the asymptotic expansion of $\omega_1(\eta)$ defined by

$$\omega_1(\eta) = \frac{1}{2}(2g_2 + 3g_3(\eta + a) + 4g_4(\eta^2 + a\eta + a^2 + 2b^2))((\eta - a)^2 - 4b^2)^{1/2},$$
$$(3.91)$$

as $\eta \to \infty$ for $\eta \notin \Omega_1 = [a - 2b, a + 2b]$, we can obtain that

$$\frac{\partial E(\rho_1)}{\partial g_2} = \int_{\Omega_1} \eta^2 \rho_1 d\eta = \frac{-1}{2\pi i} \int_{|\eta|=R} \eta^2 \omega_1 d\eta$$
$$= u^2 + v + 6g_3 u v^2 + 4g_4(6u^2 v^2 + v^3),$$

where $u = a$ and $v = b^2$. According to (3.19) and (3.20), the parameters satisfy

$$2g_2 + 6g_3 u + 12g_4(u^2 + v) - v^{-1} = 0,$$
$$g_1 + 2g_2 u + 3g_3(u^2 + 2v) + 4g_4 u(u^2 + 6v) = 0.$$

To get the concrete value of the derivative at the critical point, consider the expansions

$$u = a_c(1 + \alpha_1 \varepsilon + \alpha_2 \varepsilon^2 + \cdots), \quad (3.92)$$
$$v = b_c^2(1 + \beta_1 \varepsilon + \beta_2 \varepsilon^2 + \cdots), \quad (3.93)$$

where $\varepsilon = g_2 - g_2^c$, and other g_j's take their critical values. Substituting the ε expansions above to the restriction equations above, we can derive the following results,

$$\alpha_1 = -\frac{b_c^2}{2}, \qquad \beta_1 = -\frac{b_c^2}{2}, \qquad \alpha_2 = -\frac{b_c^4}{8}, \qquad \beta_2 = \frac{b_c^2}{16}(b_c^2 - 3a_c^2). \qquad (3.94)$$

Then, the following result holds.

Theorem 3.4 *For the ρ_1 defined by (3.43), there is*

$$\frac{\partial E(\rho_1)}{\partial g_2} = a_c^2 + 2b_c^2 - (4a_c^2 + 2b_c^2)b_c^2\varepsilon + (3a_c^2 + b_c^2)b_c^4\varepsilon^2 + O(\varepsilon^3), \qquad (3.95)$$

as $\varepsilon \to 0$, where $\varepsilon = g_2 - g_2^c > 0$.

By the results obtained above, we have that $\partial^j E(\rho_1)/\partial g_2^j$ and $\partial^j E(\rho_2)/\partial g_2^j$ are the same at the critical point g_2^c for each $j = 1$ or 2, but

$$\frac{\partial^3 E(\rho_2)}{\partial g_2^3}\bigg|_{g_2 \to g_2^c - 0} = -18a_c^2 b_c^4 < 2(3a_c^2 + b_c^2)b_c^4 = \frac{\partial^3 E(\rho_1)}{\partial g_2^3}\bigg|_{g_2 \to g_2^c + 0}, \qquad (3.96)$$

since $b_c > 0$ in any case even a_c can be 0. It can be checked that the discontinuity mainly comes from the different coefficients (3.83) and (3.94) in the expansions. Because of the bifurcation of the center and radius parameters, a becomes a_1 and a_2, and b becomes b_1 and b_2, the combination or organization of the coefficients in the expansions is changed, and the transition is then called bifurcation transition, which is a third-order phase transition in the g_2 direction. The I_0 above enhances the discontinuity when $a_c \neq 0$. In the symmetric case to be discussed in next section, there will be $I_0 = 0$, but the discontinuity still exists. The importance of I_0 appears in the Seiberg-Witten theory when the density is extended to the general density on multiple disjoint intervals as explained in the following.

The first-order derivatives of the free energy discussed above is related to the derivatives of the prepotential studied in the Seiberg-Witten theory [19], which is proportional to the logarithm of the partition function [4]. Let us consider the discussions in Sect. 2.4 in association with the multi-cut large-N limit of matrix models [4, 7]. For the eigenvalue density on multiple disjoint intervals $\Omega = \bigcup_{j=1}^l \Omega_j \equiv \bigcup_{j=1}^l [\eta_-^{(j)}, \eta_+^{(j)}]$, we have

$$\begin{cases} (P) \int_\Omega \frac{\rho(\lambda)}{\eta - \lambda} d\lambda = \frac{1}{2}W'(\eta), & \eta \in \Omega, \\ \int_\Omega \frac{\rho(\lambda)}{\eta - \lambda} d\lambda = \frac{1}{2}W'(\eta) - \omega(\eta), & \eta \in \mathbb{C} \backslash \Omega, \end{cases} \qquad (3.97)$$

where $W(\eta) = \sum_{j=0}^{2m} g_j \eta^j$. Denote

$$S_0(\eta) = \frac{1}{2}W(\eta) - \int_\Omega \ln|\eta - \lambda|\rho(\lambda)d\lambda, \qquad (3.98)$$

for η in the complex plane. By (3.97), for a point η in $(\eta_-^{(j)}, \eta_+^{(j)})$, there is $S_0(\eta) = S_0(\eta_-^{(1)}) + \sum_{k=1}^{j-1} \int_{\hat{\Omega}_k} \omega(\eta)d\eta$, where $\hat{\Omega} = \bigcup_{k=1}^{l-1} \hat{\Omega}_k \equiv \bigcup_{k=1}^{l-1} (\eta_+^{(k)}, \eta_-^{(k+1)})$. And if $\eta \in (\eta_+^{(j)}, \eta_-^{(j+1)})$, then there is $S_0(\eta) = S_0(\eta_-^{(1)}) + \operatorname{Re} \int_{\eta_-^{(1)}}^{\eta} \omega(\lambda)d\lambda$. Since $\omega(\eta)$ is real on $\hat{\Omega}$ and imaginary on Ω, there is the following general formula

$$S_0(\eta) = S_0(\eta_-^{(1)}) + \operatorname{Re} \int_{\eta_-^{(1)}}^{\eta} \omega(\lambda)d\lambda, \tag{3.99}$$

or $dS_0(\eta) = \operatorname{Re} \omega(\eta)d\eta$ in the differential form. It can be seen that if $\eta \in \Omega$, then

$$\frac{\partial S_0(\eta)}{\partial \eta} = 0, \tag{3.100}$$

since $\operatorname{Re} \omega(\eta) = 0$ on the cuts Ω. The function $S_0(\eta)$ is then a step function on $\Omega \cup \hat{\Omega}$, as discussed in the Seiberg-Witten theory, for example, see [4, 10]. The differential $dS(\eta) = \omega(\eta)d\eta$ for $\eta \in \mathbb{C} \backslash \Omega$ is corresponding to the Seiberg-Witten differential or the extended differentials defined on the Riemann surface, and free energy E is proportional to the prepotential.

Since the derivative of the free energy with respect to a potential direction g can be represented as

$$\frac{\partial}{\partial g} E = \int_{\Omega} \frac{\partial W(\eta)}{\partial g} \rho(\eta)d\eta + 2 \int_{\Omega} S_0(\eta) \frac{\partial \rho(\eta)}{\partial g} d\eta, \tag{3.101}$$

as shown before, it is then equal to

$$\int_{\Omega} \frac{\partial W(\eta)}{\partial g} \rho(\eta)d\eta + 2S_0(\eta_-^{(1)}) \int_{\Omega} \frac{\partial \rho(\eta)}{\partial g} d\eta + 2\operatorname{Re} \int_{\Omega} \left(\int_{\eta_-^{(1)}}^{\eta} \omega(\lambda) \frac{\partial \rho(\eta)}{\partial g} d\lambda \right) d\eta. \tag{3.102}$$

After simplifications, there is

$$\frac{\partial}{\partial g} E = \int_{\Omega} \frac{\partial W(\eta)}{\partial g} \rho(\eta)d\eta + 2 \sum_{j=2}^{l} \sum_{k=1}^{j-1} \int_{\hat{\Omega}_k} \omega(\lambda)d\lambda \frac{\partial}{\partial g} \int_{\Omega_j} \rho(\eta)d\eta, \tag{3.103}$$

which is consistent to the result (3.77) for the two-cut case ($l = 2$).

The Seiberg-Witten theory is developed to study the mass gap problem in the quantum Yang-Mills theory [11]. The linear combination of a and a_D [4, 10, 19], which are the integrals of the Seiberg-Witten differential over different intervals, is used to define the mass quantity according to the physical literatures. The connection between the Seiberg-Witten differential and the eigenvalue density in matrix models is important, that involves complicated mathematical and physical problems. A simple linear combination example of the integrals of the ρ and ω over different intervals is discussed in Sect. B.2 in order to experience the mathematical properties, and it is seen that the integrals are related to the Legendre's relation in the elliptic integral theory.

The relevant researches, such as instanton, integrable systems, string theory, supersymmetric Yang-Mills theory and Whitham equations can be found, for example, in [2, 5–7, 14, 17, 18] and the references therein. There are also other methods to investigate the mass gap problem, including, for example, Euler-Lagrange equation, energy-momentum operator, renormalization, propagator and much more that can be found in the publications and the archive journals. There are many challenging problems in this field, and one important problem related to our discussion is to balance the finite freedoms and the integrability. In the position models such as the soliton integrable systems with infinitely many constants, it is usually complicated to reduce the infinite dimensional space by using the periodic conditions as studied in the quantum inverse scattering method [13]. In the momentum aspect, the integrable system characterized by the infinitely many functions such as the u_n and v_n can be reduced to the physical model with finite number of parameters such as the a_j and b_j satisfying certain conditions, and the degree of the potential polynomial determines the degree of the freedoms.

The string equations are related to the density problems in the Seiberg-Witten theory as explained above. But the entire story would be complicated. As Michael R. Douglas mentioned in 2004 [8], it remains a fertile ground for mathematical discovery. There is a large field behind this fundamental problem that will connect many different researches and motivate the developments of sciences. Uncertainty and Fourier transform are fundamental in this research area, that have been typically emphasized in [11], and are associated with diverse problems.

3.5 Symmetric Cases

On the real line, denote $\Omega_1 = [-\eta_1, \eta_1]$ and $\Omega_2 = [-\eta_2, -\eta_0] \cup [\eta_0, \eta_2]$, and consider the potential $W(\eta) = g_2\eta^2 + g_4\eta^4$, where $g_4 > 0$. The densities on Ω_1 and Ω_2 are all symmetric now.

According to the discussion in last section, the density on Ω_1 now becomes

$$\rho_1(\eta) = \frac{1}{\pi}\left(g_2 + g_4(2\eta^2 + \eta_1^2)\right)\sqrt{\eta_1^2 - \eta^2}, \quad \eta \in \Omega_1, \qquad (3.104)$$

where $\eta_1 = 2b$, and the parameters satisfy the following conditions

$$g_2 + 4g_4b^2 \geq 0, \qquad (3.105)$$

$$2g_2b^2 + 12g_4b^4 = 1. \qquad (3.106)$$

It is easy to see that (3.105) implies $\rho_1(\eta) \geq 0$ for $\eta \in \Omega_1$. The conditions (3.105) and (3.106) imply $g_2b^2 + 1 = 3(g_2 + 4g_4b^2)b^2 \geq 0$, or $-g_2 \leq 1/b^2$. When g_2 is negative, there is $(-g_2)^2 \leq (1/b^2)(4g_4b^2) = 4g_4$, or

$$g_2 \geq -2\sqrt{g_4}. \qquad (3.107)$$

When g_2 is positive, (3.107) is also true. So (3.107) defines the region of the parameters for ρ_1, and the parameters for ρ_2 will stay in the region $g_2 \leq -2\sqrt{g_4}$. As a remark, the ρ_1 is an extension of the planar diagram density [3] $\rho(\eta) = \frac{1}{\pi}(\frac{1}{2} + 4gb^2 + 2g\eta^2)\sqrt{4b^2 - \eta^2}$ for potential $W(\eta) = \frac{1}{2}\eta^2 + g_4\eta^4$ since the g_2 considered here is negative around the critical point. The parameter g_4 in [3] is negative as they discuss the singular point for the free energy. Here, we consider positive g_4. The critical point $g_2/\sqrt{g_4} = -2$ was found early in 1982 by Shimamune [20], and the relevant discussions about the phase transition can also be found in [1, 9, 12], for instance.

The densities given above satisfy the normalization and the variational equation as shown before. It has been discussed in Sect. 3.1 that the free energy can be calculated by using the analytic function

$$\omega_1(\eta) = \left(g_2 + g_4(2\eta^2 + \eta_1^2)\right)\sqrt{\eta^2 - \eta_1^2}, \quad \eta \in \mathbb{C} \setminus \Omega_1, \tag{3.108}$$

which has the following asymptotics

$$\omega_1(\eta) = \frac{1}{2}\left(2g_2\eta^2 + 4g_4\eta^3\right) - \left(g_2 + 6g_4b^2\right)\frac{2b^2}{\eta} - \left(g_2 + 8g_4b^2\right)\frac{2b^4}{\eta^3} + O\left(\frac{1}{\eta^5}\right), \tag{3.109}$$

as $\eta \to \infty$ in the complex plane.

Now, let us consider

$$\rho_2(\eta) = \frac{2g_4}{\pi}\eta \operatorname{Re}\sqrt{e^{-\pi i}\left(\eta^2 - \eta_0^2\right)\left(\eta^2 - \eta_2^2\right)}, \quad \eta \in \Omega_2 = [-\eta_2, -\eta_0] \cup [\eta_0, \eta_2], \tag{3.110}$$

where $\eta_0 = b_2 - b_1$, $\eta_2 = b_1 + b_2$, and the parameters satisfy

$$4g_4b_1^2b_2^2 = 1, \tag{3.111}$$

$$g_2 + 2g_4\left(b_1^2 + b_2^2\right) = 0, \tag{3.112}$$

which imply

$$g_2 = -2g_4\left(b_1^2 + b_2^2\right) \leq -4g_4b_1b_2 = -2\sqrt{g_4}. \tag{3.113}$$

Therefore, the regions of the parameters in densities (3.104) and (3.110) are separated by the curve

$$g_2 = -2\sqrt{g_4}, \tag{3.114}$$

in the (g_2, g_4) plane. The ρ_1 is defined when $g_2 \geq -2\sqrt{g_4}$, and ρ_2 is defined when $g_2 \leq -2\sqrt{g_4}$. In addition, since

$$\sqrt{(\eta^2 - \eta_0^2)(\eta^2 - \eta_2^2)} = \begin{cases} \left|\sqrt{(\eta^2 - \eta_0^2)(\eta^2 - \eta_2^2)}\right|e^{3\pi i/2}, & -\eta_2 < \eta < -\eta_0, \\ \left|\sqrt{(\eta^2 - \eta_0^2)(\eta^2 - \eta_2^2)}\right|e^{\pi i/2}, & \eta_0 < \eta < \eta_2, \end{cases} \tag{3.115}$$

there is $\rho_2(\eta) > 0$ when $\eta \in (-\eta_2, -\eta_0) \cup (\eta_0, \eta_2)$. In fact, if $\eta \in (-\eta_2, -\eta_0)$. there are $\arg(\eta - (-\eta_2)) = 0$, and $\arg(\eta - \eta') = \pi$ for $\eta' = -\eta_0, \eta_0, \eta_2$; and if $\eta \in (\eta_0, \eta_2)$, there are $\arg(\eta - \eta') = 0$ for $\eta' = -\eta_2, -\eta_0, \eta_0$, and $\arg(\eta - \eta_2) = \pi$.

Define another analytic function

$$\omega_2(\eta) = 2g_4\eta\sqrt{(\eta^2 - \eta_0^2)(\eta^2 - \eta_2^2)}, \quad \eta \in \mathbb{C}\setminus\Omega_2, \tag{3.116}$$

which has the asymptotics

$$\omega_2(\eta) = \frac{1}{2}(2\eta + 4g_4\eta^3) - \frac{4g_4b_1^2b_2^2}{\eta} + O(\eta^{-3}), \tag{3.117}$$

as $\eta \to \infty$ in the complex plane. It can be checked that if $b_1 = b_2 = b$, then $\eta_0 = 0$ and $\eta_1 = \eta_2$, and the conditions (3.111) and (3.112) become $4g_4b^4 = 1$ and $g_2 + 4g_4b^2 = 0$ respectively, which imply (3.106). In other words, (3.111) and (3.112) can be thought as (3.106) is split to two equations as b is bifurcated to the two parameters b_1 and b_2. In the symmetric cases, we can experience the third-order phase transitions with the explicit formulations of the free energy function.

To calculate the free energy function, let us make a change for ρ_2 by the following

$$\rho_2(\eta)d\eta = \frac{1}{2}\hat{\rho}_2(\zeta)d\zeta, \tag{3.118}$$

where

$$\hat{\rho}_2(\zeta) = \frac{2g_4}{\pi}\sqrt{(\zeta_+ - \zeta)(\zeta - \zeta_-)}, \tag{3.119}$$

$\zeta_- = \eta_0^2$, $\zeta_+ = \eta_2^2$, $\zeta = \eta^2$, and $W(\eta) = \frac{1}{2}\hat{W}(\zeta)$ with $\hat{W}(\zeta) = 2g_2\zeta + 2g_4\zeta^2$ such that $\eta^{-1}\partial W(\eta)/\partial\eta = \partial\hat{W}(\zeta)/\partial\zeta$. The coefficients $1/2$ and 2 above will make the following calculations easy. Since $\rho_2(\eta)$ satisfies $\int_{-\eta_2}^{-\eta_0}\rho_2(\eta)d\eta + \int_{\eta_0}^{\eta_2}\rho_2(\eta)d\eta = 1$, and

$$(\text{P})\int_{-\eta_2}^{-\eta_0}\frac{\rho_2(\lambda')}{\eta - \lambda'}d\lambda' + (\text{P})\int_{\eta_0}^{\eta_2}\frac{\rho_2(\lambda)}{\eta - \lambda}d\lambda = \frac{1}{2}W'(\eta),$$

where $'$ for $W(\eta)$ means $\partial/\partial\eta$, we can get $\int_{\zeta_-}^{\zeta_+}\hat{\rho}_2(\zeta)d\zeta = 1$, and

$$(\text{P})\int_{\zeta_-}^{\zeta_+}\frac{\hat{\rho}_2(\xi)}{\zeta - \xi}d\xi = \frac{1}{2}\hat{W}'(\zeta), \tag{3.120}$$

by taking $\lambda' = -\lambda$, $\xi = \lambda^2$ and $\zeta = \eta^2$, where $'$ for $\hat{W}(\zeta)$ means $\partial/\partial\zeta$. Then, the free energy becomes

$$E = \frac{1}{2}\left(\hat{W}(\hat{a}) + \frac{3}{4} - \ln\hat{b}\right), \tag{3.121}$$

by using the free energy for the one-interval case discussed in Sect. 3.1 for $m = 1$, where $\hat{a} = \frac{1}{2}(\zeta_- + \zeta_+) = -\frac{g_2}{2g_4}$ and $\hat{b} = \frac{1}{4}(\zeta_+ - \zeta_-) = \frac{1}{2\sqrt{g_4}}$. One can get the same result by using the method in Sect. 3.1 and the asymptotics (3.117). The details are left to interested readers as an exercise.

The general formulas of the free energy for the potential $W(\eta) = g_2\eta^2 + g_4\eta^4$ can be summarized in the following with the parameters restricted in the different regions,

$$
E = \begin{cases}
\frac{1}{24}(2g_2v - 1)(9 - 2g_2v) - \frac{1}{2}\ln v + \frac{3}{4}, & g_2 \geq -2\sqrt{g_4}, \\
\frac{3}{8} - \frac{g_2^2}{4g_4} + \frac{1}{4}\ln(4g_4), & g_2 \leq -2\sqrt{g_4},
\end{cases} \tag{3.122}
$$

where

$$
2g_2v + 12g_4v^2 = 1. \tag{3.123}
$$

The free energy function and its first- and second-order derivatives are always continuous. but its third-order derivatives are discontinuous as the parameters pass through the Shimamune's critical curve $g_2 = -2\sqrt{g_4}$ [20], which is different from the critical point $g_2^2 + 12g_4 = 0$ in the planar diagram model [3]. In the following, we discuss the third-order discontinuities by choosing different forms of the potential $W(\eta)$.

When $W(\eta) = -\hat{g}\eta^2 + \eta^4$, we have from (3.122) that

$$
E = \begin{cases}
-\frac{1}{24}(1 + 2\hat{g}v)(9 + 2\hat{g}v) - \frac{1}{2}\ln v + \frac{3}{4}, & \hat{g} \leq 2, \\
\frac{3}{8} - \frac{\hat{g}^2}{4} + \frac{1}{2}\ln 2, & \hat{g} \geq 2,
\end{cases} \tag{3.124}
$$

where $-2\hat{g}v + 12v^2 = 1$. At the critical point $\hat{g} = 2$, there are $v = 1/2$ and $v' = -1/8$ where $' = d/d\hat{g}$. The free energy E is a continuous function of $\hat{g}$, and at the critical point $\hat{g} = 2$ there is $E|_{\hat{g}\to 2-} = -5/8 + \ln\sqrt{2} = E|_{\hat{g}\to 2+}$. The derivative of E with respect to $\hat{g}$ can be obtained by direct calculations

$$
\frac{dE}{d\hat{g}} = \begin{cases}
-v - 4v^3, & \hat{g} \leq 2; \\
-\frac{\hat{g}}{2}, & \hat{g} \geq 2.
\end{cases} \tag{3.125}
$$

Therefore, we have $\frac{dE}{d\hat{g}}|_{\hat{g}\to 2-} = -1 = \frac{dE}{d\hat{g}}|_{\hat{g}\to 2+}$, which implies the first-order derivative is continuous. The second-order derivative can be obtained as

$$
\frac{d^2E}{d\hat{g}^2} = \begin{cases}
-2v^2, & \hat{g} \leq 2, \\
-\frac{1}{2}, & \hat{g} \geq 2.
\end{cases} \tag{3.126}
$$

And then

$$
\frac{d^3E}{d\hat{g}^3} = \begin{cases}
-\frac{4v^3}{1+\hat{g}v}, & \hat{g} \leq 2, \\
0, & \hat{g} \geq 2.
\end{cases} \tag{3.127}
$$

Obviously, the third-order derivative is not continuous at the critical point $\hat{g} = 2$.

When $W(\eta) = -2\eta^2 + g\eta^4$, we have

$$E = \begin{cases} -\frac{1}{24}(1+v)(9+v) - \frac{1}{2}\ln v + \frac{3}{4}, & g \geq 1, \\ \frac{3}{8} - \frac{1}{g} + \frac{1}{4}\ln(4g), & 0 < g \leq 1, \end{cases} \tag{3.128}$$

where $-4v + 12gv^2 = 1$. At the critical point $g = 1$, there are $v = 1/2$ and $v' = -3/8$ where $' = d/dg$. We have

$$\frac{dE}{dg} = \begin{cases} (3+2v)v^2, & g \geq 1, \\ \frac{1}{g^2} + \frac{1}{4g}, & 0 < g \leq 1, \end{cases} \tag{3.129}$$

and

$$\frac{d^2E}{dg^2} = \begin{cases} -36v^4, & g \geq 1, \\ -\frac{2}{g^3} - \frac{1}{4g^2}, & 0 < g \leq 1. \end{cases} \tag{3.130}$$

Then it can be seen that the first- and second-order derivatives are all continuous, and the third-order derivative is discontinuous at $g = 1$.

When $W(\eta) = T^{-1}(-2\eta^2 + \eta^4)$, we have

$$E = \begin{cases} -\frac{v}{3T^2}(2v + 5T) - \frac{1}{2}\ln v + \frac{3}{8}, & T \geq 1, \\ \frac{3}{8} - \frac{1}{T} - \frac{1}{4}\ln\frac{T}{4}, & T \leq 1, \end{cases} \tag{3.131}$$

for a positive parameter T (temperature) with $-4v + 12v^2 = T$ such that $v = 1/2$ when $T = 1$. It can be get that $E|_{T \to 1+} = -5/8 + \ln\sqrt{2} = E|_{T \to 1-}$. Then, as a function of T, the derivative of the free energy takes the following form

$$\frac{dE}{dT} = \begin{cases} \frac{v^2(4v+5T)}{T^3(6v-1)} - \frac{1}{8v(6v-1)}, & T \geq 1, \\ \frac{1}{T^2} - \frac{1}{4T}, & T \leq 1. \end{cases} \tag{3.132}$$

At the critical point $T = 1$, dE/dT is continuous, and explicitly $\frac{dE}{dT}|_{T \to 1+} = \frac{3}{4} = \frac{dE}{dT}|_{T \to 1-}$. It is just direct calculations to check that the second-order derivative is continuous and the third-order derivative is discontinuous at $T = 1$. As a remark, there is no difference if we consider a more general potential $W(\eta) = T^{-1}(-c^2\eta^2 + \eta^4)$ for analyzing the discontinuous property. For simplicity, here we just show the results for $c = 1$.

The criticality discussed above has negative g_2^c and positive g_4^c. For potential $W(\eta) = \sum_{j=1}^4 g_j\eta^j$, the critical point $g_2 = g_2^c$ can be positive if $3a_c^2 > 2b_c^2$ according to (3.59). One can experience that for the potential $g_2\eta^2 + g_4\eta^4 + g_6\eta^6$, the critical point g_2^c will be positive. The parameter g_2 is important because in physics it is related to the mass quantity as discussed in some literatures, for example, see [9]. In the planar diagram model [3], g_4 is negative to let the parabola cover the semicircle forming a Gaussian kind distribution. By the relation (3.123), if we take g_4 as a constant and g_2 as a function of $\ln v$, then

$$\frac{dg_2}{d \ln v} = g_2 - v^{-1}. \tag{3.133}$$

So $dg_2/d \ln v = 0$ implies $g_2 = v^{-1}$ and $g_4 = -\frac{1}{12} v^{-2}$ (negative) satisfying $g_2^2/g_4 = -12$. This is the critical case for the planar diagram model different from the critical point $g_2/\sqrt{g_4} = -2$ above. The transition problem for the planar diagram model will be discussed in Sect. 4.2.2.

The parameter bifurcation introduces a preliminary knowledge about the cause of the transition. The double zero point at the critical point, which is formed in two ways as explained in the following, is a main factor for the transition. For the potential $W(\eta) = g_2 \eta^2 + g_4 \eta^4$ with the density $\rho_c = \frac{2g_4}{\pi} \eta^2 \sqrt{4b^2 - \eta^2}$ at the critical point, we have experienced that the double zero $\eta = 0$ is from the parabola pushing the semicircle to give a gap in the density when we consider the density on one interval. For the density on two intervals, one zero is from the outside of the square root in the density model, and another zero is from the inside of the square root meeting with the outside zero to form a double zero. The density functions have behaviors $O(|\eta - \eta_0|^2)$ at the bifurcation point $\eta_0 = 0$ at the critical point, while normally at the end points the behaviors are of square root order. It is the same behaviors for the higher degree potentials except more double zero points. This would indicate that a matter is transferred during the transitions with the splitting or merging deformations at the middle points of the densities based on the bifurcations. The transition can be also caused by the deformation of the density at the largest (or smallest) end point of the eigenvalues in the distribution, in which case there will be a small system released or added at the largest eigenvalue point, that will be discussed in Sect. 4.3.3 by using double scaling method.

References

1. Bleher, P., Eynard, B.: Double scaling limit in random matrix models and a nonlinear hierarchy of differential equations. J. Phys. A **36**, 3085–3106 (2003)
2. Braden, H.W., Krichever, I.M. (eds.): Integrability: The Seiberg-Witten and Whitham Equations. Gordon & Breach, Amsterdam (2000)
3. Brézin, E., Itzykson, C., Parisi, G., Zuber, J.B.: Planar diagrams. Commun. Math. Phys. **59**, 35–51 (1978)
4. Chekhov, L., Mironov, A.: Matrix models vs. Seiberg–Witten/Whitham theories. Phys. Lett. B **552**, 293–302 (2003)
5. Chekhov, L., Marshakov, A., Mironov, A., Vasiliev, D.: Complex geometry of matrix models. Proc. Steklov Inst. Math. **251**, 254–292 (2005)
6. Dijkgraaf, R., Vafa, C.: Matrix models, topological strings, and supersymmetric gauge theories. Nucl. Phys. B **644**, 3–20 (2002)
7. Dijkgraaf, R., Moore, G.W., Plesser, R.: The partition function of 2-D string theory. Nucl. Phys. B **394**, 356–382 (1993)
8. Douglas, M.R.: Report on the status of the Yang-Mills millenium prize problem (2004). http://www.claymath.org/millenium

9. Fuji, H., Mizoguchi, S.: Remarks on phase transitions in matrix models and $N = 1$ supersymmetric gauge theory. Phys. Lett. B **578**, 432–442 (2004)

10. Gorsky, A., Krichever, I., Marshakov, A., Mironov, A., Morozov, A.: Integrability and Seiberg-Witten exact solution. Phys. Lett. B **355**, 466–477 (1995)

11. Jaffe, A., Witten, E.: Quantum Yang-Mills theory. In: Carlson, J., Jaffe, A., Wiles, A. (eds.) The Millennium Prize Problems, pp. 129–152. AMS, Providence (2006)

12. Jurkiewicz, J.: Regularization of one-matrix models. Phys. Lett. B **235**, 178–184 (1990)

13. Korepin, V.E., Bogoliubov, N.M., Izergin, A.G.: Quantum Inverse Scattering Method and Correlation Functions. Cambridge University Press, Cambridge (1993)

14. Marshakov, A., Nekrasov, N.: Extended Seiberg-Witten theory and integrable hierarchy. J. High Energy Phys. **0701**, 104 (2007)

15. McLeod, J.B., Wang, C.B.: Eigenvalue density in Hermitian matrix models by the Lax pair method. J. Phys. A, Math. Theor. **42**, 205205 (2009)

16. Mehta, M.L.: Random Matrices, 3rd edn. Academic Press, New York (2004)

17. Nakatsu, T., Takasaki, K.: Whitham-Toda hierarchy and $N = 2$ supersymmetric Yang-Mills theory. Mod. Phys. Lett. A **11**, 157–168 (1996)

18. Nekrasov, N.: Seiberg-Witten prepotential from instanton counting. Adv. Theor. Math. Phys. **7**, 831–864 (2004)

19. Seiberg, N., Witten, E.: Electric-magnetic duality, monopole condensation, and confinement in $N = 2$ supersymmetric Yang-Mills theory. Nucl. Phys. B **426**, 19–52 (1994). Erratum, ibid. B **430**, 485–486 (1994)

20. Shimamune, Y.: On the phase structure of large N matrix models and gauge models. Phys. Lett. B **108**, 407–410 (1982)

Chapter 4
Large-N Transitions and Critical Phenomena

The bifurcation transition models discussed in the last chapter can be extended to large-N transitions, which will be explained in this chapter based on hypergeometric-type differential equations and the double scaling method. The singular values of the hypergeometric-type differential equation are related to the elliptic functions that are the fundamental mathematical tools for studying the vertex models in statistical physics. The double scaling method can connect the string system to the soliton system. Different transitions, or discontinuities, will be discussed in this chapter, especially the odd-order transitions, such as first-, third- and fifth-order transitions, which can be formulated by using the density models. The second-order divergences (critical phenomena) that are usually discussed in physics by using renormalization methods can be obtained by considering the derivatives of the logarithm of the partition function in the original potential parameter direction and using the Toda lattice. The third-order divergence for the planar diagram model is investigated in association with the critical phenomenon and double scaling. The fourth-order discontinuity is studied by using the analytic properties of the integrable system.

4.1 Cubic Potential

4.1.1 Models in Large-N Asymptotics

In last chapter, we have discussed the transition models based on the asymptotics $\sqrt{-\det \hat{A}_n} = V'(z)/2 - n/z + O(z^{-2})$ as $z \to \infty$ for the matrix $\hat{A}_n$ which is obtained from A_n. It is known that $\sqrt{-\det A_n}$ itself also has such an asymptotics. Then it is natural to ask whether there is a transition directly from the density model $\frac{1}{n\pi}\sqrt{\det A_n}$. In this chapter, we are going to discuss such transitions and some new transitions such as the first- and fifth-order transitions different from the third-order transitions discussed in last chapter. First, in this subsection, we introduce some preliminary formulas for the later discussions.

C.B. Wang, *Application of Integrable Systems to Phase Transitions*,
DOI 10.1007/978-3-642-38565-0_4, © Springer-Verlag Berlin Heidelberg 2013

We have obtained before that the orthogonal polynomials p_n's satisfy

$$p_{n,z} = -v_n\big[3t_3 + 4t_4(u_n + u_{n-1} + z)\big]p_n$$
$$+ v_n\big[2t_2 + 3t_3 u_n + 4t_4(u_n^2 + v_n + v_{n+1}) + (3t_3 + 4t_4 u_n)z + 4t_4 z^2\big]p_{n-1},$$
$$(4.1)$$

for the potential $V(z) = \sum_{j=1}^{4} t_j z^j$. If we take the limit case $t_2 \to 0$ and $t_4 \to 0$, then the equation above becomes

$$p_{n,z} = -3t_3 v_n p_n + 3t_3 v_n (z + u_n) p_{n-1},\qquad (4.2)$$

in which case the polynomials are not orthogonal any more since the weight function is not finite at infinity that leads the integral for the orthogonality of the polynomials to be divergent. But the Lax pair structure still exists, that can be applied to obtain the continuum differential equation and the singularity of the second-order derivative of the free energy in large-N asymptotics.

Let us consider the potential

$$V(z) = tz + \frac{2}{3}z^3.\qquad (4.3)$$

The string equation now becomes the following set of two equations,

$$2(u_n + u_{n-1})v_n = n,\qquad (4.4)$$
$$t + 2(u_n^2 + v_n + v_{n+1}) = 0.\qquad (4.5)$$

Denote $\Phi_n = e^{-\frac{1}{2}V(z)}(p_n, p_{n-1})^T$. By using the recursion formula, (4.2) can be changed to the following equation,

$$\Phi_{n,z} = A_n \Phi_n,\qquad (4.6)$$

where

$$A_n = \begin{pmatrix} -z^2 - \frac{t}{2} - 2v_n & 2v_n(z + u_n) \\ -2(z + u_{n-1}) & z^2 + \frac{t}{2} + 2v_n \end{pmatrix}.\qquad (4.7)$$

It follows that

$$-\det A_n = z^4 + tz^2 - 2nz + \left(2v_n + \frac{t}{2}\right)^2 - \frac{n^2}{4v_n} + \frac{(v_n')^2}{v_n},\qquad (4.8)$$

where $' = d/dt$. If we make the following scalings for large n,

$$\frac{t}{n^{2/3}} = -g + \frac{\xi}{n},\qquad z = n^{1/3}\eta,\qquad v_n = n^{2/3}v,\qquad (4.9)$$

then

$$-\det A_n = n^{4/3}\big(\eta^4 - g\eta^2 - 2\eta - X + O(n^{-1})\big),\qquad (4.10)$$

Fig. 4.1 Three X curves in the Hermitian model

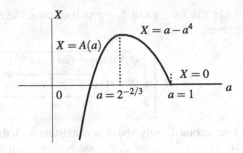

and

$$X = \frac{1}{4v} - \left(2v - \frac{g}{2}\right)^2 - \frac{v_\xi^2}{v}, \tag{4.11}$$

with $v_\xi = dv/d\xi$. The transitions obtained from the densities reduced from the above large-N asymptotics will be called large-N transitions.

It is discussed in Sect. D.1 that if we make a change of variable for the g parameter

$$g = 2a^2 + a^{-1}, \tag{4.12}$$

based on the relation between t and u_n given by (4.4) and (4.5) above in order to factorize the determinant of the coefficient matrix of the hypergeometric-type differential equation, $\det C$, discussed in Sect. D.1, then the X variable is found to have four different singular values. These four singular cases can be expressed as four functions $X(a)$ in terms of a, that can be reorganized according to the intersections of the function curves to formulate new X function curves to achieve the transition processes with the intersection as critical point. One may try all the possible combinations for the X functions based on the roots of the determinant of the coefficient matrix talked in Sect. D.1. Here, we choose the following cases,

$$X = X(a) = \begin{cases} A(a), & 0 < a \le 2^{-2/3}, \\ a - a^4, & 2^{-2/3} \le a \le 1, \\ 0, & a \ge 1, \end{cases} \tag{4.13}$$

where $A = A(a)$ is given in Sect. D.1,

$$2A(a) = \left(a^2 + 2a^{-1}\right)\sqrt{a^4 + 2a} - \left(a^4 + 3a + \frac{1}{2}a^{-2}\right), \tag{4.14}$$

shown in Fig. 4.1. It can be checked that at $a = 2^{-2/3}$, both dA/da and $d(a - a^4)/da$ are equal to 0.

In the next section, we will discuss the transitions among the following three phases,

$$\rho(\eta) = \begin{cases} \frac{1}{\pi}\sqrt{A + 2\eta + (2a^2 + a^{-1})\eta^2 - \eta^4}, & 0 < a \le 2^{-2/3}, \\ \frac{1}{\pi}(\eta + a)\sqrt{a^{-1} - (\eta - a)^2}, & 2^{-2/3} \le a \le 1, \\ \frac{1}{\pi}\sqrt{2\eta + (2a^2 + a^{-1})\eta^2 - \eta^4}, & a \ge 1. \end{cases} \qquad (4.15)$$

The second density above is consistent with the one-interval density for the potential $W(\eta) = \sum_{j=1}^{4} g_j \eta^j$ discussed before when the potential is degenerated to the current case, and $X(a) = a - a^4$ is corresponding to $v = 1/(4a)$ in (4.11) with $v_\xi = 0$. It is then seen that the method to derive the eigenvalue density by the hypergeometric-type differential equation has extended the index folding technique discussed in the previous chapters. The index folding assumes an even kind pressure in the system as explained Sect. 2.2, while the X variable in the hypergeometric-type differential equation is related to many other parameters or functions in the model that characterize the correlations between the different factors in the system. The large-N transitions include the uneven pressure cases that extend the model obtained from the index folding.

4.1.2 First-Order Discontinuity

The density models for $a \ge 2^{-2/3}$ can be changed to be in term of g with $g \ge 3/2^{1/3}$ since $g = 2a^2 + a^{-1}$ is monotonically increasing when $a \ge 2^{-2/3}$ and we want to discuss the transition in the g direction.

Lemma 4.1 *If $g > 3$, the algebraic equation $(\eta^3 - g\eta - 2)\eta = 0$ has four roots*

$$\eta_-^{(1)} \le \eta_+^{(1)} < \eta_-^{(2)} < \eta_+^{(2)}, \qquad (4.16)$$

where $\eta_-^{(2)} = 0$. As $g \to 3+$, there are the following expansions

$$g = 3 + 3\varepsilon^2, \qquad a = 1 + \varepsilon^2 + \cdots, \qquad (4.17)$$

$$\eta_-^{(1)} = -1 - \varepsilon + \cdots, \qquad (4.18)$$

$$\eta_+^{(1)} = -1 + \varepsilon + \cdots, \qquad (4.19)$$

$$\eta_+^{(2)} = 2 + \varepsilon^2 + \cdots. \qquad (4.20)$$

Proof Let $\mu(\eta) = \eta^3 - (2a^2 + a^{-1})\eta - 2$, where $g = 2a^2 + a^{-1}$ with $a > 1$. It is easy to see that $\mu(-\infty) = -\infty$, $\mu(-a) = a^3 - 1 > 0$, $\mu(0) = -2 < 0$ and $\mu(\infty) = \infty$. Then, the equation $\mu(\eta) = 0$ has three roots $\eta_-^{(1)}$, $\eta_+^{(1)}$ and $\eta_+^{(2)}$ satisfying

$$\eta_-^{(1)} < -a < \eta_+^{(1)} < 0 < \eta_+^{(2)}.$$

The expansions can be obtained by using the equations $\mu(\eta_-^{(1)}) = 0$, $\mu(\eta_+^{(1)}) = 0$ and $\mu(\eta_+^{(2)}) = 0$. $\qquad\square$

Now, we define ρ as

$$\rho(\eta) = \begin{cases} \frac{1}{\pi}(\eta+a)\sqrt{a^{-1}-(\eta-a)^2}, & \eta \in [a-a^{-1/2}, a+a^{-1/2}], \\ \quad 3/2^{1/3} \le g \le 3, \\ \frac{1}{\pi}\sqrt{2\eta + g\eta^2 - \eta^4}, & \eta \in \Omega, \\ \quad g \ge 3, \end{cases} \qquad (4.21)$$

where $\Omega = [\eta_-^{(1)}, \eta_+^{(1)}] \cup [\eta_-^{(2)}, \eta_+^{(2)}]$, and the parameter a is related to g by the equation $g = 2a^2 + a^{-1}$. At the critical point $g = 3$, there is $\rho_c = \frac{1}{\pi}(\eta+1)\sqrt{\eta(2-\eta)}$. Define

$$\omega(\eta) = \begin{cases} (\eta+a)\sqrt{(\eta-a)^2 - a^{-1}}, & \eta \in \mathbb{C}\setminus[a-a^{-1/2}, a+a^{-1/2}], \\ \quad 3/2^{1/3} \le g \le 3, \\ \sqrt{\eta^4 - g\eta^2 - 2\eta}, & \eta \in \mathbb{C}\setminus\Omega, \quad g \ge 3. \end{cases} \qquad (4.22)$$

When $g = 3$, we have $a = 1$ and $\eta_-^{(1)} = \eta_+^{(1)} = -1$, that means as g decreases to the critical point $g = 3$, the left interval of Ω does not approach to the right interval, but keep a distance away from the right interval until it shrinks to a point. The relation between ρ and ω is

$$\rho(\eta) = \begin{cases} \frac{1}{\pi i}\omega(\eta)|_{[a-a^{-1/2}, a+a^{-1/2}]^+}, & 3/2^{1/3} \le g \le 3, \\ \frac{1}{\pi i}\omega(\eta)|_{\Omega^+}, & g \ge 3. \end{cases} \qquad (4.23)$$

As before, we have

$$\begin{cases} (P)\int_\Omega \frac{\rho(\lambda)}{\eta-\lambda}d\lambda = \frac{1}{2}W'(\eta), & \eta \in (\eta_-^{(1)}, \eta_+^{(1)}) \cup (\eta_-^{(2)}, \eta_+^{(2)}), \\ \int_\Omega \frac{\rho(\lambda)}{\eta-\lambda}d\lambda = \frac{1}{2}W'(\eta) - \omega(\eta), & \eta \in (\eta_+^{(1)}, \eta_-^{(2)}). \end{cases} \qquad (4.24)$$

When $3/2^{1/3} \le g \le 3$, by using the asymptotics $\omega(\eta) = \frac{1}{2}W'(\eta) - \eta^{-1} + O(\eta^{-2})$ as $\eta \to \infty$, we can get $\int_\Omega \rho(\eta)d\eta = 1$ and

$$(P)\int_\Omega \frac{\rho(\lambda)}{\eta-\lambda}d\lambda = \frac{1}{2}W'(\eta), \quad \eta \in (a-a^{-1/2}, a+a^{-1/2}), \qquad (4.25)$$

where Ω stands for $[a-a^{-1/2}, a+a^{-1/2}]$.
Based on the above results,

$$E = \int_\Omega W(\eta)\rho(\eta)d\eta - \int_\Omega\int_\Omega \ln|\lambda-\eta|\rho(\lambda)\rho(\eta)d\lambda d\eta \qquad (4.26)$$

has the following first-order derivative formulas,

$$\frac{\partial}{\partial g}E(\rho) = \begin{cases} -\int_{a-a^{-1/2}}^{a+a^{-1/2}} \eta\rho(\eta)d\eta, & 3/2^{1/3} \leq g \leq 3, \\ -\int_{\Omega} \eta\rho(\eta)d\eta - 2\int_{\eta_+^{(1)}}^{\eta_-^{(2)}} \omega(\eta)d\eta \frac{d}{dg}\int_{\eta_-^{(1)}}^{\eta_+^{(1)}} \rho(\eta)d\eta, & g \geq 3, \end{cases}$$

(4.27)

where $\int_{\eta_-^{(1)}}^{\eta_+^{(1)}} \rho + \int_{\eta_-^{(2)}}^{\eta_+^{(2)}} \rho = 1$, that implies $\frac{d}{dg}\int_{\eta_-^{(2)}}^{\eta_+^{(2)}} \rho = -\frac{d}{dg}\int_{\eta_-^{(1)}}^{\eta_+^{(1)}} \rho$.

Lemma 4.2 As $g \to 3+$, there is

$$\int_{\eta_-^{(1)}}^{\eta_+^{(1)}} \sqrt{2\eta + g\eta^2 - \eta^4}d\eta = \frac{\sqrt{3}}{2}\pi\varepsilon^2 + O(\varepsilon^3),$$

(4.28)

where $g = 3 + 3\varepsilon^2$.

Proof If we make a change of variable $\eta = -1 + \zeta\varepsilon$ for $\eta \in [\eta_-^{(1)}, \eta_+^{(1)}]$, then

$$2\eta + g\eta^2 - \eta^4 = 3(1 - \zeta^2)\varepsilon^2 + O(\varepsilon^3).$$

According to the asymptotic expansions discussed in Lemma 4.1, we have

$$\int_{\eta_-^{(1)}}^{\eta_+^{(1)}} \sqrt{2\eta + g\eta^2 - \eta^4}d\eta = \sqrt{3}\varepsilon^2\int_{-1}^{1}\sqrt{1 - \zeta^2}d\zeta + O(\varepsilon^3) = \frac{\sqrt{3}}{2}\pi\varepsilon^2 + O(\varepsilon^3).$$

Then this lemma is proved.   □

Lemma 4.3 As $g \to 3+$, there is

$$\int_{\eta_+^{(1)}}^{\eta_-^{(2)}} \sqrt{\eta^4 - g\eta^2 - 2\eta}d\eta = \sqrt{3} + \ln(2 - \sqrt{3}) + O(\varepsilon).$$

(4.29)

Proof It can be checked that when $g = 3$ there is

$$\int_{\eta_+^{(1)}}^{\eta_-^{(2)}} \sqrt{\eta^4 - g\eta^2 - 2\eta}d\eta = \int_{-1}^{0}(\eta + 1)\sqrt{\eta(\eta - 2)}d\eta.$$

Then the rest of this lemma can be verified by the integral calculations.   □

Theorem 4.1 As $g \to 3$, the first-order derivative of E with respect to g is discontinuous at $g = 3$,

$$E'(3 - 0) - E'(3 + 0) = 1 + \frac{1}{\sqrt{3}}\ln(2 - \sqrt{3}),$$

(4.30)

where $' = d/dg$.

Proof By Lemma 4.2, we have that

$$\frac{d}{dg} \int_{\eta_-^{(1)}}^{\eta_+^{(1)}} \rho(\eta)d\eta = \frac{1}{2\sqrt{3}} + o(1),$$

where $g = 3 + 3\varepsilon^2$. Then, Lemma 4.3 implies

$$2 \int_{\eta_+^{(1)}}^{\eta_-^{(2)}} \omega(\eta)d\eta \frac{d}{dg} \int_{\eta_-^{(1)}}^{\eta_+^{(1)}} \rho(\eta)d\eta = 1 + \frac{1}{\sqrt{3}} \ln(2 - \sqrt{3}) + o(1).$$

It is not hard to see that as $g \to 3-$, $\int_{a-a^{-1/2}}^{a+a^{-1/2}} \eta\rho(\eta)d\eta$ approaches to $\int_\Omega \eta\rho(\eta)d\eta$ same as $g \to 3+$. So we conclude the first-order derivative is discontinuous at $g = 3$. $\qquad\square$

The first-order discontinuity is caused by the term I_0 shown above. In the third-order transition models discussed in last chapter, there is a similar term I_0 which is $O(|g - g_c|^2)$ at the critical point. In the first-order discontinuity model above, the two intervals where ρ is defined on do not merge together, that implies the corresponding I_0 is $O(1)$ and the first-order derivative is then not continuous.

It can be seen that when $a > 1$ the algebraic equation $g = 2a^2 + a^{-1}$ and the differential equation $v_\xi^2 = \frac{1}{4} - v(2v - \frac{g}{2})^2$ where $v = 1/(4a)$ imply $v_\xi^2 = \frac{1}{4}(1 - a^3)$, which indicates that ξ should be imaginary. When $2^{-2/3} < a < 1$ with $X = a - a^4$, there is $v_\xi^2 = 0$, that means v is independent of ξ. The eigenvalue scale parameter $v = 1/(4a)$ satisfies $v \geq \frac{1}{4}$ when $0 < a < 1$. If $a > 1$, the eigenvalue scale becomes a wave in the ξ direction. As a remark, it is given in Sect. 3.2 that $\frac{d^2}{dt^2} \ln Z_n = v_n$. By the scalings $t = -n^{2/3}g + n^{-1/3}\xi$ and $v_n = n^{2/3}v$, it becomes $\frac{d^2}{d\xi^2} \ln \tau_0(\xi) = v$, where the partition function is replaced by a new function τ_0. Such a formula often appears in the soliton theories and the correlation function theories. By the differential equation for v above, we can get $w_x^2 = 4w^3 - \frac{g^2}{12}w + \frac{g^3}{6^3} - \frac{1}{4}$, where $w = v - g/6$ and $\xi = ix$, that implies w is the Weierstrass elliptic $\wp$-function, and then v has infinitely many poles. The calculations of the elliptic functions would give a result involving the logarithmic function as studied in the physical theories such as lattice models or correlation function theories.

It must be noted that when $g > 3$ ($X = 0$), the ρ defined on the two intervals takes negative values on the left interval and positive values on the right interval. But the summation of the integrals of ρ over these two parts satisfies the normalization condition. This can be checked by using the numerical computations or the complex number arguments discussed in Sect. 3.3. Therefore in this case the ρ is not a direct physical model. But this negative-positive density function defined on two intervals can be changed to a non-negative density defined on two edges of the cut by referring the idea of two coupled strings in physics.

If we make a change of variable $\eta = \sqrt{\zeta}$ with the positive real line R_+ as the cut for this square root, then on the upper edge R_+^+ there is $\rho(\eta)d\eta = \rho(\sqrt{\zeta})\frac{d\zeta}{2\sqrt{\zeta}} \equiv$

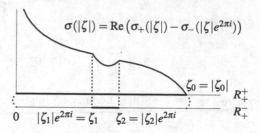

$$\sigma(|\zeta|) = \mathrm{Re}\left(\sigma_+(|\zeta|) - \sigma_-(|\zeta|e^{2\pi i})\right)$$

$\zeta_0 = |\zeta_0|$ $R_+^{\pm}$

0 $|\zeta_1|e^{2\pi i} = \zeta_1$ $\zeta_2 = |\zeta_2|e^{2\pi i}$ R_+^-

$\sigma_+(\zeta)d\zeta$ where $d\zeta$ is from left to right; and on the lower edge R_+^- there is $\rho(\eta)d\eta = \rho(\sqrt{\zeta})\frac{d\zeta}{2\sqrt{\zeta}} \equiv \sigma_-(\zeta)d\zeta$ where $d\zeta$ is from right to left, or explicitly

$$\sigma_+(\zeta) = \sigma_+(|\zeta|) = \frac{1}{2\pi}\sqrt{2|\zeta|^{-1/2} + g - |\zeta|}, \tag{4.31}$$

and

$$\sigma_-(\zeta) = \sigma_-\left(|\zeta|e^{2\pi i}\right) = \frac{1}{2\pi}\sqrt{-2|\zeta|^{-1/2} + g - |\zeta|}. \tag{4.32}$$

If we denote $\zeta_0 = (\eta_+^{(2)})^2 = |\eta_+^{(2)}|^2$, $\zeta_1 = (\eta_+^{(1)})^2 = |\eta_+^{(1)}|^2 e^{2\pi i}$ and $\zeta_2 = (\eta_-^{(1)})^2 = |\eta_-^{(1)}|^2 e^{2\pi i}$, it can be checked that the interval $[|\zeta_1|, |\zeta_2|]$ is in $(0, |\zeta_0|)$. Also $\sigma_+(\zeta) = \mathrm{Re}\,\sigma_+(\zeta) \geq 0$ for $\zeta \in (0, \zeta_0]$, and $\sigma_-(\zeta) = \mathrm{Re}\,\sigma_-(\zeta) \geq 0$ for $\zeta \in [\zeta_1, \zeta_2]$. Then we get a non-negative density model

$$\sigma(|\zeta|) = \mathrm{Re}\left(\sigma_+(|\zeta|) - \sigma_-\left(|\zeta|e^{2\pi i}\right)\right), \tag{4.33}$$

for $\zeta \in (0, |\zeta_0|]$ satisfying $\int_0^{|\zeta_0|} \sigma(|\zeta|)d|\zeta| = 1$ (see Fig. 4.2). The σ is defined in terms of $|\zeta|$ in order to combine the two parts on the two edges.

The variational equation becomes

$$(\mathrm{P}) \int_{\zeta_2}^{\zeta_1} \frac{\sigma_-(\tau)}{\sqrt{\zeta} - \sqrt{\tau}}d\tau + (\mathrm{P}) \int_0^{\zeta_0} \frac{\sigma_+(\tau)}{\sqrt{\zeta} - \sqrt{\tau}}d\tau = \frac{1}{2}\frac{d}{d\sqrt{\zeta}}\hat{W}(\zeta), \tag{4.34}$$

for $\zeta \in (\zeta_1, \zeta_2)$ (lower edge) or $\zeta \in (0, \zeta_0)$ (upper edge), obtained from (4.24), where $\hat{W}(\zeta) = -g\zeta^{1/2} + \frac{2}{3}\zeta^{3/2}$. The point ζ is not on the cut now, differing from the previous cases, but on the edges of the cut. The complementary equation becomes

$$\int_{\zeta_2}^{\zeta_1} \frac{\sigma_-(\tau)}{\sqrt{\zeta} - \sqrt{\tau}}d\tau + \int_0^{\zeta_0} \frac{\sigma_+(\tau)}{\sqrt{\zeta} - \sqrt{\tau}}d\tau = \frac{1}{2}\frac{d}{d\sqrt{\zeta}}\hat{W}(\zeta) - \omega(\sqrt{\zeta}), \tag{4.35}$$

for $\zeta \in (0, \zeta_1)$. Then $E(\rho)$ can be expressed in terms of σ_+ and σ_-, but the details are not discussed here since in this book we want to keep the discussions in terms of ρ for consistency.

4.1.3 Fifth-Order Phase Transition

For the fifth-order discontinuity discussed in the following, it is convenient to use the parameter a as the transition variable since $g = 2a^2 + a^{-1}$ is not monotonic on $a \in (0, 1)$. The potential is now written as $W(\eta) = -(2a^2 + a^{-1})\eta + \frac{2}{3}\eta^3$.

Consider the densities in terms of the parameter a,

$$\rho(\eta) = \begin{cases} \frac{1}{\pi}\sqrt{A + 2\eta + (2a^2 + a^{-1})\eta^2 - \eta^4}, & 0 < a \leq 2^{-2/3}, \\ \frac{1}{\pi}(\eta + a)\sqrt{a^{-1} - (\eta - a)^2}, & 2^{-2/3} \leq a \leq 1, \end{cases} \tag{4.36}$$

where the first density is defined for η on Ω to be given next, and A is given by (4.14).

Lemma 4.4 *The density defined by* (4.36) *satisfies*

$$\rho(\eta) = \begin{cases} \frac{1}{\pi}(\eta + \eta_0)\sqrt{2a^2 + 2a\eta_0 - (\eta - \eta_0)^2}, & 0 < a \leq 2^{-2/3}, \\ \frac{1}{\pi}(\eta + a)\sqrt{a^{-1} - (\eta - a)^2}, & 2^{-2/3} \leq a < 1, \end{cases} \tag{4.37}$$

for $\eta \in \Omega$, where $\Omega = [\eta_0 - \sqrt{2a^2 + 2a\eta_0}, \eta_0 + \sqrt{2a^2 + 2a\eta_0}]$ when $0 < a \leq 2^{-2/3}$ with

$$\eta_0 = \left(\sqrt{a^2 + 2a^{-1}} - a\right)/2 \to 2^{2/3}, \quad (\text{as } a \to 2^{-2/3}) \tag{4.38}$$

satisfying $2a\eta_0^2 + 2a^2\eta_0 - 1 = 0$, $\eta_0^4 - \eta_0 + A = 0$ and $-\eta_0 \leq -a \leq \eta_0 - \sqrt{2a^2 + 2a\eta_0}$; and $\Omega = [a - a^{-1/2}, a + a^{-1/2}]$ when $2^{-2/3} \leq a < 1$ with the parameter a satisfying $-a \leq a - a^{-1/2}$.

Proof It can be checked that when $0 < a \leq 2^{-2/3}$ there is

$$\eta^4 - (2a^2 + a^{-1})\eta^2 - 2\eta - A = (\eta + \eta_0)^2\left((\eta - \eta_0)^2 - 2a^2 - 2a\eta_0\right).$$

Here, $-\eta_0$ is on the left side of the interval where the eigenvalue density ρ is defined on. The rest of the lemma can be verified by the elementary calculations. $\qquad\square$

Define

$$\omega(\eta) = \begin{cases} (\eta + \eta_0)\sqrt{(\eta - \eta_0)^2 - 2a^2 - 2a\eta_0}, & 0 < a \leq 2^{-2/3}, \\ (\eta + a)\sqrt{(\eta - a)^2 - a^{-1}}, & 2^{-2/3} \leq a \leq 1, \end{cases} \tag{4.39}$$

in the complex plane outside the cut Ω. We see that $\omega(\eta)$ satisfies

$$\omega(\eta) = \frac{1}{2}W'(\eta) - \eta^{-1} + O(\eta^{-2}), \tag{4.40}$$

as $\eta \to \infty$, and then

$$(\text{P})\int_\Omega \frac{\rho(\lambda)}{\eta - \lambda}d\lambda = \frac{1}{2}W'(\eta), \quad \eta \in \Omega, \tag{4.41}$$

where $W(\eta) = -(2a^2 + a^{-1})\eta + \frac{2}{3}\eta^3$ and $' = \partial/\partial\eta$. Now, consider

$$E(a) = \int_\Omega \left(\frac{2}{3}\eta^3 - (2a^2 + a^{-1})\eta\right)\rho(\eta)d\eta - \int_\Omega \int_\Omega \ln|\lambda - \eta|\rho(\lambda)\rho(\eta)d\lambda d\eta.$$
(4.42)

Since ρ is defined on one interval in both $a \le 2^{-2/3}$ and $a \ge 2^{-2/3}$ cases, the first-order derivatives in these two cases have the following formulation,

$$\frac{d}{da}E(a) = -(4a - a^{-2})\int_\Omega \eta\rho(\eta)d\eta,$$
(4.43)

that implies

$$\frac{d}{da}E(a) = \begin{cases} -a^5 + \frac{5}{4}a^2 - \frac{1}{4}a^{-1} - (a^5 - \frac{1}{4}a^2)(1 + 2a^{-3})^{3/2}, & 0 < a \le 2^{-2/3}, \\ -4a^2 + \frac{1}{2}a^{-1} + \frac{1}{8}a^{-4}, & 2^{-2/3} \le a < 1. \end{cases}$$
(4.44)

By direct calculations, one can find that the E function has continuous derivatives up to the fourth order with the values 0, $-9 \times 2^{1/3}$, 36 and $-252 \times 2^{2/3}$ of $d^j E/da^j$ for $j = 1, 2, 3, 4$ respectively at $a = 2^{-2/3}$. But the fifth-order derivative is not continuous. Then we get a fifth-order transition model summarized in the following.

Theorem 4.2 *When $0 < a < 1$, the E function (4.42) has continuous derivatives $d^j E/da^j$ for $j = 1, 2, 3, 4$, but the fifth-order derivative is discontinuous at the critical point $a = 2^{-2/3}$.*

However, when $a > 2^{-2/3}$ if we consider the transition in the g direction,

$$g = 2a^2 + a^{-1},$$
(4.45)

a power-law divergence occurs at the third-order derivative at the critical point $g^c = 3 \times 2^{-1/3}$ which is corresponding to $a = 2^{-2/3}$. By (4.44) for $a > 2^{-2/3}$, it is not hard to get that

$$\frac{dE(g)}{dg} = -a - \frac{1}{8}a^{-2},$$
(4.46)

and

$$\frac{d^2 E(g)}{dg^2} = -\frac{1}{4}a^{-1}.$$
(4.47)

Since (4.45) permits the following expansions

$$g = g^c + \varepsilon^2, \qquad a = 2^{-2/3} + 6^{-1/2}\varepsilon + \cdots,$$
(4.48)

we then have the power-law divergence for the third-order derivative in the g direction

$$\frac{d^3 E(g)}{dg^3} = O\left(\frac{1}{|g - g^c|^{1/2}}\right),$$
(4.49)

as $g \to g^c + 0$.

The above discussion has changed the fifth-order discontinuity to the third-order power-law divergence. Can we further change the third-order divergence to a first-order divergence? If we make the following change of variable

$$x = \frac{2}{3}a^3 - 3 \times 2^{-1/3}a + \ln a, \quad a > 2^{-2/3}, \tag{4.50}$$

which is an monotonic increasing function for $a > 2^{-2/3}$, then since

$$\frac{dx}{da} = 2(1 + 2^{1/3}a^{-1})(a - 2^{-2/3})^2, \tag{4.51}$$

the formula (4.44) for $a > 2^{-2/3}$ can be changed to

$$\frac{dE(x)}{dx} = \frac{(8a^3 + 1)(a^2 + 2^{-2/3}a + 2^{-4/3})}{4a^3(a + 2^{1/3})(a - 2^{-2/3})}. \tag{4.52}$$

It can be checked that as $a \to 2^{-2/3} + 0$, there is

$$x - x^c = 2(a - 2^{-2/3})^3 + O((a - 2^{-2/3})^4), \tag{4.53}$$

where $x^c = -\frac{4}{3} - \frac{2}{3}\ln 2$. Therefore, we have

$$\frac{dE(x)}{dx} = O\left(\frac{1}{|x - x^c|^{1/3}}\right), \tag{4.54}$$

as $x \to x^c + 0$. So we get a first-order divergence.

Also, the critical point can be determined by considering the vanishing case of

$$\frac{dg}{da} = 4a - a^{-2}, \tag{4.55}$$

based on the relation $g = 2a^2 + a^{-1}$ given above that also implies

$$g = 2(a - 2^{-2/3})^2 + (2^{2/3} - a^{-1/2})^2 + 3 \times 2^{-1/3} \geq 3 \times 2^{-1/3}, \tag{4.56}$$

for $a > 0$.

The above examples give a indication that in some cases the high order transition models can be reduced to low order transitions by properly changing the transition variable. In this situation, the string equations can play an important role to investigate the first-order transition problems. We have seen that the nonlinear relations obtained from the string equations can give the fractional power-law divergence for the first-order transition which is generally believed in statistical mechanics to be harder than the second-order transition problems. The above discussions are all about the odd order transitions. We will discuss in the next section that the second-order divergence of the free energy (critical phenomenon) can be obtained by using both string equation and Toda lattice.

4.2 Quartic Potential

Now, let us consider the model with the potential $V(z) = t_2 z^2 + t_4 z^4$. By the orthogonality, we have

$$p_{n,z} = -v_n z p_n + v_n [2t_2 + 4t_4 (v_n + v_{n+1} + z^2)] p_{n-1}, \qquad (4.57)$$

and the v_n's satisfy the string equation,

$$(2t_2 + 4t_4 (v_n + v_{n-1} + v_{n+1})) v_n = n. \qquad (4.58)$$

Denote $\Phi_n = e^{-\frac{1}{2}V(z)} (p_n, p_{n-1})^T$. Equation (4.57) is then changed to [11]

$$\Phi_{n,z} = A_n \Phi_n, \qquad (4.59)$$

where

$$A_n = \begin{pmatrix} -2t_4 z^3 - t_2 z - 4t_4 v_n z & v_n (4t_4 z^2 + 2t_2 + 4t_4 (v_n + v_{n+1})) \\ -4t_4 z^2 - 2t_2 - 4t_4 (v_n + v_{n-1}) & 2t_4 z^3 + t_2 z + 4t_4 v_n z \end{pmatrix}. \qquad (4.60)$$

Consequently, there is

$$-\det A_n = 4t_4^2 z^6 + 4t_4 t_2 z^4 + (t_2^2 - 4t_4 n) z^2 - v_n \left(t + 2t_4 v_n + \frac{n}{2v_n} \right)^2 + 4t_4^2 \frac{(v_n')^2}{v_n}. \qquad (4.61)$$

If we make the following scalings

$$\frac{t}{n^{1/2}} = g_2 + \frac{\xi}{n}, \qquad z = n^{1/4} \eta, \qquad v_n = n^{1/2} v, \qquad (4.62)$$

then

$$-\det A_n = n^{3/2} \left(\frac{1}{4} \eta^6 + g_2 \eta^4 + (g_2^2 - 1) \eta^2 - X + O(n^{-1}) \right), \qquad (4.63)$$

where

$$X = 4v(g_2 + 4g_4 v)^2 - 4g_4^2 \frac{v_\xi^2}{v}, \qquad (4.64)$$

where

$$2g_2 v + 12g_4 v^2 = 1. \qquad (4.65)$$

One may use the method discussed in Appendix D to find the singular values for X. The corresponding coefficient matrix C would be a 6×6 matrix. Here, we just consider three possible singular cases

$$X = X(g_2) = \begin{cases} 4g_2(9g_2^2 - 1), \\ 4v(g_2 + 4g_4 v)^2, \\ 0, \end{cases} \qquad (4.66)$$

in order to give some second-order transition examples. The formula $X = 4g_2(9g_2^2 - 1)$ is obtained such that the right hand side of (4.63) can be so factorized as shown next when n goes to infinity. Without loss of generality in the case $g_4 > 0$, we are going to discuss the transition between $X = 4g_2(9g_2^2 - 1)$ and $X = 4v(g_2 + 4g_4v)^2$ next by taking $g_4 = 1/4$ for convenience in discussion. The transition between $X = 4v(g_2 + 4g_4v)^2$ with $g_4 > 0$ and $X = 0$ has been studied in Sect. 3.5.

4.2.1 Second-Order Transition

Consider

$$\rho(\eta) = \begin{cases} \frac{1}{\pi}\sqrt{(4g_2 - \eta^2)((\frac{1}{2}\eta^2 + 2g_2)^2 + 5g_2^2 - 1)}, & 1/3 < g_2 \le 1/\sqrt{5}, \\ \frac{1}{\pi}(g_2 + v + \frac{1}{2}\eta^2)\sqrt{4v - \eta^2}, & g_2 \ge 1/\sqrt{5}, \end{cases} \tag{4.67}$$

where the equation $(\frac{1}{2}\eta^2 + 2g_2)^2 + 5g_2^2 - 1 = 0$ has four pure imaginary roots $\pm i\eta_1$ and $\pm i\eta_2$ because $1 - 5g_2^2 < (2g_2)^2$, and they become to $\pm i2/5^{1/4}$ as $g_2 \to 1/\sqrt{5} - 0$. When $g_2 \ge 1/\sqrt{5}$, g_2 and v are related by $2g_2v + 3v^2 = 1$ as given before.

When $g_2 \ge 1/\sqrt{5}$, the derivative of the free energy has the following formula

$$\frac{dE}{dg_2} = \int_{-2\sqrt{g_2}}^{2\sqrt{g_2}} \eta^2 \rho(\eta) d\eta. \tag{4.68}$$

When $g_2 \le 1/\sqrt{5}$, the eigenvalues are still distributed on one interval on the real line, but the phase includes two cuts in the complex plane, that is a case $l_1 = 1$ and $l_2 = 2$ discussed in Sect. 2.4. Since the variational equation only involves one interval on the real line, the calculation for the first-order derivative of the free energy is still like the $g_2 \ge 1/\sqrt{5}$ case. So the above formula is true for the two cases. By the asymptotics

$$\sqrt{\frac{1}{4}\eta^6 + g_2\eta^4 + (g_2^2 - 1)\eta^2 - X}$$

$$= \frac{1}{2}\eta^3 + g_2\eta - \eta^{-1} - 12g_2(6g_2^2 - 1)\eta^{-3} + O(\eta^{-5}), \tag{4.69}$$

as $\eta \to \infty$, where $X = 4g_2(9g_2^2 - 1)$, we then get

$$\frac{dE}{dg_2} = 6g_2(6g_2^2 - 1), \quad 1/3 < g_2 \le 1/\sqrt{5}. \tag{4.70}$$

Note that in the contour integral calculations, the cuts in the complex plane are involves, but these parts are canceled when we take the real or imaginary parts due to the complex conjugates. Hence, there is

$$\frac{d^2E}{dg_2^2} = 6(18g_2^2 - 1), \quad 1/3 < g_2 \leq 1/\sqrt{5}. \tag{4.71}$$

It has been obtained in Sect. 3.5 that

$$\frac{dE}{dg_2} = v(v^2 + 1), \quad g_2 \geq 1/\sqrt{5}, \tag{4.72}$$

and

$$\frac{d^2E}{dg_2^2} = 2v^2, \quad g_2 \geq 1/\sqrt{5}, \tag{4.73}$$

where $2g_2v + 3v^2 = 1$. Therefore dE/dg_2 is continuous and d^2E/dg_2^2 is discontinuous at the critical point $g_2 = 1/\sqrt{5}$ with $v = 1/\sqrt{5}$,

$$E'(1/\sqrt{5} - 0) = \frac{6}{5\sqrt{5}} = E'(1/\sqrt{5} + 0),$$
$$E''(1/\sqrt{5} - 0) = \frac{78}{5} > \frac{2}{5} = E''(1/\sqrt{5} + 0). \tag{4.74}$$

This example shows that the string equations can also create second-order transition models while the discussions before are all about first- or third-order transition models. The difference between this model and others is the four complex roots talked above. In the first-order discontinuity model in Sect. 4.1, there is no complex root.

4.2.2 Critical Phenomenon

The planar diagram model [3] is the case $X = 4v(g_2 + 4g_4v)^2$ with $g_4 < 0$, and the density has been discussed in Sect. 2.5. We want to study whether the planar diagram model has a transition or divergence at the critical point $g_2^2 + 12g_4 = 0$, which was originally given as $g_2 = 1/2$ and $g_4 = -1/48$ in [3]. One can try to use the hypergeometric-type equation to find a transition for the planar diagram model. As talked above, the coefficient matrix for the hypergeometric-type equation would be a 6×6 matrix, and the calculations will be complicated. The discussion in the following based on the double scaling will involve the parameter n in the transition, which is not directly included in the X variable talked before. The method is then different from the hypergeometric-type differential equation. We will first discuss the divergence of the third-order derivative of the free energy at $g_2 = g_2^c$, that can be obtained based on the algebraic equation derived from the string equation by using the ε-expansion method. Then in the t_2 direction, the second-order derivative is divergent at the critical point $t_2 = \sqrt{n}g_2^c$, that is a critical phenomenon. This subsection is planned to discuss the main steps in deriving the divergences for the planar diagram model. The details will be given in next section for the generalized planar diagram model.

Based on (4.65), denote

$$R_c = 2g_2^c + 12g_4^c v - v^{-1}, \qquad (4.75)$$

where $g_2^c = v_c^{-1}$ and $g_4^c = -\frac{1}{12}v_c^{-2}$ with a constant number $v_c > 0$. Let

$$v = v_c(1 + \alpha\varepsilon + \cdots). \qquad (4.76)$$

We have

$$R_c = -\alpha^2 v_c^{-1}\varepsilon^2 + \cdots. \qquad (4.77)$$

According to the restriction condition $2g_2 + 12g_4 v - v^{-1} = 0$ and (4.77), if g_4 is a constant $g_4 = g_4^c$, then we need to choose $g_2 = g_2^c + \varepsilon^2$ to have the coefficient of ε^2 in the restriction condition equal to 0. Then we get $\alpha^2 = 2v_c$ that implies $dv/dg_2 = O(\varepsilon^{-1})$. Since $d^2 E/dg_2^2 = -2v$ as discussed in last chapter, there is

$$\frac{d^3 E(g_2)}{dg_2^3} = -2\frac{dv}{dg_2} = O\big(|g_2 - g_2^c|^{-1/2}\big), \qquad (4.78)$$

as $g_2 \to g_2^c + 0$. If we consider $g_2 = g_2^c$ and $g_4 = g_4^c + \varepsilon^2$, there is $\alpha^2 = 12v_c$. Since $d^2 E/dg_4^2 = -36v^4$, there is a corresponding power-law divergence $d^3 E(g_4)/dg_4^3 = O(|g_4 - g_4^c|^{-1/2})$ as $g_4 \to g_4^c + 0$. The critical point in the above third-order divergence can be directly obtained from the relation $2g_2 + 12g_4 v - v^{-1} = 0$. If we take g_4 as a negative constant and consider the vanishing point of

$$\frac{dg_2}{dv} = -6g_4 - \frac{1}{2v^2}, \qquad (4.79)$$

then the critical point is obtained as $g_2^c = 2\sqrt{-3g_4}$ when $v = 1/\sqrt{-12g_4}$. And g_2^c is the lower bound of g_2 since $g_2 = \frac{1}{2}(v^{-1} + 12(-g_4)v) \geq 2\sqrt{-3g_4} = g_2^c$, that indicates we need to use different method to discuss the case $g_2 < g_2^c$. We can investigate the density model when $g_2 < g_2^c$ based on the general density model $\frac{1}{n\pi}\sqrt{\det A_n(z)}dz$, in which case the g_j parameters need to be changed to the t_j parameters. It will be discussed in Sect. 4.3.1 that the second-order derivative of the free energy function is consistent with the formula $\partial^2 \ln Z_n/\partial t_2^2 = v_n(v_{n+1} + v_{n-1})$ when the density above is on one interval with some cuts in the complex plane. Then the first- and second-order derivatives of the free energy function are continuous at the critical point for the planar diagram model, even the models on the two sides of the critical point look quite different. This discussion at least brings the information that the planar diagram model is like the Gross-Witten model that they both have third-order discontinuity or divergence. The interesting thing is that these models and their generalizations, to be discussed in next section, have second-order divergences if we slightly change the direction of the transition as explain in the following.

The formula above can be changed to a new expression

$$\frac{\partial^2}{\partial t_2^2} \ln Z_n = 2 v_n v_{n-1} - \frac{d v_n}{d t_2}, \tag{4.80}$$

by using the relation $v_n v_{n+1} = v_n v_{n-1} - d v_n / d t_2$ in order to eliminate the index $n+1$ in the formula and get a closed system since in the formula of Z_n there are just n random variables z_j ($1 \le j \le n$). This change more clearly shows the dynamics of the model in the t_2 direction, and we can study what can happen in the t_2 direction in large-N asymptotics. It will be discussed in Sect. 4.3.3 by using double scaling that the string equation $(2t_2 + 12t_4(v_{n-1} + v_n + v_{n+1}))v_n = n$ can be reduced to a nonlinear differential equation in the t_2 direction with the asymptotics $t_2 = t_2^c(1 + O(n^{-4/5}))$ and $v_n = v_n^c(1 + O(n^{-2/5}))$. So we have $d v_n / d t_2 = O(|t_2 - t_2^c|^{-1/2})$ as $t_2 \to t_2^c - 0$, that implies the critical phenomenon (second-order divergence)

$$\frac{\partial^2 \ln Z_n}{\partial t_2^2} = O\left(\frac{1}{|t_2 - t_2^c|^{1/2}}\right), \tag{4.81}$$

as $t_2 \to t_2^c - 0$, where $t_2^c = \sqrt{n} g_2^c$ to be discussed later in the double scaling. When the parameter passes the critical point, the v_n and v_{n+1} will be scaled in the same way to get back to the planar diagram model in the region $g_2 > g_2^c$, where the divergence occurs at the third-order derivatives of the free energy function as discussed above.

The third-order and second-order divergences are mutually related each other, that show the singularities in the different directions, g_2 or t_2. The t_2 direction seems local since the t_2^c involves parameter n, while in the g_2 direction, the parameter n completely disappears. If we compare the divergences with the discontinuities, it may be found that the divergence comes from the slight change of the integrable system, and the discontinuities are due to relatively heavy changes, although all the reductions are just tiny deformations from the integrable systems. The third-order transition models discussed in last chapter are based on the periodic reduction on the Lax pair by replacing $u_{n-lq+s-1}$ and v_{n-lq+s} by $x_n^{(s)}$ and $y_n^{(s)}$ respectively. The above divergences are obtained by considering the singularities of the X variable in large-N asymptotics.

The density models for the second-order transitions in the discussion for the t_2 direction above and Sect. 4.2.1 have cuts in complex plane that are different from other density models we have talked before. This would indicate that the second-order transition models do not completely stay in the momentum aspect. In fact, the Toda lattice involved in the discussion is a dynamic model that is researched to find the position properties. And the large-N asymptotics to be discussed in Sect. 4.3.3 transforms the string system to a differential equation system which is equivalent to the KdV system [11], indicating a relation to the position aspect at the critical point. In the second-order transition model considered in Sect. 4.2.1, there are complex roots in the first density in (4.67), which become $\pm i2/5^{1/4}$ as $g_2 \to 1/\sqrt{5}$. The large-N reduction of the discrete integrable system to the KdV system discussed in [11] is at a complex point $z = iy_0$, say, for the 2D quantum gravity model. The corresponding density problems with cuts in the complex plane might be related to the different types of large-N asymptotics with complex coefficients. These properties raise new problems such as the correspondences between the complex numbers in the different methods, and introduce a possibility to understand the singularities in the second-order transition problems.

Furthermore, it is known that the KdV system can be reduced to a second-order transition model by the q-periodicity constraints $k_{j+M} = qk_j$ and $\theta_{j+M}^{(0)} = \theta_j^{(0)}$ for the parameters in the τ-function of the KdV equation discussed in [16–18] by Loutsenko and Spiridonov. It is unknown whether there is a relation between the periodic constraint and the periodic reduction in the string equation discussed above. Also, the string system can give the critical phenomenon as discussed in this section, that is usually studied in the renormalization group theory. The integrable systems provide a structure to balance the invariance and the renormalization by properly working on the reduction in the discrete or continuum direction(s). Interested readers can find more discussions in the literatures, for example, [1, 2, 4, 7–10, 12, 13, 15, 22] (matrix models and phase transitions), [5, 14] (phase transition in quantum chromodynamics), [6, 19, 21] (Wilson loops) and much more. In the next section, we will discuss the general quartic potential to extend the planar diagram model and analyze what a difference it will make.

4.3 General Quartic Potential

4.3.1 Density Model with Discrete Parameter

Now, we are going to discuss the transition for a generalized planar diagram model for the potential

$$V(z) = t_1 z + t_2 z^2 + t_3 z^3 + t_4 z^4, \tag{4.82}$$

with the technical details that have been omitted in Sect. 4.2.2. The following results are based on the Lax pair discussed in Chap. 2.

Consider the orthogonal polynomials $p_n(z) = z^n + \cdots$ defined by $\langle p_n, p_{n'} \rangle \equiv \int_{-\infty}^{\infty} p_n(z) p_{n'}(z) e^{-V(z)} dz = h_n \delta_{nn'}$. The recursion formula

$$p_{n+1}(z) + u_n p_n(z) + v_n p_{n-1}(z) = z p_n(z), \tag{4.83}$$

where $v_n = h_n/h_{n-1} > 0$ can be written as

$$\Phi_{n+1} = L_n(z)\Phi_n, \tag{4.84}$$

where

$$L_n(z) = \begin{pmatrix} z - u_n & -v_n \\ 1 & 0 \end{pmatrix}. \tag{4.85}$$

And $\Phi_n(z) = e^{-\frac{1}{2}V(z)}(p_n(z), p_{n-1}(z))^T$ also satisfies

$$\frac{\partial}{\partial z}\Phi_n = A_n(z)\Phi_n, \quad n \geq 2, \tag{4.86}$$

where

$$A_n(z) = \begin{pmatrix} \gamma_n & v_n\delta_n \\ -\delta_{n-1} & -\gamma_n \end{pmatrix},$$

$$\gamma_n = -v_n(3t_3 + 4t_4(u_n + u_{n-1} + z)) - \frac{1}{2}V'(z), \tag{4.87}$$

$$\delta_n = 2t_2 + 3t_3 u_n + 4t_4(u_n^2 + v_n + v_{n+1}) + (3t_3 + 4t_4 u_n)z + 4t_4 z^2.$$

The coefficients u_n and v_n are functions of the potential parameters t_j, $j = 1, 2, 3, 4$. Initially, $u_0 = \langle z, 1 \rangle / h_0$ and $v_0 = 0$ corresponding to $p_1 + u_0 p_0 = z p_0$, where $p_0 = 1$. Also, $u_1 = \langle z p_1, p_1 \rangle / h_1$ and $v_1 = \langle z p_1, 1 \rangle / h_0$. When $n \geq 2$, u_n and v_n can be obtained from the recursion relations

$$\left[2t_2 + 3t_3(u_n + u_{n-1}) + 4t_4(u_n^2 + u_{n-1}^2 + u_n u_{n-1} + v_{n+1} + v_n + v_{n-1})\right]v_n = n, \tag{4.88}$$

$$t_1 + 2t_2 u_n + 3t_3(u_n^2 + v_{n+1} + v_n) + 4t_4(u_n^3 + (u_{n+1} + 2u_n)v_{n+1}$$

$$+ (2u_n + u_{n-1})v_n) = 0, \tag{4.89}$$

which are derived from the following relations based on the orthogonality of the polynomials

$$\langle p_n(z), V'(z)p_{n-1}(z) \rangle = nh_{n-1},$$

$$\langle p_n(z), V'(z)p_n(z) \rangle = 0,$$

where $n \geq 1$. The set of the two discrete equations (4.88) and (4.89) for u_n and v_n is called string equation. The consistency condition for (4.86) and (4.84) is of the form $\partial L_n / \partial z = A_{n+1}L_n - L_n A_n$. It can be verified by direct calculations that this consistency condition is equivalent to (4.88) and (4.89). So (4.86) and (4.84) are called the Lax pair for the string equation, and then we get a discrete integrable system. In Sect. 4.3.3, we will discuss in what situation the discrete integrable system can be reduced to a continuum integrable system by double scaling in order to find what causes the discontinuity of the free energy. But first, let us work on the basic properties to get ready for discussing the transition problem.

Since the roots of $\det A_n = 0$ may be distributed on the real line or in the complex plane as the parameter n changes, it is uncertain to conclude what a density model

$\sqrt{\det A_n}$ can create. If $\det A_n = 0$ has four or six real roots, that are the cases like the model discussed in Sect. 4.1.2. In the following, let us consider a case of two real roots z_1 and z_2 with four complex roots z_3, z_4 and their complex conjugates $\bar{z}_3$ and $\bar{z}_4$, where z_3 and z_4 are in the upper half plane, say. Then the density is like a case discussed in Sect. 2.4 with $l = l_1 + l_2$ where $l_1 = 1$ and $l_2 = 2$ with the formula

$$\rho(z, n) = \frac{1}{n\pi} \operatorname{Re} \sqrt{\det A_n(z)}, \quad -\infty < z < \infty, \tag{4.90}$$

or $\rho(z, n) = \frac{1}{n\pi} \sqrt{\det A_n(z)}$ for $z_1 \le z \le z_2$, satisfying

$$\int_{z_1}^{z_2} \rho(z, n)dz = 1, \tag{4.91}$$

and

$$(\mathrm{P}) \int_{z_1}^{z_2} \frac{\rho(\zeta, n)}{z - \zeta} d\zeta = \frac{1}{2n} V'(z), \tag{4.92}$$

according to the discussions in Sects. 2.3 and 2.4, where $' = \partial/\partial z$ and z is an inner point in $[z_1, z_2]$, that implies

$$\int_{z_1}^{z_2} \ln|z - \zeta| \rho(\zeta, n)d\zeta - \frac{1}{2n} V'(z) = \int_{z_1}^{z_2} \ln|z_1 - \zeta| \rho(\zeta, n)d\zeta - \frac{1}{2n} V'(z_1).$$

If the free energy function is defined as

$$E = \frac{1}{n} \int_{z_1}^{z_2} V(z)\rho(z, n)dz - \int_{z_1}^{z_2} \int_{z_1}^{z_2} \ln|z - \zeta| \rho(z, n)\rho(\zeta, n)dzd\zeta,$$

then the first-order derivative in the t_2 direction is

$$\frac{\partial E}{\partial t_2} = \frac{1}{n^2\pi} \operatorname{Re} \int_{z_1}^{z_2} z^2 \sqrt{\det A_n} dz, \tag{4.93}$$

by the similar discussion as in Sect. 3.4. To further compute the derivative of this integral, we need the consistency condition

$$A_{n,t_2} = M_{n,z} + M_n A_n - A_n M_n, \tag{4.94}$$

where $\partial \Phi_n/\partial t_2 = M_n \Phi_n$, which together with (4.84) form the Lax pair for the Toda lattice in the t_2 direction, to reduce the complexity by the coefficient matrix M_n for the linear equation in the t_2 direction, where

$$M_n = \begin{pmatrix} -\frac{1}{2}z^2 - v_n & v_n(z + u_n) \\ -z - u_{n-1} & \frac{1}{2}z^2 - u_{n-1}^2 - v_{n-1} \end{pmatrix}. \tag{4.95}$$

Since $A_n^2 - (\operatorname{tr} A_n)A_n + (\det A_n)I = 0$ where $\operatorname{tr} A_n = 0$, we have $\det A_n = -\frac{1}{2} \operatorname{tr} A_n^2$, that implies

$$(\det A_n)_{t_2} = -\operatorname{tr}(A_n A_{n,t_2})$$
$$= -\operatorname{tr}\Big(A_n(M_{n,z} + M_n A_n - A_n M_n)\Big)$$
$$= -\operatorname{tr}(A_n M_{n,z})$$
$$= n + v_n\big(2t_2 + 4t_4\big(v_n - u_n u_{n-1} - (u_n + u_{n-1})z\big)\big) - zV'(z),$$

by using the string equation. Since

$$\frac{\partial^2 E}{\partial t_2^2} = \frac{1}{4n^2\pi}\operatorname{Re}\frac{1}{i}\int_{\Omega^*} z^2 \frac{(\det A_n)_{t_2}}{\sqrt{-\det A_n(z)}}dz, \qquad (4.96)$$

where Ω^* is the contour around $[z_1, z_2]$, and $\sqrt{-\det A_n} = \frac{1}{2}V'(z) - nz^{-1} + O(z^{-2})$
as $z \to \infty$, there is

$$\frac{\partial^2 E}{\partial t_2^2} = -\frac{1}{n^2}v_n\big(v_{n+1} + v_{n-1} + (u_n + u_{n-1})^2\big), \qquad (4.97)$$

by noting that the contour integrals around the cuts between z_3 and z_4 and between $\bar{z}_3$ and $\bar{z}_4$ are canceled by taking the real part because all the parameters are real. We see that the result above is consistent with the formula $\partial^2 \ln Z_n/\partial t_2^2 = v_n(v_{n+1} + v_{n-1} + (u_n + u_{n-1})^2)$ obtained in Sect. 3.2.

4.3.2 Expansion for the Generalized Model

We are going to discuss the transition between the $\rho(z, n)$ given above and the generalized planar diagram density

$$\rho(\eta) = \frac{1}{2\pi}\big(2g_2 + 3g_3(\eta + a) + 4g_4(\eta^2 + a\eta + a^2 + 2b^2)\big)\sqrt{4b^2 - (\eta - a)^2}, \quad (4.98)$$

given in Sect. 2.5, with the parameters satisfying the following conditions

$$2g_2 + 3g_3(\eta + a) + 4g_4(\eta^2 + a\eta + a^2 + 2b^2) \geq 0, \quad \eta \in [\eta_-, \eta_+], \quad (4.99)$$
$$2g_2b^2 + 6g_3ab^2 + 12g_4(a^2 + b^2)b^2 = 1, \qquad (4.100)$$
$$g_1 + 2g_2a + 3g_3(a^2 + 2b^2) + 4g_4a(a^2 + 6b^2) = 0, \qquad (4.101)$$

where $\eta_\pm = a \pm 2b$. It has been discussed that this one-interval density model satisfies the following properties

$$\int_{\eta_-}^{\eta_+} \rho(\eta)d\eta = 1, \qquad (4.102)$$

and

$$(P) \int_{\eta_-}^{\eta_+} \frac{\rho(\lambda)}{\eta - \lambda} d\lambda = \frac{1}{2} W'(\eta), \tag{4.103}$$

where $W(\eta) = \sum_{j=0}^{4} g_j \eta^j$ and $' = \partial/\partial \eta$. These equations do not have parameter n, while the equations (4.91) and (4.92) for $\rho(z, n)$ involve the parameter n. The parameter n can provide the large-N asymptotics to meet with the singularity from the $\rho(\eta)$ model as explained in the following.

Differing from the bifurcation transition discussed in Chap. 3, the transition discussed in this section is a case that at the critical point the parameters satisfy the condition

$$2g_2 + 3g_3(\eta + a) + 4g_4(\eta^2 + a\eta + a^2 + 2b^2) = 4g_4((\eta - a)^2 - 4b^2). \tag{4.104}$$

That means the factor outside the square root in the formula of the $\rho(\eta)$ approaches to the phase in the square root at the critical point. By comparing the coefficients on both sides of the equation above, it is not hard to see that the parameter values at the critical point satisfy

$$g_1^c = -\frac{5}{3b_c}, \qquad 2g_2^c = \frac{1}{b_c^2}, \qquad 3g_3^c = \frac{1}{b_c^3}, \qquad 4g_4^c = -\frac{1}{3b_c^4}, \qquad a_c = b_c. \tag{4.105}$$

And the density $\rho(\eta)$ becomes $\frac{1}{6\pi b_c^4}(4b_c^2 - (\eta - b_c)^2)^{3/2}$. Consider (4.100) and (4.101). If we take a and b as functions of g_2 and all other parameters are constants with the critical values, then it can be checked that the condition in the implicit function theorem is not satisfied at the critical point. That leads us to investigate the divergence or critical phenomenon around the critical point.

Let us start from the free energy for the density $\rho(\eta)$. It has been obtained in Sect. 3.1 that the free energy function for the density $\rho(\eta)$ has the following explicit formula

$$E = W(a) + \frac{3}{4} - \ln b - 4g_4 b^4 - 6(g_3 + 4g_4 a)^2 b^6 - 6g_4^2 b^8. \tag{4.106}$$

We are going to use the ε-expansion method to investigate whether a derivative of the free energy has a discontinuity or divergence.

For convenience in discussion, denote

$$R_1^c \equiv 2g_2^c + 6g_3^c a + 12g_4^c(a^2 + v) - v^{-1}, \tag{4.107}$$

$$R_2^c \equiv g_1^c + 2g_2^c a + 3g_3^c(a^2 + 2v) + 4g_4^c a(a^2 + 6v), \tag{4.108}$$

where $v = b^2$. Now, let us expand a and v in terms of a small parameter ε,

$$a = b_c(1 + \alpha_1 \varepsilon + \alpha_2 \varepsilon^2 + \cdots), \tag{4.109}$$

$$v = b_c^2(1 + \beta_1 \varepsilon + \beta_2 \varepsilon^2 + \cdots). \tag{4.110}$$

Substituting the expansions (4.109) and (4.110) into (4.107) and (4.108) and comparing the coefficients of the ε terms, we first have the vanishing $O(1)$ terms,

$$2g_2^c + 6g_3^c b_c + 24g_4^c b_c^2 - v_c^{-1} = 0,$$

$$g_1^c + 2g_2^c b_c + 3g_3^c b_c^2 + 4g_4^c b_c^3 + 2\left(3g_3^c + 12g_4^c b_c\right)b_c^2 = 0,$$

according to the critical values given above. For the $O(\varepsilon)$ terms, the coefficients of ε for both R_1 and R_2 also vanish,

$$\left(6g_3^c b_c + 24g_4^c b_c^2\right)\alpha_1 + \left(12g_4^c b_c^2 + b_c^{-2}\right)\beta_1 = 0,$$

$$\left(2g_2^c b_c + 6g_3^c b_c^2 + 36g_4^c b_c^3\right)\alpha_1 + \left(6g_3^c b_c^2 + 24g_4^c b_c^3\right)\beta_1 = 0,$$

without any condition for α_1 or β_1. This has indicated a different style if we compare with the expansions in Sect. 3.4. For the $O(\varepsilon^2)$ terms, the coefficients can be combined and simplified as the following,

$$\left(6g_3^c b_c + 24g_4^c b_c^2\right)\alpha_2 + \left(12g_4^c b_c^2 + b_c^{-2}\right)\beta_2 + 12g_4^c b_c^2\alpha_1^2 - b_c^{-2}\beta_1^2 = -\frac{1}{b_c^2}\left(\alpha_1^2 + \beta_1^2\right),$$

$$\left(2g_2^c b_c + 6g_3^c b_c^2 + 36g_4^c b_c^3\right)b_c\alpha_2 + \left(3g_3^c b_c^2 + 12g_4^c b_c^3\right)\alpha_1^2 + \left(6g_3^c b_c^2 + 24g_4^c b_c^3\right)\beta_2$$

$$+ 24g_4^c b_c^3\alpha_1\beta_1 = -\frac{2}{b_c}\alpha_1\beta_1.$$

We will use these terms to find the expansion of g_2.

Since a and v are restricted by the conditions

$$R_1 \equiv 2g_2 + 6g_3 a + 12g_4\left(a^2 + v\right) - v^{-1} = 0, \tag{4.111}$$

$$R_2 \equiv g_1 + 2g_2 a + 3g_3\left(a^2 + 2v\right) + 4g_4 a\left(a^2 + 6v\right) = 0, \tag{4.112}$$

we can use $g_j = g_j^c + \varepsilon^\mu$ to create new terms to match with the $O(\varepsilon^2)$ terms discussed above so that the $O(\varepsilon^2)$ terms all disappear. If we consider $g_1 = g_1^c + \varepsilon^\mu$, it can be found that it does not add any thing to the R_1 equation to vanish the $O(\varepsilon^2)$ term. So the simplest case is the g_2 direction. This is one of the reasons that in our discussions we always consider the g_2 direction. Let

$$g_2 = g_2^c + \varepsilon^2, \qquad g_j = g_j^c, \qquad j \neq 2. \tag{4.113}$$

The above discussions yield

$$R_1 = R_1^c + 2\varepsilon^2 = \left[2 - \frac{1}{b_c^2}\left(\alpha_1^2 + \beta_1^2\right)\right]\varepsilon^2 + O\left(\varepsilon^3\right), \tag{4.114}$$

$$R_2 = R_2^c + 2a\varepsilon^2 = \left[2b_c - \frac{2}{b_c}\alpha_1\beta_1\right]\varepsilon^2 + O\left(\varepsilon^3\right). \tag{4.115}$$

Since $R_1 = R_2 = 0$, we have

$$\alpha_1^2 + \beta_1^2 = 2b_c^2, \tag{4.116}$$

$$\alpha_1\beta_1 = b_c^2. \tag{4.117}$$

The solution is $\alpha_1 = b_c$ and $\beta_1 = b_c$. Then we get the ε-expansions for a and v,

$$a = b_c(1 + b_c\varepsilon + \cdots), \tag{4.118}$$

$$v = b_c^2(1 + b_c\varepsilon + \cdots). \tag{4.119}$$

Now, come back to the free energy (4.106). By the above expansions, we have

$$E = \frac{1}{8} - \ln b_c + O(\varepsilon^2), \tag{4.120}$$

that means the first-order derivative with respect to g_2 does not have singularity at the critical point. Interested readers can experience the higher order terms in the expansions. The result should be $E = \frac{1}{8} - \ln b_c + O(\varepsilon^2) + O(\varepsilon^4) + O(\varepsilon^5)$, that is different from the expansion discussed in Sect. 3.4 for $g_4 > 0$. The expansion results above are enough to discuss the transition problem as explained in the following. The continuity of the first-order derivative can be seen from the integral formulas. The second-order derivative has the following formula

$$\frac{\partial^2 E}{\partial g_2^2} = -2v(2a^2 + v), \tag{4.121}$$

as discussed before, where the singularity is canceled if we use the free energy formula and the parameter conditions. We then have

$$\frac{\partial^2 E}{\partial g_2^2} = -4b_c^4 - 10b_c^5\varepsilon + O(\varepsilon^2), \tag{4.122}$$

where $g_2 - g_2^c = \varepsilon^2$, which implies

$$\frac{\partial^3 E}{\partial g_2^3} = \pm\frac{5b_c^5}{|g_2 - g_2^c|^{1/2}}(1 + o(1)), \tag{4.123}$$

where the sign depends we choose $\varepsilon = -|g_2 - g_2^c|^{1/2}$ or $\varepsilon = |g_2 - g_2^c|^{1/2}$, or simply $\partial^3 E / \partial g_2^3 = O(|g_2 - g_2^c|^{-1/2})$ as $g \to g_2^c + 0$.

The continuity of the second-order derivative of the free energy in the g_2 direction can be seen as we connect (4.121) with (4.97). The details should be discussed by using the double scaling as explained next to see how the two sets of parameters are transferred at the critical point. Locally, the divergence singularity is already hidden in the t_2 direction with the parameter n, that can be seen from the second-order derivative formula

$$\frac{\partial^2}{\partial t_2^2} \ln Z_n = v_n \big(v_{n+1} + v_{n-1} + (u_n + u_{n-1})^2 \big), \qquad (4.124)$$

when the v_{n+1} is changed to dv_n/dt_2 as discussed in Sect. 4.2.2. The discontinuity does not appear in the g_2 direction for the second-order derivative because the large-N factor removed such singularity in a global scale. Generally, there would be various critical phenomena $\frac{\partial^2 \ln Z_n}{\partial t_j^2} = O(|t_j - t_j^c|^{-\gamma_j})$ in the different t_j directions. That could be discussed by using the string equations and Toda lattice, and left for further investigations.

4.3.3 Double Scaling at the Critical Point

The nonlinear differential equation

$$y_{xx} = 6y^2 + x, \qquad (4.125)$$

has a Lax pair of the form [11]

$$\psi_{xx} = 2(y + \lambda)\psi, \qquad (4.126)$$

$$\psi_\lambda = 2y_x \psi + 4(2\lambda - y)\psi_x, \qquad (4.127)$$

that is obtained from the Lax pair of the KdV equation as discussed in [11] where (4.126) is the Schrödinger equation [11]. As a remark, the nonlinear differential equation (4.125) is called Painlevé I equation in [11]. It is discussed in [11] that when $V(z) = t_2 z^2 + t_4 z^4$, the equation (4.88) for v_n with the vanishing u_n's can be reduced to (4.125) by using a double scaling. Since their double scaling involves the complex coefficients that could be related to the nonlinear Schrödinger equation according to some literatures for a purpose different from here and we expect a double scaling with real coefficients for discussing the density and free energy function, we need to search a different double scaling with real coefficients. In the following discussions, v_n is coupled with u_{n-1}, instead of u_n, because the J_n matrix in Sect. 2.2 is defined by u_{n-1} and v_n. When the coefficients in the u_{n-1}'s expansion are equal to 0 in the following discussions, the double scaling is for the potential $V(z) = t_2 z^2 + t_4 z^4$.

We want to find real parameters c_k and the exponents r_k in the following double scaling in large-N asymptotics,

$$t_2 = n^{\frac{1}{2}} \left(g_2^c + \frac{c_1}{n^{r_1}} x \right), \qquad t_j = n^{1 - \frac{j}{4}} g_j^c, \quad j = 1, 3, 4, \qquad (4.128)$$

$$u_{n-1} = n^{\frac{1}{4}} \left(u_c + \frac{c_2}{n^{r_2}} y(x) \right), \qquad v_n = n^{\frac{1}{2}} \left(v_c + \frac{c_3}{n^{r_2}} y(x) \right), \qquad (4.129)$$

$$z = n^{\frac{1}{4}}(z_c + c_0\lambda h^2), \tag{4.130}$$

$$\psi_n(z) = \psi(x, \lambda)(1 + o(1)), \tag{4.131}$$

with

$$u_n = u_{(n-1)+1} = n^{\frac{1}{4}}\left(u_c + \frac{c_2}{n^{r_2}}\left(y + y_x h + \frac{1}{2}y_{xx}h^2 + \cdots\right)\right), \tag{4.132}$$

$$v_{n\pm 1} = n^{\frac{1}{2}}\left(v_c + \frac{c_3}{n^{r_2}}\left(y \pm y_x h + \frac{1}{2}y_{xx}h^2 + \cdots\right)\right), \tag{4.133}$$

$$\psi_{n\pm 1} = \psi \pm \psi_x h + \frac{1}{2}\psi_{xx}h^2 + O(h^3), \tag{4.134}$$

to reduce the discrete system to the continuum system given above, where $\psi_n = e^{-\frac{1}{2}V(z)}p_n(z)/h_n^{1/2}$, and h and other parameter values will be given below.

To get (4.126), rewrite (4.83) as

$$\psi_{n+1} + \psi_{n-1} - 2\psi_n + \left(\frac{u_n}{v_{n+1}^{1/2}} + 2\right)\psi_n + \left(\frac{v_n^{1/2}}{v_{n+1}^{1/2}} - 1\right)\psi_{n-1} = \frac{z}{v_{n+1}^{1/2}}\psi_n,$$

and substitute the double scaling formulas above into it. After some asymptotic calculations for the leading terms, there is

$$\psi_{xx}h^2 + \left[\frac{1}{\sqrt{v_c}}\left(u_c + \frac{c_2 y}{n^{r_2}} - \frac{c_3 u_c y}{2v_c n^{r_2}}\right) + 2\right]\psi = \frac{1}{\sqrt{v_c}}\left[z_c + c_0\lambda h^2 - \frac{c_3 z_c y}{2v_c n^{r_2}}\right]\psi.$$

If we choose

$$z_c = u_c + 2\sqrt{v_c}, \tag{4.135}$$

$$c_0 = 2\sqrt{v_c}, \tag{4.136}$$

$$c_2\sqrt{v_c} + c_3 = -2\gamma^2 v_c, \quad \gamma = hn^{r_2/2}, \tag{4.137}$$

then the above equation becomes (4.126), where γ is a constant, that means $h = O(n^{-r_2/2})$.

To get (4.127), write the first equation in (4.86) as

$$\frac{\partial}{\partial z}\psi_n = -\left[\frac{1}{2}V'(z) + v_n(3t_3 + 4t_4(u_n + u_{n-1} + z))\right]\psi_n$$

$$+ \left[2t_2 + 3t_3 u_n + 4t_4(u_n^2 + v_n + v_{n+1}) + (3t_3 + 4t_4 u_n)z\right.$$

$$\left. + 4t_4 z^2\right]v_n^{1/2}\psi_{n-1}.$$

By the asymptotic calculations, it can be verified that if the parameters additionally satisfy

$$r_2 = \frac{2}{5}, \quad \gamma^5 = \frac{3}{2}, \tag{4.138}$$

$$g_3^c + 4g_4^c u_c = 0, \qquad 12g_4^c v_c^2 = -1, \qquad (4.139)$$

and

$$2g_2^c + 3g_3^c u_c + 4g_4^c(u_c^2 + 2v_c) + (3g_3^c + 4g_4^c u_c)z_c + 4g_4^c z_c^2 = 0, \quad (4.140)$$

$$g_1^c + 2g_2^c z_c + 3g_3^c z_c^2 + 4g_4^c z_c^3 + 2v_c(3g_3^c + 4g_4^c(2u_c + z_c)) = 0, \quad (4.141)$$

then the above equation is reduced to (4.127).

If we substitute the double scaling formulas above with

$$r_1 = 2r_2, \qquad (4.142)$$

into (4.88) and (4.89), then we get

$$\gamma^2 y_{xx} = \frac{c_3^2 + v_c c_2^2}{4g_4^c c_3 v_c^3} y^2 - \frac{c_1}{2g_4^c c_3} x, \qquad (4.143)$$

$$c_2 \gamma^2 y_{xx} = -\frac{6c_2 c_3}{v_c} y^2 - \frac{c_1 u_c}{2g_4^c v_c} x. \qquad (4.144)$$

In the case $u_c \neq 0$ and $c_2 \neq 0$, if the parameters satisfy

$$c_1 = -\frac{\gamma^4}{6v_c}, \qquad c_2 = -u_c \gamma^2, \qquad c_3 = -v_c \gamma^2, \qquad u_c = \sqrt{v_c}, \qquad (4.145)$$

then the above two equations (4.143) and (4.144) both become to (4.125). If $u_c = c_2 = 0$ and

$$c_1 = -\frac{\gamma^4}{3v_c}, \qquad c_3 = -2v_c \gamma^2, \qquad v_c > 0, \qquad (4.146)$$

then (4.144) becomes $0 = 0$, and (4.143) becomes (4.125). In the first case, we have the critical point

$$g_1^c = -\frac{5}{3\sqrt{v_c}}, \qquad 2g_2^c = \frac{1}{v_c}, \qquad 3g_3^c = \frac{1}{v_c\sqrt{v_c}}, \qquad 4g_4^c = -\frac{1}{3v_c^2}, \qquad (4.147)$$

which is consistent to (4.105). In the second case, the critical point is

$$2g_2^c = \frac{2}{v_c}, \qquad 4g_4^c = -\frac{1}{3v_c^2}, \qquad (4.148)$$

with $g_1^c = g_3^c = 0$, which is the critical point for the planar diagram model.

If we work on the higher degree potential, then the recursion formula still becomes the Schrödinger equation, and the z equation $\Psi_{n,z} = A_n \Psi_n$ becomes a linear equation coupled with the Schrödinger equation to form a Lax pair for the higher order equation in the continuum hierarchy which is corresponding to the KdV hierarchy. However, the t_j equations $\Psi_{n,t_j} = M_n^{(j)} \Psi_n$ can not join with the double scaling because the hierarchy does not have the corresponding equations. This property affects the whole system, and the critical phenomena are researched based on this singular property. The critical exponent in the critical phenomenon can be studied based on the expansions including the important relation $r_1 = 2r_2$ discussed above.

The double scaling reveals that a small system at the point $z = z_c$ is separated at the critical point to reduce the discrete integrable system to a physical model such as the planar diagram model. The release of the small system from the integrable system would help us to think about the cause of the transition. The double scaling method can also reduce the Hermitian matrix model to another continuum integrable system as researched in [2] for the corresponding critical points. In the two-matrix models [8–10], the large-N scaling method is applied to study the correlation function by using the loop equation or Schwinger-Dyson equation, which plays a role as the string equation, in order to get the Yang-Baxter equation. The scalings for the whole system as discussed in [2, 8–11, 20] and this section, for instance, provide a new interpretation for the discontinuity of the transition phenomena, so that we have more tools to study these challenging problems.

4.4 Searching for Fourth-Order Discontinuity

We have discussed the discontinuities of the first-, second-, third- and fifth-order derivatives of the free energy before. In this section, we are going to search a fourth-order discontinuity by considering the potential

$$W(\eta) = g_1\eta + g_2\eta^2 + g_3\eta^3. \tag{4.149}$$

We will choose $g_3 > 0$ in the following discussion. The discussion for $g_3 < 0$ is similar to the $g_3 > 0$ case. By choosing $g_4 = 0$ in (4.98) and the corresponding conditions, we have

$$\rho_2(\eta) = \frac{1}{2\pi}(2g_2 + 3g_3(\eta + u))\sqrt{4v - (\eta - u)^2}, \tag{4.150}$$

and

$$R_1 \equiv 2g_2 + 6g_3u - v^{-1} = 0, \tag{4.151}$$

$$R_2 \equiv g_1 + 2g_2u + 3g_3(u^2 + 2v) = 0. \tag{4.152}$$

The critical point of the fourth-order transition is when the parameters satisfy the condition

$$2g_2 + 3g_3(\eta + u) = 3g_3(\eta - u + 2\sqrt{v}). \tag{4.153}$$

Combining this condition with (4.151) and (4.152), we get the values or the relations of the parameters for the critical point,

$$g_1 = -3/(2u_c), \qquad g_2^c = 0, \qquad 6g_3^c u_c v_c = 1, \qquad u_c = b_c = \sqrt{v_c}. \tag{4.154}$$

If one wants to discuss the case $u_c = -\sqrt{v_c}$, the straight line should be put to the right side of the semicircle for the density. It can be seen that for the model (4.150), the parameter g_2 is positive ($g_2 \geq g_2^c$).

Denote

$$R_1^c \equiv 2g_2^c + 6g_3^c u - v^{-1}, \tag{4.155}$$

$$R_2^c \equiv g_1^c + 2g_2^c u + 3g_3^c(u^2 + 2v), \tag{4.156}$$

with the expansions

$$u = b_c(1 + \alpha_1\varepsilon + \alpha_2\varepsilon^2 + \alpha_3\varepsilon^3 + \cdots), \tag{4.157}$$

$$v = v_c(1 + \beta_1\varepsilon + \beta_2\varepsilon^2 + \beta_3\varepsilon^3 + \cdots), \tag{4.158}$$

where $\varepsilon = g_2 - g_2^c$. We have for the $O(1)$ terms,

$$O(1): \quad 2g_2^c + 6g_3^c b_c - v_c^{-1} = 0,$$

$$g_1^c + 2g_2^c b_c + 3g_3^c(b_c^2 + 2v_c) = 0,$$

by the relations (4.154). For the $O(\varepsilon)$ terms, there are

$$O(\varepsilon): \quad 6g_3^c b_c \alpha_1 + v_c^{-1}\beta_1,$$

$$(2g_2^c b_c + 6g_3^c b_c^2)\alpha_1 + 6g_3^c v_c \beta_1.$$

One can experience that if we do not choose $\beta_1 = 0$, then there will be a contradiction in the $O(\varepsilon^2)$ terms. Therefore, we get $\alpha = \beta_1 = 0$, consequently implying easier calculation in the rest analysis. For the $O(\varepsilon^2)$ terms, we have

$$O(\varepsilon^2): \quad 2 + 6g_2^c b_c \alpha_2 + v_c^{-1}\beta_2,$$

$$2b_c + 2g_2^c b_c \alpha_2 + 3g_3(2b_c^2\alpha_2 + 2v_c\beta_2).$$

By choosing $g_2 = g_2^c + \varepsilon$ in order to finally have $R_1 = R_1^c + 2\varepsilon = O(\varepsilon^3)$ and $R_2 = R_2^c + 2b_c\alpha_1\varepsilon^2 = O(\varepsilon^3)$, we get the following equations for the coefficients,

$$\alpha_1 + \beta_1 = -2b_c^2,$$

$$\alpha_2 + \beta_2 = \beta_1^2,$$

$$\alpha_2 + \beta_2 = -\frac{1}{2}\alpha_1^2 - 2b_c^2\alpha_1.$$

It can be solved that

$$\alpha_1 = \left(-2 \pm \frac{2}{\sqrt{3}}\right)b_c^2, \qquad \beta_1 = \mp\frac{2}{\sqrt{3}}b_c^2, \qquad \alpha_2 + \beta_2 = \frac{3}{4}b_c^4. \tag{4.159}$$

Since $\partial^2 E/\partial g_2^2 = -2v(v + 2u^2)$, it can be derived by the expansions above that

$$\frac{\partial^2}{\partial g_2^2} E = -2b_c^2 \left(3b_c^2 - 8b_c^4 \varepsilon + 12b_c^6 \varepsilon^2 + \cdots \right), \tag{4.160}$$

which implies that

$$\frac{\partial^3}{\partial g_2^3} E(0+) = 16b_c^6, \qquad \frac{\partial^4}{\partial g_2^4} E(0+) = -48b_c^8. \tag{4.161}$$

Now, consider

$$A_n(z) = \begin{pmatrix} \alpha_n & v_n \beta_n \\ -\beta_{n-1} & -\alpha_n \end{pmatrix}, \tag{4.162}$$

with

$$\alpha_n = -3t_3 v_n - \frac{1}{2}\left(t_1 + 2t_2 z + 3t_3 z^2\right),$$

$$\beta_n = 2t_2 + 3t_3(u_n + z),$$

followed by

$$\sqrt{\det A_n(z)} = \sqrt{v_n \beta_n \beta_{n-1} - \alpha_n^2}, \tag{4.163}$$

and

$$\frac{1}{n\pi} \operatorname{Re} \sqrt{\det A_n(z)} dz = \rho(\eta) d\eta, \tag{4.164}$$

with the scaling $t_1 n^{-2/3}$, $t_2 n^{-1/3}$, t_3, $u_n n^{-1/3}$, $u_{n-1} n^{-1/3}$, $v_n n^{-2/3}$ and $z n^{-1/3}$ replaced by g_1, g_2, g_3, u, $\hat{u}$, v and η respectively. Since

$$\frac{1}{n^{5/3}\pi} \int z^2 \operatorname{Re} \sqrt{\det A_n(z)} dz = \int \eta^2 \rho(\eta) d\eta, \tag{4.165}$$

if $\det A_n = 0$ has two real roots and two complex roots as $n \to \infty$, then the first-order derivative $\partial E/\partial g_2$ is continuous according to the discussion for the one-interval density models. If it is not the case, then it is a first-order discontinuity.

We have obtained in Sect. 3.2 that the second-order derivative of $\ln Z_n$ has the following formula

$$\frac{\partial^2}{\partial t_2^2} \ln Z_n = v_n \left(v_{n+1} + v_{n-1} + (u_n + u_{n-1})^2\right), \tag{4.166}$$

where Z_n is the partition function. To investigate the continuity of the higher order derivatives, the derivatives of the u_n and v_n are needed since the expansion method does not work when the model keeps the n parameter. It is not hard to get

$$2t_2 + 3t_3(u_n + u_{n-1}) = \frac{n}{v_n}, \tag{4.167}$$

$$t_1 + 2t_2 u_n + 3t_3 (u_n^2 + v_n + v_{n+1}) = 0, \tag{4.168}$$

and

$$\frac{\partial u_n}{\partial t_2} = -(u_{n+1} + u_n)v_{n+1} + (u_n + u_{n-1})v_n, \tag{4.169}$$

$$\frac{\partial v_n}{\partial t_2} = -v_n (v_{n+1} - v_{n-1} + u_n^2 - u_{n-1}^2). \tag{4.170}$$

The parameters at the critical point satisfy

$$t_2 = 0, \qquad 3t_3(u_n + u_{n-1})v_n = n. \tag{4.171}$$

By these relations, it can be derived that

$$\frac{\partial u_n}{\partial t_2} = -\frac{1}{3t_3} + \frac{2t_2}{3t_3}(v_{n+1} - v_n), \tag{4.172}$$

$$\frac{\partial v_n}{\partial t_2} = \frac{2t_2}{3t_3} v_n (u_n - u_{n-1}), \tag{4.173}$$

which imply at the critical point,

$$\frac{\partial u_n}{\partial t_2} = -\frac{1}{3t_3}, \qquad \frac{\partial v_n}{\partial t_2} = 0. \tag{4.174}$$

Denote

$$S_n = v_n(v_{n+1} + v_{n-1}), \qquad T_n = v_n(u_n + u_{n-1})^2,$$

for the right hand side of (4.166), and change them to the following by using (4.167) and (4.168),

$$S_n = -\frac{2t_2}{3t_3}v_n - \frac{2t_2}{(3t_3)^2}(n - 2t_2 v_n) - (u_n^2 + u_{n-1}^2)v_n - 2v_n^2,$$

$$T_n = \frac{v_n}{(3t_3)^2}\left(\frac{n}{v_n} - 2t_2\right)^2.$$

The derivatives of S_n and T_n with respect to t_2 have complicated formulas. But at the critical point $t_2 = 0$ with $u_n \sim u_{n-1}$, $v_{n+1} \sim v_n$ and $u_n \sim \sqrt{v_n}$, there are the following based on the results obtained above,

$$\frac{\partial}{\partial t_2}S_n = 0, \qquad \frac{\partial^2}{\partial t_2^2}S_n = 0, \tag{4.175}$$

and

$$\frac{\partial}{\partial t_2}T_n = -\frac{4n}{(3t_3)^2}, \qquad \frac{\partial^2}{\partial t_2^2}T_n = \frac{8v_n}{(3t_3)^2}. \tag{4.176}$$

Then, at the critical point there are

$$\frac{\partial^3}{\partial t_2^3} \ln Z_n = -\frac{4n}{(3t_3)^2}, \qquad \frac{\partial^4}{\partial t_2^4} \ln Z_n = \frac{8v_n}{(3t_3)^2}, \qquad (4.177)$$

which imply that $E = \lim_{n\to\infty} \frac{-1}{n^2} \ln Z_n$ satisfies

$$\frac{\partial^3}{\partial g_2^3} E(0-) = 16b_c^6, \qquad \frac{\partial^4}{\partial g_2^4} E(0-) = -32b_c^8. \qquad (4.178)$$

Summarizing the discussions above, we have that if $b_c \neq 0$ then in the $g = g_2$ direction there are

$$\frac{\partial^j}{\partial g_2^j} E(0-) = \frac{\partial^j}{\partial g_2^j} E(0+), \quad j = 2, 3, \qquad (4.179)$$

$$\frac{\partial^4}{\partial g_2^4} E(0-) = -32b_c^8 > -48b_c^8 = \frac{\partial^4}{\partial g_2^4} E(0+). \qquad (4.180)$$

If the first-order derivative is continuous as explained above, then this is a fourth-order discontinuity model with a $O(1)$ discontinuity at the critical point $g_2 = 0$.

References

1. Bertola, M., Marchal, O.: The partition of the two-matrix models as an isomonodromic τ function. J. Math. Phys. **50**, 013529 (2009)
2. Bleher, P., Eynard, B.: Double scaling limit in random matrix models and a nonlinear hierarchy of differential equations. J. Phys. A **36**, 3085–3106 (2003)
3. Brézin, E., Itzykson, C., Parisi, G., Zuber, J.B.: Planar diagrams. Commun. Math. Phys. **59**, 35–51 (1978)
4. Chekhov, L., Mironov, A.: Matrix models vs. Seiberg–Witten/Whitham theories. Phys. Lett. B **552**, 293–302 (2003)
5. Douglas, M.R., Kazakov, V.A.: Large N phase transition in continuum QCD in two-dimensions. Phys. Lett. B **319**, 219–230 (1993)
6. Douglas, M.R., Shenker, S.H.: Strings in less than one dimension. Rutgers preprint RU-89-34 (1989)
7. Eynard, B.: Topological expansion for the 1-Hermitian matrix model correlation functions. J. High Energy Phys. **0411**, 031 (2004)
8. Eynard, B., Orantin, N.: Topological expansion of the 2-matrix model correlation functions: diagrammatic rules for a residue formula. J. High Energy Phys. **12**, 034 (2005)
9. Eynard, B., Orantin, N.: Mixed correlation functions in the 2-matrix model, and the Bethe ansatz. J. High Energy Phys. **08**, 028 (2005)
10. Eynard, B., Orantin, N.: Invariants of algebraic curves and topological expansion. Commun. Number Theory Phys. **1**, 347–452 (2007)
11. Fokas, A.S., Its, A.R., Kitaev, A.V.: Discrete Painlevé equations and their appearance in quantum gravity. Commun. Math. Phys. **142**, 313–344 (1991)
12. Gerasimov, A., Marshakov, A., Mironov, A., Morozov, A., Orlov, A.: Matrix models of 2D gravity and Toda theory. Nucl. Phys. B **357**, 565–618 (1991)

13. Gorsky, A., Krichever, I., Marshakov, A., Mironov, A., Morozov, A.: Integrability and Seiberg-Witten exact solution. Phys. Lett. B **355**, 466–477 (1995)
14. Gross, D.J., Matytsin, A.: Instanton induced large-N phase transitions in two-dimensional and four-dimensional QCD. Nucl. Phys. B **429**, 50–74 (1994)
15. Jurkiewicz, J.: Regularization of one-matrix models. Phys. Lett. B **235**, 178–184 (1990)
16. Loutsenko, I.M., Spiridonov, V.P.: Self-similar potentials and Ising models. JETP Lett. **66**, 747–753 (1997)
17. Loutsenko, I.M., Spiridonov, V.P.: Spectral self-similarity, one-dimensional Ising chains and random matrixes. Nucl. Phys. B **538**, 731–758 (1999)
18. Loutsenko, I.M., Spiridonov, V.P.: A critical phenomenon in solitonic Ising chains. SIGMA **3**, 059 (2007)
19. Passerini, F., Zarembo, K.: Wilson loops in $N = 2$ super-Yang-Mills from matrix model. J. High Energy Phys. **1109**, 102 (2011)
20. Pastur, L., Figotin, A.: Spectra of Random and Almost Periodic Operators. Springer, Berlin (1992)
21. Rey, S.-J., Suyama, T.: Exact results and holography of Wilson loops in $N = 2$ superconformal (quiver) gauge theories. J. High Energy Phys. **1101**, 136 (2011)
22. Shimamune, Y.: On the phase structure of large N matrix models and gauge models. Phys. Lett. B **108**, 407–410 (1982)

Chapter 5
Densities in Unitary Matrix Models

The unitary matrix model is another important topic in quantum chromodynamics (QCD) and lattice gauge theory. The Gross-Witten weak and strong coupling densities are the most popular density models in QCD for studying the third-order phase transition problems, which are related to asymptotic freedom and confinement. For the Gross-Witten weak and strong coupling densities and the generalizations to be discussed in this chapter, it should be noted that the densities are defined on the complement of the cuts in the unit circle, and there are two essential singularities, which are different from the Hermitian models. The orthogonal polynomials on the unit circle are applied to study these problems by using the string equation. The recursion formula now becomes the discrete AKNS-ZS system, and the reduction of the eigenvalue density is now based on new linear systems of equations satisfied by the orthogonal polynomials on the unit circle. The integrable systems and string equation discussed in this chapter provide a structure for finding the generalized density models and parameter relations that will be used as the mathematical foundation to investigate the transition problems discussed in next chapter.

5.1 Variational Equation

According to the discussion in [4], an eigenvalue density $\rho(\theta)$ on an interval or multiple disjoint intervals Ω_θ in $[-\pi, \pi]$ for the potential $2\sum_{j=1}^{m} g_j \cos(j\theta)$ should be defined to satisfy $\rho(\theta) \geq 0$, $\int_{\Omega_\theta} \rho(\theta)d\theta = 1$ and the variational equation

$$(\text{P}) \int_{\Omega_\theta} \cot\frac{\theta - \theta'}{2}\rho(\theta')d\theta' = 2\sum_{j=1}^{m} jg_j \sin(j\theta). \tag{5.1}$$

Convert the interval $[-\pi, \pi]$ to the unit circle by $z = e^{i\theta}$, then Ω_θ becomes Ω to be as an arc or multiple disjoint arcs on the unit circle, and $\rho(\theta)$ is changed to $\dot\sigma(z)$ by $\rho(\theta)d\theta = \sigma(z)dz$. For the potential $U(z) = \sum_{j=1}^{m} g_j(z^j + z^{-j})$, $\sigma(z)$ needs to

C.B. Wang, *Application of Integrable Systems to Phase Transitions*,
DOI 10.1007/978-3-642-38565-0_5, © Springer-Verlag Berlin Heidelberg 2013

satisfies $\int_\Omega \sigma(z)dz = 1$ and the variational equation

$$(P) \int_\Omega \frac{\zeta\sigma(\zeta)}{\zeta - z}d\zeta = \frac{1}{2}zU'(z) + \frac{1}{2}, \tag{5.2}$$

where z is an inner point of Ω, $' = d/dz$, $zU'(z) = z(U_0'(z) + U_0'(z^{-1})) = \sum_{j=1}^m jg_j(z^j - z^{-j})$ and $U_0(z) = \sum_{j=1}^m g_j z^j$. Note that $\sigma(z)$ is not positive, or it is even not a real function. This transformation will make the discussions much easier to apply the properties of the analytic functions. If $U(z) = T^{-1}(z + z^{-1})$, then the variational equation becomes

$$(P) \int_{\Omega_\theta} \cot\frac{\theta - \theta'}{2}\rho(\theta')d\theta' = \frac{2}{T}\sin\theta, \tag{5.3}$$

or

$$(P) \int_\Omega \frac{\zeta\sigma(\zeta)}{\zeta - z}d\zeta = \frac{1}{2T}(z - z^{-1}) + \frac{1}{2}, \tag{5.4}$$

where $z = e^{i\theta} \in \Omega$, $\theta \in \Omega_\theta$, Ω_θ is corresponding to Ω, and $\sigma(\zeta)d\zeta = \rho(\theta')d\theta'$ with $\zeta = e^{i\theta'}$. Equation (5.3) is the original variational equation for the unitary matrix model [4], and (5.4) or the general form (5.2) is derived from it in order to use the asymptotics in the complex plane to study the density problems.

We have seen in the Hermitian matrix models that the eigenvalue density can be obtained from the coefficient matrix in the Lax pair by discussing the associated orthogonal polynomials. For the unitary matrix models, it is the same idea, but the process is much complicated because now there are two essential singular points $z = 0$ and $z = \infty$. The details will be discussed in the following sections. Let us still use the Gross-Witten density models to explain how to change an eigenvalue density problem to an asymptotics problem for an analytic function $\omega(z)$. The $z\omega(z)$ needs to have an asymptotics $\frac{1}{2}z(U_0'(z) - U_0'(z^{-1})) + \frac{1}{2}$ as $z \to 0$ or $z \to \infty$. Note that $\frac{1}{2}z(U_0'(z) - U_0'(z^{-1})) + \frac{1}{2}$ is not $\frac{1}{2}zU'(z) + \frac{1}{2}$, that is an important difference from the Hermitian matrix models. Consider

$$\sigma(z)dz = \begin{cases} (1 + \frac{1}{T}(z + z^{-1}))\frac{dz}{2\pi iz}, & T \geq 2 \\ \frac{1}{T}(z^{1/2} + z^{-1/2})\sqrt{(z^{1/2} + z^{-1/2})^2 + 2(T-2)}\frac{dz}{2\pi iz}, & T \leq 2, \end{cases} \tag{5.5}$$

where $\sigma(z)$ is defined on $\Omega = \{z||z| = 1\}$ when $T \geq 2$, and defined on $\Omega = \{z||z| = 1\}\backslash\hat{\Omega}$ when $T \leq 2$, where

$$\hat{\Omega} = \{z = e^{i\theta}||\cos(\theta/2)| \leq \sqrt{1 - (T/2)}, |\theta| \leq \pi\}. \tag{5.6}$$

Note that $\hat{\Omega}$ is a cut around $z = -1$, and Ω is the complement of $\hat{\Omega}$ in the unit circle. Define $\omega(z)$ by

$$z\omega(z) = \begin{cases} \frac{1}{2}(1 + \frac{1}{T}(z + z^{-1})), z \in \mathbb{C}, & T \geq 2 \\ \frac{1}{2T}(z^{1/2} + z^{-1/2})\sqrt{(z^{1/2} + z^{-1/2})^2 + 2(T-2)}, z \in \mathbb{C}\backslash\hat{\Omega}, & T \leq 2. \end{cases}$$
$$\tag{5.7}$$

It can be checked that $\sigma(z)dz = \frac{1}{2\pi}(1 + \frac{2}{T}\cos\theta)d\theta$ for $T \geq 2$, and

$$\sigma(z)dz = \frac{2}{\pi T}\cos\frac{\theta}{2}\sqrt{\frac{T}{2} - \sin^2\frac{\theta}{2}}d\theta$$

for $T \leq 2$. So the $\sigma(z)$ is the complex variable form of the Gross-Witten density models [4]. In both cases, there is $\omega(z) = \pi i \sigma(z)$ for $z \in \Omega$. This property is also different from the Hermitian matrix models since Ω is now in the domain where $\omega(z)$ is defined. There is

$$z\omega(z) = \frac{1}{2T}(z + z^{-1}) + \frac{1}{2} + O((z^{1/2} + z^{-1/2})^{-1}), \qquad (5.8)$$

as $|z^{1/2} + z^{-1/2}| \to \infty$ for both $T \leq 2$ and $T \geq 2$. Note that if $z^{1/2} + z^{-1/2} = 0$, then $z = -1$ which is a point in the cut $\hat{\Omega}$, excluded from the domain of the $\omega(z)$. Therefore, we have obtained an analytic function with the expected asymptotics at 0 and ∞ as mentioned above.

To see why the $\sigma(z)$ satisfies the variational equation, let us consider a simple case: $T \geq 2$. For a small $\varepsilon > 0$, make a small circle of radius ε with center z ($|z| = 1$). Denote the "semicircle" inside the unit circle as γ_ε^-, and the outside "semicircle" as γ_ε^+. Remove the small arc on the unit circle around z enclosed by the ε circle above from the unit circle, and the remaining arc is denoted as $\Omega(\varepsilon)$. As $\varepsilon \to 0$, $\Omega(\varepsilon)$ tends to Ω, so that $\int_{\Omega(\varepsilon)} \to (P)\int_\Omega$, that will simplify the discussions. Denote $\Omega^- = \Omega(\varepsilon) \cup \gamma_\varepsilon^-$ and $\Omega^+ = \Omega(\varepsilon) \cup \gamma_\varepsilon^+$. All the closed contours are oriented counterclockwise. Then, we have

$$\frac{1}{2\pi i}\int_{\Omega(\varepsilon)} \frac{1}{\zeta - z}(1 + T^{-1}(\zeta + \zeta^{-1}))d\zeta$$

$$= \frac{1}{2\pi i}\int_{\Omega^-} \frac{1}{\zeta - z}d\zeta + \frac{1}{2T\pi i}\int_{\Omega^-} \frac{1}{\zeta - z}\zeta d\zeta + \frac{1}{2T\pi i}\int_{\Omega^+} \frac{1}{\zeta - z}\zeta^{-1}d\zeta$$

$$- \frac{1}{2\pi i}\int_{\gamma_\varepsilon^-} \frac{1}{\zeta - z}d\zeta - \frac{1}{2T\pi i}\int_{\gamma_\varepsilon^-} \frac{1}{\zeta - z}\zeta d\zeta - \frac{1}{2T\pi i}\int_{\gamma_\varepsilon^+} \frac{1}{\zeta - z}\zeta^{-1}d\zeta$$

$$\to \frac{1}{2} + \frac{1}{2T}(z - z^{-1}),$$

as $\varepsilon \to 0$, where all the three integrals along Ω^+ or Ω^- vanish. And then the variational equation for $T \geq 2$ is satisfied. The variational equation for $T \leq 2$ can be obtained by using the asymptotics (5.8), that will be discussed with other complicated densities in the later sections based on the asymptotics.

The free energy function in the unitary matrix model [4]

$$E = -\frac{2}{T}\int_{\Omega_\theta} \cos\theta\rho(\theta)d\theta - \int_{\Omega_\theta}\int_{\Omega_\theta} \ln\left|\sin\frac{\theta - \theta'}{2}\right|\rho(\theta)\rho(\theta')d\theta d\theta' - \ln 2, \quad (5.9)$$

can be changed or generalized to

$$E = \int_{\Omega} (-U(z)) \sigma(z) dz - \int_{\Omega} \int_{\Omega} \ln |z - \zeta| \sigma(z) \sigma(\zeta) dz d\zeta, \qquad (5.10)$$

where Ω is the arc(s) in the z plane corresponding to Ω_θ. For the general potential $U(z)$, the free energy will have similar results as the Gross-Witten model [4], and the phase transition problems will be discussed in detail in the next chapter. In the next section, we first consider the orthogonal polynomials in order to get an analytic function with the expected asymptotics $\frac{1}{2}z(U_0'(z) - U_0'(z^{-1})) + \frac{1}{2}$.

5.2 Recursion and Discrete AKNS-ZS System

The orthogonal polynomials on the unit circle $|z| = 1$ in the complex plane, specially for the weight $\exp\{s(z + z^{-1})\}$, have been studied many years ago to extend the orthogonal polynomials on the real line such as the Hermite polynomials, see [11]. Many literatures have been published about the orthogonal polynomials on the unit circle in association with the unitary matrix models or the circular ensembles in the random matrix theory. Interested readers can also find the recent developments in this field in [9, 10]. In this book, we consider the orthogonal polynomials on the unit circle in association with the string equation and the phase transition problems in the unitary matrix models. In this chapter, we consider the orthogonal polynomials $p_n(z) = z^n + \cdots$ on the unit circle with a general weight function $\exp V(z) = \exp\{\sum_{j=1}^{m} s_j(z^j + z^{-j})\}$, satisfying

$$\langle p_n(z), p_m(z) \rangle \equiv \oint p_n(z) \bar{p}_m(z) e^{V(z)} \frac{dz}{2\pi i z} = h_n \delta_{nm}, \qquad (5.11)$$

where s_j are real parameters, $\bar{p}_m(z)$ is the complex conjugate of $p_m(z)$, and the integral is along the unit circle $|z| = 1$ counterclockwise. For convenience in discussion, denote $V(z) = V_0(z) + V_0(z^{-1})$, where

$$V_0(z) = \sum_{j=1}^{m} s_j z^j. \qquad (5.12)$$

There are many interesting properties for the orthogonal polynomials on the unit circle, that will be explained in the later discussions. In this section, we talk about the recursion formula

$$z(p_n + v_n p_{n-1}) = p_{n+1} + u_n p_n, \qquad (5.13)$$

obtained from the Szegö's equation [12]. The recursion formula is fundamental in our discussions as we have seen in the previous chapters.

On the unit circle $z = e^{i\theta}$, $V(e^{i\theta}) = \sum_{j=1}^{m} 2s_j \cos(j\theta)$ is an even function of θ, and $\exp\{V(e^{i\theta})\}$ is a positive function of period 2π. Then the coefficients of p_n are real [11]. The orthogonal polynomials satisfy the recursion formula [11],

$$p_{n+1} = zp_n + p_{n+1}(0)p_n^*, \tag{5.14}$$

where

$$p_n^* = z^n \bar{p}_n(z).$$

The discussions in [6, 7], for instance, are based on this recursion formula. By taking complex conjugates on both sides of (5.14), and restricting z on the unit circle, we have $\bar{p}_{n+1}(z) = (1/z)\bar{p}_n(z) + p_{n+1}(0)z^{-n}p_n(z)$, which implies $p_{n+1}^* = p_n^* + p_{n+1}(0)zp_n$. Combining this equation with (5.14), we have

$$\begin{pmatrix} p_{n+1} \\ p_{n+1}^* \end{pmatrix} = \begin{pmatrix} z & p_{n+1}(0) \\ zp_{n+1}(0) & 1 \end{pmatrix} \begin{pmatrix} p_n \\ p_n^* \end{pmatrix}. \tag{5.15}$$

This is the recursion formula given in [11], called Szegö's equation.

In the following, we consider a more general form of (5.15) as discussed in [5]. The result is independent of the orthonormal polynomials. The system (iii) below is called the discrete AKNS-ZS system. The $\tilde{p}_n$ in the theorem is a generalization of the p_n^* above, and $\hat{p}_n$ is a generalization of $\bar{p}_n$. And we will take $\gamma_0 = -1/2$ in the later discussions, in which case $\tilde{p}_n$ becomes p_n^*, and $\hat{p}_n$ becomes $\bar{p}_n$.

Theorem 5.1 *Suppose r_n and q_n are arbitrary functions of n, and that $p_n(z)$ and $\hat{p}_n(z)$ are complex functions of z, depending on n. Then the following three systems are equivalent in the sense that any one can be obtained from any other.*

(i)

$$\begin{cases} z(p_n + v_n p_{n-1}) = p_{n+1} + u_n p_n, \\ zp_n = p_{n+1} - q_n \tilde{p}_n, \end{cases} \tag{5.16}$$

where $\tilde{p}_n = z^{n+\gamma_0+1/2}\hat{p}_n$, γ_0 is a constant, and

$$u_n = -\frac{q_n}{q_{n-1}}, \tag{5.17}$$

$$v_n = -\frac{q_n}{q_{n-1}}(1 - r_{n-1}q_{n-1}). \tag{5.18}$$

(ii)

$$\begin{pmatrix} p_{n+1} \\ \tilde{p}_{n+1} \end{pmatrix} = \begin{pmatrix} z & q_n \\ zr_n & 1 \end{pmatrix} \begin{pmatrix} p_n \\ \tilde{p}_n \end{pmatrix}. \tag{5.19}$$

(iii)

$$\begin{pmatrix} \chi_{n+1} \\ \hat{\chi}_{n+1} \end{pmatrix} = \begin{pmatrix} \eta & q_n \\ r_n & \eta^{-1} \end{pmatrix} \begin{pmatrix} \chi_n \\ \hat{\chi}_n \end{pmatrix}. \tag{5.20}$$

where $\eta = z^{1/2}$, *and*

$$
\begin{cases}
\chi_n(\eta) = z^{-\frac{n+\gamma_0}{2}} p_n(z), \\
\hat{\chi}_n(\eta) = z^{\frac{n+\gamma_0}{2}} \hat{p}_n(z)
\end{cases}
\tag{5.21}
$$

Proof To derive the system (i) from (ii), we just need to substitute $\tilde{p}_n = (p_{n+1} - zp_n)/q_n$ (the first equation of (ii)) into the second equation of (ii). After simplifications, the first equation of (i) is obtained. Conversely, the first equation of (i) can be converted to the second equation of (ii) by using the second equation of (i). So (i) and (ii) are equivalent.

To get (ii) from (iii), we substitute (5.21) into (5.20). Then we have

$$
\begin{pmatrix} p_{n+1} \\ \hat{p}_{n+1} \end{pmatrix} = \begin{pmatrix} z^{\frac{n+\gamma_0+1}{2}} & 0 \\ 0 & z^{-\frac{n+\gamma_0+1}{2}} \end{pmatrix} \begin{pmatrix} z^{1/2} & q_n \\ r_n & z^{-1/2} \end{pmatrix} \begin{pmatrix} z^{-\frac{n+\gamma_0}{2}} & 0 \\ 0 & z^{\frac{n+\gamma_0}{2}} \end{pmatrix} \begin{pmatrix} p_n \\ \hat{p}_n \end{pmatrix},
$$

or

$$
\begin{pmatrix} 1 & 0 \\ 0 & z^{(n+\gamma_0+1)+\frac{1}{2}} \end{pmatrix} \begin{pmatrix} p_{n+1} \\ \hat{p}_{n+1} \end{pmatrix} = \begin{pmatrix} z & z^{n+\gamma_0+\frac{1}{2}} q_n \\ zr_n & z^{n+\gamma_0+\frac{1}{2}} \end{pmatrix} \begin{pmatrix} p_n \\ \hat{p}_n \end{pmatrix}.
$$

Thus (ii) is obtained. And obviously (iii) can also be derived from (ii). Then the theorem is proved. □

In the orthonormal case, the original Szegö's equation is the following recursion formulas

$$
z\kappa_{n-1} P_{n-1}(z) = \kappa_n P_n(z) - P_n(0) P_n^*(z),
\tag{5.22}
$$

$$
\kappa_n P_{n+1}(z) = z\kappa_{n+1} P_n(z) + P_{n+1}(0) P_n^*(z),
\tag{5.23}
$$

where $P_n(z) = \kappa_n p_n(z)$ and $\kappa_n^2 = 1/h_n$. By eliminating $P_n^*(z)$, we have

$$
z\left(p_n(z) - \frac{p_{n+1}(0)}{p_n(0)} \frac{\kappa_{n-1}^2}{\kappa_n^2} p_{n-1}(z) \right) = p_{n+1} - \frac{p_{n+1}(0)}{p_n(0)} p_n(z).
\tag{5.24}
$$

Therefore, we have the following relation [12]

$$
\frac{h_n}{h_{n-1}} = \frac{v_n}{u_n} = 1 - x_n^2.
\tag{5.25}
$$

The later discussions will be based on the orthogonal polynomials, in which case the u_n and v_n are real. But the results in the above theorem is independent of the orthogonal polynomials, that means the u_n and v_n (or r_n and q_n) can be complex functions, in which case the integrable system and the string equation will play an important role because the integrability is not limited to the orthogonal system. The complex function cases can appear in the transition problems as we will see in Sect. 6.4 when discussing the power-law divergence. Here, let us give the following remarks about the more general forms of the string equation, so that as we further investigate the application problems there is a background to extend the basic models.

The u_n and v_n will be discussed to satisfy the coupled discrete equations (5.35) and (5.36). The set of these coupled equations is the string equation in the unitary matrix model in terms of the two functions u_n and v_n. By (5.17) and (5.18), these coupled equations can be changed to

$$\frac{n}{s} = -\frac{q_n + q_{n-2}}{q_{n-1}}(1 - r_{n-1}q_{n-1}), \tag{5.26}$$

$$\frac{n}{s} = -\frac{r_n + r_{n-2}}{r_{n-1}}(1 - r_{n-1}q_{n-1}). \tag{5.27}$$

In particular, if we choose $r_{n-1} = q_{n-1} = x_n$ (real), then both (5.26) and (5.27) become the string equation in terms of x_n,

$$\frac{n}{s}x_n = -(1 - x_n^2)(x_{n+1} + x_{n-1}). \tag{5.28}$$

If we choose $r_{n-1} = \bar{q}_{n-1}$(complex conjugate of q_{n-1}), then (5.26) and (5.27) become

$$\frac{n}{s}q_{n-1} = -(1 - |q_{n-1}|^2)(q_n + q_{n-2}). \tag{5.29}$$

It should be noted that the orthogonality is an important case of the consistency of the integrable system, but does not cover the general consistency. When the application problems are not in the scope of the orthogonality, the integrable systems can extend the model to discuss a wider range of problems.

5.3 Lax Pair and String Equation

5.3.1 Special Potential

By using the orthogonality of the polynomials with the special potential $V(z) = s(z + z^{-1})$ ($m = 1$ case), we will show in this section that the orthogonal polynomials $p_n = p_n(z, s)$ satisfy the following equations:

$$z(p_n + v_n p_{n-1}) = p_{n+1} + u_n p_n, \tag{5.30}$$

$$\frac{\partial p_n}{\partial z} = E_n p_{n-1} + F_n p_{n-2}, \tag{5.31}$$

where

$$u_n = -\frac{x_{n+1}}{x_n}, \tag{5.32}$$

$$v_n = -\frac{x_{n+1}}{x_n}\left(1 - x_n^2\right), \tag{5.33}$$

with $x_n = p_n(0, s)$, and

$$E_n = n, \qquad F_n = s\frac{v_n v_{n-1}}{u_n u_{n-1}}. \tag{5.34}$$

We are going to show that the consistency condition for (5.30) and (5.30) is the following string equation in terms of u_n and v_n (coupled discrete equations),

$$\frac{n}{s} = v_n + \frac{v_n}{u_n u_{n-1}}, \tag{5.35}$$

$$\frac{n}{s} = v_n\frac{u_{n-1} - v_{n-1}}{u_n - v_n} + \frac{v_n}{u_n u_{n+1}}\frac{u_{n+1} - v_{n+1}}{u_n - v_n}. \tag{5.36}$$

In the following, we first discuss how to derive these equations and explain what the consistency condition means for the discrete model.

The recursion formula (5.30) is obtained from the Szegö's equation derived from the Christoffel-Darboux formula [11] (Theorem 11.4.2) as discussed in last section, and simplified into the current form [5, 12]. To derive (5.31), write $\partial p_n/\partial z$ as a linear combination of $p_{n-1}, p_{n-2}, \ldots, p_0$:

$$\frac{\partial p_n}{\partial z} = \sum_{k=0}^{n-1} a_k p_k.$$

Then $a_k h_k = \langle\frac{\partial p_n}{\partial z}, p_k\rangle$. It is not hard to see that $a_{n-1}h_{n-1} = nh_{n-1}$. For $k < n - 1$, since on the unit circle ($|z| = 1$) there is $1/z = \bar{z}$, we have

$$a_k h_k = \oint \frac{\partial}{\partial z} p_n \bar{p}_k e^{s(z+1/z)}\frac{dz}{2\pi i z}$$

$$= -\oint p_n\left(-\frac{1}{z^2}\frac{\partial}{\partial\bar{z}}\bar{p}_k - \frac{1}{z}\bar{p}_k + s\left(1 - \frac{1}{z^2}\right)\bar{p}_k\right)e^{s(z+1/z)}\frac{dz}{2\pi i z}$$

$$= sh_n\delta_{n-2,k},$$

which implies $a_k = 0$ when $k < n - 2$, and $a_{n-2} = sh_n/h_{n-2} = F_n$, where we have used the relation $h_n/h_{n-1} = v_n/u_n$ given by (5.25). This gives (5.31).

In the following, we derive the string equation in terms of u_n and v_n by considering the leading coefficient and the second coefficient of $p_{n,z}$. Equation (5.35) is derived from

$$\langle zp_{n,z}, p_{n-1}\rangle = \langle E_n zp_{n-1} + F_n zp_{n-2}, p_{n-1}\rangle, \tag{5.37}$$

and (5.36) is obtained from

$$\langle p_{n+1,z}, p_{n-1}\rangle = \langle E_{n+1}p_n + F_{n+1}p_{n-1}, p_{n-1}\rangle. \tag{5.38}$$

It can be seen that, by the recursion formula (5.30), the right hand side of (5.37) is equal to $E_n(u_{n-1} - v_{n-1})h_{n-1} + F_n h_{n-1}$. For the left hand side of (5.37), consider $p_n = z^n + \beta_n z^{n-1} + \cdots$. Then

$$z \frac{\partial p_n}{\partial z} = nz^n + (n-1)\beta_n z^{n-1} + \cdots$$

$$= np_n - \beta_n p_{n-1} + \cdots.$$

Equation (5.37) then becomes

$$-\beta_n = n(u_{n-1} - v_{n-1}) + s \frac{v_n v_{n-1}}{u_n u_{n-1}}.$$

By using the Lemma 5 in [12], we have

$$v_n(u_{n-1} - v_{n-1}) + \frac{v_n}{u_n} = \frac{n}{s}(u_{n-1} - v_{n-1}) + \frac{v_n v_{n-1}}{u_n u_{n-1}},$$

which is (5.35).

To compute the left hand side of (5.38), consider $p_{n+1} = z^{n+1} + \beta_{n+1} z^n + \cdots$, and then

$$\frac{\partial p_{n+1}}{\partial z} = (n+1)\left[z^n + \beta_n z^{n-1} + \left(\frac{n}{n+1} \beta_{n+1} - \beta_n \right) z^{n-1} + \cdots \right]$$

$$= (n+1)p_n + (n\beta_{n+1} - (n+1)\beta_n)p_{n-1} + \cdots.$$

Equation (5.38) then becomes

$$(n\beta_{n+1} - (n+1)\beta_n) = s \frac{v_n v_{n+1}}{u_n u_{n+1}}.$$

By Lemma 5 in [12] again, we get

$$-n(u_n - v_n) + s\left(v_n(u_{n-1} - v_{n-1}) + \frac{v_n}{u_n} \right) = s \frac{v_n v_{n+1}}{u_n u_{n+1}}.$$

Thus (5.36) is obtained.

The string equation (5.28) in terms of x_n can be obtained as follows. Since p_n satisfy the recursion formula, $\langle p_{n,z}, p_{n-1} \rangle$ can be calculated by using the recursion formula. We have that by integration by parts

$$nh_{n-1} = \oint \frac{\partial}{\partial z} p_n \bar{p}_{n-1} e^{s(z+1/z)} \frac{dz}{2\pi i z}$$

$$= -\oint p_n \left[\frac{-1}{z^2} \frac{\partial \bar{p}_{n-1}}{\partial z} - \frac{1}{z} \bar{p}_{n-1} + s\left(1 - \frac{1}{z^2} \right) \bar{p}_{n-1} \right] e^{s(z+1/z)} \frac{dz}{2\pi i z}$$

$$= nh_n + s(u_n - v_n + u_{n-1} - v_{n-1})h_n.$$

Using $h_{n-1}/h_n = u_n/v_n$, we obtain the string equation (5.28). These equations can also be derived from the Szegö's equation. Relevant discussions can be found, for example, in [6, 8]. To avoid the $*$ polynomial p_n^* in the Szegö's equation, we use the recursion formula (5.30) here.

In the following, we show that (5.30) and (5.31) form a Lax pair for the coupled discrete equations (5.35) and (5.36). The results obtained in this section are independent of the orthonormal polynomials. We only assume that $p_n(z)$, as a function of z, satisfies (5.30) and (5.31), and the functions u_n and v_n satisfy (5.35) and (5.36).

Lemma 5.1 *If* (5.31) *holds, then the recursion formula* (5.30) *can be changed to*

$$z(E_n p_{n-1} + X_n p_{n-2}) = E_n p_n + Y_n p_{n-1}, \tag{5.39}$$

where

$$X_n = F_n + v_n E_{n-1} - \frac{u_n}{u_{n-2}} F_n, \tag{5.40}$$

$$Y_n = F_{n+1} + u_n E_n - v_n - \frac{u_n}{u_{n-2}} F_n. \tag{5.41}$$

Proof Taking $\partial/\partial z$ on both sides of (5.30) and applying (5.31), we get

$$p_n + v_n p_{n-1} + z\big(E_n p_{n-1} + (F_n + v_n E_{n-1})p_{n-2}\big) + z v_n F_{n-1} p_{n-3}$$
$$= E_{n+1} p_n + (F_{n+1} + u_n E_n)p_{n-1} + u_n F_n p_{n-2}.$$

This equation represents a recursion relation for the polynomials p_n, which should be consistent with (5.30). To see that, we express $v_n F_{n-1} z p_{n-3} - u_n F_n p_{n-2}$ in terms of p_{n-1} and $z p_{n-2}$, such that the equation above comes back to the format of (5.30). In fact, it is not hard to see that

$$\frac{v_n}{v_{n-2}} F_{n-1} = \frac{u_n}{u_{n-2}} F_n,$$

by (5.34). Then we have

$$v_n F_{n-1} z p_{n-3} - u_n F_n p_{n-2} = \frac{u_n}{u_{n-2}} F_n(v_{n-2} z p_{n-3} - u_{n-2} p_{n-2})$$

$$= \frac{u_n}{u_{n-2}} F_n(p_{n-1} - z p_{n-2}).$$

The recursion formula above then becomes

$$z E_n p_{n-1} + z(F_n + v_n E_{n-1})p_{n-2} - z\frac{u_n}{u_{n-2}} F_n p_{n-2}$$

$$= E_n p_n + (F_{n+1} + u_n E_n - v_n)p_{n-1} - \frac{u_n}{u_{n-2}} F_n p_{n-1}.$$

Then this lemma is proved.   □

Lemma 5.2 *If u_n and v_n satisfy the discrete equations (5.35) and (5.36), then the E_n and F_n defined by (5.34) satisfy the relation*

$$F_{n+1} + (u_n - v_n)E_n = F_n + (u_{n-1} - v_{n-1})E_n. \tag{5.42}$$

Proof Since u_n, v_n satisfy (5.35), by multiplying both sides of (5.35) by $s(u_{n-1} - v_{n-1})$, we have

$$(u_{n-1} - v_{n-1})E_n = s(u_{n-1} - v_{n-1})v_n + s\frac{v_n}{u_n} - F_n.$$

By (5.36), we have

$$(u_n - v_n)E_n = s(u_{n-1} - v_{n-1})v_n + s\frac{v_n}{u_n} - F_{n+1}.$$

Eliminating $s(u_{n-1} - v_{n-1})v_n$ in these two equations, we get (5.42).   □

Lemma 5.3 *If u_n and v_n satisfy the discrete equations (5.35) and (5.36), then X_n and Y_n defined by (5.40) and (5.41) respectively satisfy*

$$X_n = v_{n-1}E_n, \tag{5.43}$$

$$Y_n = u_{n-1}E_n. \tag{5.44}$$

Proof Write (5.35) in the form

$$s\frac{v_n}{u_n u_{n-1}} = E_n - s v_n,$$

and apply it to $F_n = s\frac{v_n v_{n-1}}{u_n u_{n-1}}$ in (5.40). We then obtain

$$X_n = v_{n-1}(E_n - s v_n) + v_n E_{n-1} - v_n(E_{n-1} - s v_{n-1}).$$

Simplification of this equation yields (5.43).

By eliminating X_n in (5.40) and (5.43), we get

$$\frac{u_n}{u_{n-2}}F_n = F_n + v_n E_{n-1} - v_{n-1}E_n.$$

Then by applying this formula to the last term in (5.41), we obtain

$$Y_n = F_{n+1} - F_n - v_n E_{n-1} + v_{n-1}E_n + u_n E_n - v_n.$$

Since $E_{n-1} + 1 = E_n$, the last lemma implies (5.44).   □

Combining these three lemmas, we have proved the following result.

Theorem 5.2 *The equations*

$$z(p_n + v_n p_{n-1}) = p_{n+1} + u_n p_n, \tag{5.45}$$

$$\frac{\partial p_n}{\partial z} = n p_{n-1} + s \frac{v_n}{u_n} \frac{v_{n-1}}{u_{n-1}} p_{n-2}, \tag{5.46}$$

form a Lax pair for the coupled discrete equations (5.35) *and* (5.36).

On the unit circle $|z| = 1$, we can write the above equations of the Lax pair into the following matrix forms,

$$\Phi_{n+1} = L_n(z) \Phi_n, \tag{5.47}$$

$$\frac{\partial}{\partial z} \Phi_n = A_n(z) \Phi_n, \tag{5.48}$$

where $\Phi_n(z) = e^{\frac{s}{2}(z+1/z)}(z^{-n/2} p_n(z), z^{n/2} \overline{p_n(z)})^T$,

$$L_n = \begin{pmatrix} z^{1/2} & x_{n+1} z^{-1/2} \\ x_{n+1} z^{1/2} & z^{-1/2} \end{pmatrix}, \tag{5.49}$$

$$A_n(z) = \begin{pmatrix} \frac{s}{2} + \frac{s}{2z^2} + \frac{n-2sx_n x_{n+1}}{2z} & s(x_{n+1} - \frac{x_n}{z})z^{-1} \\ s(x_n - \frac{x_{n+1}}{z}) & -\frac{s}{2} - \frac{s}{2z^2} - \frac{n-2sx_n x_{n+1}}{2z} \end{pmatrix}, \tag{5.50}$$

and x_n satisfies the string equation

$$\frac{n}{s} x_n = -\left(1 - x_n^2\right)(x_{n+1} + x_{n-1}), \tag{5.51}$$

with $x_n \in [-1, 1]$. Note that (5.47) is equivalent to the Szegö's equation, and it is an equation for $\Phi_n(z)$ here.

Because of the $z^{1/2}$ term in the Lax pair above, in some literatures, the Lax pair for the string equation is also discussed by using the variable $\eta = z^{1/2}$. Based on Theorem 5.1 and Theorem 5.2, we can get a more general result for the Lax pair for (5.26) and (5.27), which consists of the following two linear systems of equations [5]

$$\begin{pmatrix} \psi_{n+1} \\ \hat{\psi}_{n+1} \end{pmatrix} = \bar{L}_n \begin{pmatrix} \psi_n \\ \hat{\psi}_n \end{pmatrix}, \tag{5.52}$$

$$\frac{\partial}{\partial \eta} \begin{pmatrix} \psi_n \\ \hat{\psi}_n \end{pmatrix} = \bar{A}_n \begin{pmatrix} \psi_n \\ \hat{\psi}_n \end{pmatrix}, \tag{5.53}$$

where $(\psi_n, \hat{\psi}_n)^T = e^{\frac{s}{2}(\eta^2 + \eta^{-2})}(\chi_n, \hat{\chi}_n)^T$ with the $(\chi_n, \hat{\chi}_n)$ given in last section (Theorem 5.1),

$$\bar{L}_n = \begin{pmatrix} \eta & q_n \\ r_n & \eta^{-1} \end{pmatrix}, \tag{5.54}$$

$$\bar{A}_n = \begin{pmatrix} a_n & b_n \\ c_n & d_n \end{pmatrix}, \tag{5.55}$$

and

$$a_n = s\eta + \frac{s}{\eta^3} + \frac{n - \gamma_0 - 2sq_n r_{n-1}}{\eta},$$

$$b_n = 2s\left(q_n - \frac{q_{n-1}}{\eta^2}\right),$$

$$c_n = 2s\left(r_{n-1} - \frac{r_n}{\eta^2}\right),$$

$$d_n = -s\eta - \frac{s}{\eta^3} - \frac{n + \gamma_0 + 1 - 2sq_{n-1}r_n}{\eta}.$$

One can directly check that the consistency condition $\partial \bar{L}_n / \partial \eta = \bar{A}_{n+1}\bar{L}_n - \bar{L}_n \bar{A}_n$ is equivalent to (5.26) and (5.27). When $r_{n-1} = q_{n-1} = x_n$ and $\gamma_0 = -1/2$, the above Lax pair (5.52) and (5.53) becomes the Lax pair (5.47) and (5.48). Also, the Lax pairs shown above are consistent with the results obtained in [1, 2], for instance. For the Lax pair of the discrete Painlevé II equation discussed in [2], if we replace n by $n + 1$, and let $\nu = 1/2$ and $\kappa = s$, then their Lax pair becomes the discrete AKNS-ZS system (5.52) and (5.53) with $r_{n-1} = q_{n-1} = x_n$ and $\gamma_0 = -1/2$.

As a remark, the coupled equations (5.35) and (5.36) can be converted into the alternate discrete Painlevé II equation [3]. Write (5.35) and (5.36) as follows,

$$\frac{n+1}{s}v_n = v_n v_{n+1} + \frac{v_n v_{n+1}}{u_n u_{n+1}},$$

$$\frac{n}{s}(u_n - v_n) = v_n(u_{n-1} - v_{n-1}) + \frac{v_n}{u_n} - \frac{v_n v_{n+1}}{u_n u_{n+1}}.$$

Adding these two equations, we have

$$\frac{n}{s}u_n + \frac{v_n}{s} = v_n(u_{n-1} - v_{n-1}) + \frac{v_n}{u_n} + v_n v_{n+1}.$$

Since $\frac{n}{s}u_n = v_n u_n + v_n/u_{n-1}$, the equation above then becomes

$$\left(v_{n+1} + v_n - u_n + \frac{1}{u_n}\right) - \left(v_n + v_{n-1} - u_{n-1} + \frac{1}{u_{n-1}}\right) = \frac{1}{s},$$

or

$$v_{n+1} + v_n - u_n + \frac{1}{u_n} = c_0 + \frac{n}{s},$$

where $c_0 = v_1 + v_0 - u_0 + 1/u_0$. This equation together with (5.35) is the alternate string equation in [3]. So the set of the coupled equations (5.35) and (5.36) is equivalent to the alternate string equation.

5.3.2 General Potential

Let us consider a general potential $V(z) = V_0(z) + V_0(1/z)$ where

$$V_0(z) = s \sum_{j=1}^{m} c_j z^j, \qquad (5.56)$$

and discuss how to get the Lax pair when $m \geq 1$, where c_j are constants. We are going to show that the orthogonal polynomials $p_n = p_n(z, s)$ now satisfy the following linear equations:

$$z\big(p_n(z) + v_n p_{n-1}(z)\big) = p_{n+1}(z) + u_n p_n(z); \qquad (5.57)$$

$$z\frac{\partial p_n(z)}{\partial z} = \sum_{k=n-m}^{n+m} a_k p_k(z) - z V_0'(z) p_n(z), \qquad (5.58)$$

where if $k < n$,

$$a_k h_k = -\big\langle z V_0'(1/z) p_n(z), p_k(z)\big\rangle,$$

and if $k \geq n$,

$$a_k h_k = n h_n \delta_{k,n} + \big\langle z V_0'(z) p_n(z), p_k(z)\big\rangle;$$

and

$$z^{-1}\frac{\partial}{\partial s}\left(\frac{p_n(z)}{x_n}\right) = \sum_{k=n-m-1}^{n+m-1} b_k p_k(z) - \frac{1}{s x_n} z^{-1} V_0(z) p_n(z), \qquad (5.59)$$

where if $k \leq n - 1$,

$$b_k h_k = \frac{\partial}{\partial s}\left(\frac{h_n}{x_n}\right)\delta_{k,n-1} - \frac{1}{s x_n}\big\langle z V_0(z) p_k(z), p_n(z)\big\rangle,$$

and if $k > n - 1$,

$$b_k h_k = \frac{1}{s x_n}\big\langle z^{-1} V_0(z) p_n(z), p_k(z)\big\rangle.$$

If $V_0(z) = \sum_{j=1}^{m} s c_j z^j$ is changed to $V_0(z) = \sum_{j=1}^{m} s_j z^j$, we need to consider the partial derivatives for each s_j, and the method is similar to (5.59). For simplicity, let us just think one s parameter s in $V_0(z)$ here.

Note that the left hand side of (5.58) is $z\frac{\partial p_n(z)}{\partial z}$, whereas the left hand side of (5.46) is $\frac{\partial p_n(z)}{\partial z}$. In fact, when $m = 1$, (5.58) becomes

$$z\frac{\partial p_n(z)}{\partial z} = s p_{n+1} + \big(n + s(u_n - v_n)\big)p_n + s\frac{v_n}{u_n}p_{n-1} - s z p_n, \qquad (5.60)$$

with $c_1 = 1$ and $s_1 = s$, which can be changed to (5.46) by using the recursion formula and string equation. However, the above equation is obviously not convenient for discussing the consistency condition. Equation (5.46) can directly and easily provide an example about the consistency condition. The form of (5.58) is so chosen for the general potential that the matrix form of the Lax pair can be easily formulated in order to study the eigenvalue density problems. It is left to interested readers to experience that the equation in the form of (5.46) is not easy to derive the matrix equation. Also, the different forms of the Lax pair are presented here to connect the different results of the Lax pair theory obtained other literatures for a wider impact.

The recursion formula $z(p_n + v_n p_{n-1}) = p_{n+1} + u_n p_n$ is still satisfied as discussed before. The z equation (5.58) can be derived by considering

$$z \frac{\partial p_n(z)}{\partial z} = \sum_{k=0}^{n+m} a_k p_k(z) - z V_0'(z) p_n(z),$$

where the term $z V_0'(z) p_n(z)$ on the right hand side is so chosen based on the following calculations. Using integration by parts and orthogonality for $k < n$, we have

$$a_k h_k = \oint \frac{\partial}{\partial z} p_n(z) \bar{p}_k(z) e^{V(z)} \frac{dz}{2\pi i} + \langle z V_0'(z) p_n, p_k \rangle$$

$$= -\oint p_n \left[-\frac{1}{z^2} \frac{\partial \bar{p}_k(z)}{\partial \bar{z}} + \left(V_0'(z) + V_0'(1/z) \right) \bar{p}_k(z) \right] z d\mu$$

$$+ \langle z V_0'(z) p_n(z), p_k(z) \rangle$$

$$= -\langle z V_0'(1/z) p_n(z), p_k(z) \rangle,$$

where $d\mu = e^{V(z)} \frac{dz}{2\pi i z}$. It is not hard to see that $a_k = 0$ when $k < n - m$ since $\deg(z V_0'(1/z) \bar{p}_k) < n$, where $z V_0'(1/z) \bar{p}_k(z)$ is a polynomial in $\bar{z}$ of degree $k + m$ since $|z| = 1$. When $k \geq n$, a_k can be easily found by taking $\langle \cdot, p_k \rangle$ on both sides of the equation.

The s equation can be obtained by considering

$$z^{-1} \frac{\partial}{\partial s} \left(\frac{p_n}{x_n} \right) = \sum_{k=0}^{n+m-1} b_k p_k - \frac{1}{s x_n} z^{-1} V_0(z) p_n(z).$$

The left side of the equation is a polynomial in z of degree $n - 1$, and the factor $\frac{1}{s x_n} z^{-1}$ on the right hand side is chosen based on the following calculations. When $k > n - 1$, taking $\langle \cdot, p_k \rangle$ on both sides of this equation, we get the formula for b_k. When $k \leq n - 1$, b_j can be found by using the product rule and orthogonality of the polynomials. Still taking $\langle \cdot, p_k \rangle$ on both sides of this equation, we have

$$b_k h_k = \oint z^{-1} \frac{\partial}{\partial s} \left(\frac{p_n}{x_n} \right) \bar{p}_k d\mu + \frac{1}{s x_n} \langle z^{-1} V_0(z) p_n(z), p_k(z) \rangle$$

$$= \frac{\partial}{\partial s} \oint z^{-1} \frac{p_n}{x_n} \bar{p}_k d\mu - \oint z^{-1} \frac{p_n}{x_n} \frac{\partial \bar{p}_k}{\partial s} d\mu$$

$$- \oint z^{-1} \frac{p_n}{x_n} \bar{p}_k \frac{1}{s} V(z) d\mu + \frac{1}{s x_n} \langle z^{-1} V_0(z) p_n(z), p_k(z) \rangle$$

$$= \frac{\partial}{\partial s} \left(\frac{h_n}{x_n} \right) \delta_{k,n-1} - \frac{1}{s x_n} \oint p_n z^{-1} V_0(1/z) \bar{p}_k d\mu$$

$$= \frac{\partial}{\partial s} \left(\frac{h_n}{x_n} \right) \delta_{k,n-1} - \frac{1}{s x_n} \langle z V_0(z) p_k(z), p_n(z) \rangle.$$

It can be seen that when $k < n - m - 1$ there are $b_k = 0$.

We have shown that the orthogonal polynomials p_n satisfy the three linear equations simultaneously. We are interested in finding the consistency conditions for these linear equations. If we choose x_n as the function variable, the consistency condition for (5.57) and (5.58) is the string equation hierarchy, and the consistency condition for (5.57) and (5.59) is the Toda lattice for the unitary matrix models. The consistency condition for (5.58) and (5.59) is the continuum Painlevé III or V hierarchy, which is a more complicated case, and we will discuss the special cases in Sect. C.3.

First, the string equation hierarchy has the following general formula [6, 7]

$$n(h_n - h_{n-1}) = \langle V'(z) p_n(z), p_{n-1}(z) \rangle. \tag{5.61}$$

In fact, by integration by parts, we have

$$n h_{n-1} = \oint p_{n,z} \bar{p}_{n-1} e^{V(z)} \frac{dz}{2\pi i z}$$

$$= - \oint p_n \left(\frac{-1}{z^2} \frac{\partial \bar{p}_{n-1}}{\partial z} - \frac{1}{z} \bar{p}_{n-1} + V'(z) \bar{p}_{n-1} \right) d\mu$$

$$= n h_n - \langle V'(z) p_n(z), p_{n-1}(z) \rangle.$$

When $m = 1$, $V'(z) = s_1(1 - z^{-2})$, (5.61) is (5.28).
When $m = 2$, $V'(z) = s_1(1 - z^{-2}) + 2s_2(z - z^{-3})$ with $s_j = s c_j$, (5.61) becomes

$$n(h_{n-1} - h_n)$$

$$= s_1 \langle z^2 p_{n-1}(z), p_n(z) \rangle - 2s_2 \langle z p_n(z), p_{n-1}(z) \rangle + 2s_2 \langle z^3 p_{n-1}(z), p_n(z) \rangle.$$

By repeatedly using the recursion formula, we can find

$$z p_n = p_{n+1} + u_n p_n - v_n (p_n + u_{n-1} p_{n-1} - v_{n-1} z p_{n-2}),$$

$$z^2 p_{n-1} = p_{n+1} + u_n p_n + (u_{n-1} - v_{n-1} - v_n)(p_n + u_{n-1} p_{n-1} - v_{n-1} z p_{n-2})$$

$$- v_{n-1} z (u_{n-2} p_{n-2} - v_{n-2} z p_{n-3}),$$

$$z^3 p_{n-1} = p_{n+2} + u_{n+1} p_{n+1}$$

$$+ (u_n - v_n + u_{n-1} - v_{n-1} - v_{n+1})(p_{n+1} + u_n p_n - v_n z p_{n-1})$$

$$+ (u_{n-1} - v_{n-1} - v_n)z(u_{n-1}p_{n-1} - v_{n-1}zp_{n-2})$$
$$- v_{n-1}z^2(u_{n-2}p_{n-2} - v_{n-2}zp_{n-3}).$$

The equation can be written as

$$n(h_{n-1} - h_n) = s_1(u_n - v_n + u_{n-1} - v_{n-1})h_n + 2s_2v_n(u_{n-1} - v_{n-1})h_{n-1}$$
$$+ 2s_2\big[(u_n - v_n + u_{n-1} - v_{n-1} - v_{n+1})(u_n - v_n)$$
$$+ (u_{n-1} - v_{n-1} - v_n)(u_{n-1} - v_{n-1}) - v_{n-1}(u_{n-2} - v_{n-2})\big]h_n.$$

And it can be further simplified to

$$\frac{nx_n}{1 - x_n^2} = -s_1(x_{n+1} + x_{n-1}) + 2s_2x_n(x_{n+1} + x_{n-1})^2$$
$$- 2s_2x_{n+2}\big(1 - x_{n+1}^2\big) - 2s_2x_{n-2}\big(1 - x_{n-1}^2\big), \tag{5.62}$$

by using (5.32) and (5.33). This is (3.7) in [1].

The consistency conditions can be also formulated in terms of the u_n and v_n functions. As in the $m = 1$ case, we need to consider $\langle p_{n,z}(z), p_{n-2}(z)\rangle$ and $\langle zp_{n,z}(z), p_{n-1}(z)\rangle$. Let $p_n = z^n + \beta_n z^{n-1} + \cdots$. Then we have

$$\langle p_{n,z}(z), p_{n-2}(z)\rangle = \big[n(\beta_n - \beta_{n-1}) - \beta_n\big]h_{n-2}, \tag{5.63}$$
$$\langle zp_{n,z}(z), p_{n-1}(z)\rangle = -\beta_n h_{n-1}. \tag{5.64}$$

The β_n satisfy the following relations

$$\beta_n - \beta_{n-1} = -(u_{n-1} - v_{n-1}),$$
$$\beta_n h_{n-1} = -\langle zV'(z)p_n(z), p_{n-1}(z)\rangle.$$

The first relation holds because $p_n - zp_{n-1} = (\beta_n - \beta_{n-1})z^{n-1} + \cdots = -u_{n-1}p_{n-1} + v_{n-1}zp_{n-2}$. For the second relation, consider $J = \langle zp_n(z), p_{n-1}(z)\rangle$. By integration by parts, we have

$$J = \frac{1}{s}\oint p_n\bar{p}_{n-1}e^{V(z)-sz}\frac{de^{sz}}{2\pi i}$$
$$= \frac{-1}{s}\oint\left[\frac{\partial p_n}{\partial z}\bar{p}_{n-1} + p_n\frac{-1}{z^2}\frac{\partial\bar{p}_{n-1}}{\partial\bar{z}} + p_n\bar{p}_{n-1}(V'(z) - s)\right]e^{V(z)}\frac{dz}{2\pi i}$$
$$= \frac{-1}{s} - \beta_n h_{n-1} - \frac{1}{s}\langle zV'(z)p_n(z), p_{n-1}(z)\rangle + \langle zp_n(z), p_{n-1}(z)\rangle.$$

Thus we get $\beta_n h_{n-1} = -\langle zV'(z)p_n(z), p_{n-1}(z)\rangle$.

By the s equation (5.59), we can get the following Toda lattice,

$$x_n'(h_n - h_{n-1}) = \frac{x_n}{s}\langle V(z)(p_n(z) - zp_{n-1}(z)), p_n(z)\rangle. \tag{5.65}$$

In fact, the left hand side of (5.59) is a polynomial in z of degree $n - 1$. If we take the inner product $\langle \cdot, p_{n-1}(z) \rangle$ on both sides of (5.59), then after some simplifications there is

$$x_n'(h_n - h_{n-1}) = x_n h_n' - \frac{x_n}{s}\langle zV(z)p_{n-1}(z), p_n(z)\rangle.$$

By the definition of h_n, we have

$$h_n' = \frac{1}{s}\langle V(z)p_n(z), p_n(z)\rangle.$$

By eliminating h_n' in these two equations above, we get (5.65).

5.4 Densities Reduced from the Lax Pair

Now, let us come back to the general potential $V(z) = V_0(z) + V_0(z^{-1})$, where $V_0(z) = \sum_{j=1}^m s_j z^j$. We want to obtained the expected analytic function $\omega(z)$ for solving the eigenvalue density problem by using the coefficient matrix A_n in the Lax pair based on the discussions in the last two sections using the similar method we have done in Chap. 2.

Based on (5.58) obtained in last section

$$z\frac{\partial p_n(z)}{\partial z} = \sum_{j=-m}^m a_{n+j} p_{n+j}(z) - zV_0'(z)p_n(z), \qquad (5.66)$$

we have the following matrix equation on the unit circle

$$z\frac{\partial}{\partial z}\begin{pmatrix} p_n \\ \bar{p}_n \end{pmatrix} = \sum_{j=-m}^m a_{n+j}\sigma_3\begin{pmatrix} p_{n+j} \\ \bar{p}_{n+j} \end{pmatrix} - \begin{pmatrix} zV_0'(z) & 0 \\ 0 & zV_0'(z^{-1}) \end{pmatrix}\begin{pmatrix} p_n \\ \bar{p}_n \end{pmatrix}, \qquad (5.67)$$

where $\sigma_3 = \mathrm{diag}(1, -1)$ and $V_0' = \frac{\partial}{\partial z}V_0$. The recursion formula (5.47) with $r_{n-1} = q_{n-1} = x_n$ and $\gamma_0 = -1/2$ or the Szegö's equation implies

$$\begin{pmatrix} p_{n+1} \\ z^{n+1}\bar{p}_{n+1} \end{pmatrix} = T_{n+1}\hat{L}_n T_n^{-1}\begin{pmatrix} p_n \\ z^n \bar{p}_n \end{pmatrix}, \qquad (5.68)$$

where

$$\hat{L}_n = \begin{pmatrix} z - u_{n+1} & zv_{n+1} \\ 1 & 0 \end{pmatrix}, \qquad T_n = \begin{pmatrix} 0 & 1 \\ \frac{1}{x_{n+1}} & \frac{-z}{x_{n+1}} \end{pmatrix}. \qquad (5.69)$$

Let $\Phi_n(z) = e^{\frac{1}{2}(V_0(z)+V_0(1/z))}(z^{-n/2}p_n(z), z^{n/2}\overline{p_n(z)})^T$. Then there is

$$\frac{\partial}{\partial z}\Phi_n(z) = A_n(z)\Phi_n(z), \qquad (5.70)$$

where

$$z\sigma_3 A_n(z) = \sum_{j=1}^{m} a_{n+j} \begin{pmatrix} 1 & 0 \\ 0 & z^{-j} \end{pmatrix} T_{n+j} \hat{L}_{n+j-1} \cdots \hat{L}_n T_n^{-1}$$

$$+ \sum_{j=1}^{m} a_{n-j} \begin{pmatrix} 1 & 0 \\ 0 & z^{j} \end{pmatrix} T_{n-j} \hat{L}_{n-j} \cdots \hat{L}_{n-1} T_n^{-1}$$

$$+ \left(a_n - \frac{n}{2} \right) I - \frac{1}{2} z (V_0'(z) - V_0'(z^{-1})) I. \qquad (5.71)$$

The eigenvalue density $\rho(\theta)$ will be obtained from the Lax pair in the unitary matrix models following a process, roughly expressed as

$$\frac{1}{n\pi} \sqrt{\det A_n(z)} dz \sim \sigma(z) dz = \rho(\theta) d\theta, \qquad (5.72)$$

for $z = e^{i\theta}$ on the complement of the arc cuts in the unit circle. We also need

$$\frac{1}{n} \sqrt{-\det A_n(z)} \sim \omega(z), \qquad (5.73)$$

where $z \neq 0$, $z \neq \infty$ and z is outside the cuts, to analyze the asymptotics as $z \to 0$ or $z \to \infty$. Note that the eigenvalues are now on the complement of the cuts, and it is easy to see that

$$\sigma(z) = \frac{1}{\pi i} \omega(z), \qquad (5.74)$$

for z on the complement of the cuts. The details are much more complicated than the Hermitian matrix models. In order to get similar results as in the Hermitian models discussed in Chap. 2, we need to apply the different forms of the recursion formula discussed in Sect. 5.2 such that the determinant of the matrix in the corresponding Cayley-Hamilton theorem is independent of z, and the factor $(-1)^n$ or the similar factors will play an important role in the matrix transformations. The factor $(-1)^n$ is fundamental in the reduction of the density models for the unitary matrix models. In the Hermitian matrix models, a fundamental consideration is about the index change from n to $n+1$, say, and the eigenvalues are distributed on the cut(s) on the real line. In the unitary matrix models, we typically need to consider the factor $(-1)^n$ for constructing the eigenvalue densities which are defined on the complement of the cut(s) in the unit circle. Interested readers can first try the case $m = 2$ or 3. To reduce the pages of the book and pay more attention to the phase transition problems, the details for the reduction and factorization processes are omitted here. In the following, we introduce some simple examples for readers to see the key steps to obtain the density formulas. The general formulations are not discussed here since this book focus on the application problems by considering the special models. The terminologies, strong and weak couplings, are used in the following discussions just for reader's convenience to associating with the strong and weak couplings in the Gross-Witten model. Here, we only talk about the mathematical formulations.

5.4.1 Strong Couplings: General Case

The strong coupling densities are derived by assuming u_n and v_n both approach to v as $n \to \infty$, so that $(\operatorname{tr} \hat{L}_n)^2 - 4 \det \hat{L}_n$ becomes a complete square and the formula of the density finally does not have square root. In this case, x_n will approach to 0, $\hat{L}_{n \pm k}$ will be replaced by $\hat{L}_n$, and $T_{n \pm j}$ will be replaced by $\operatorname{diag}(1, (-v)^{\mp j}) T_n$. Also by the orthogonality of the polynomials, there are $a_{n \pm j} = j s_j$ and $a_n = n$. The $\sqrt{-\det A_n}$ can be rescaled and reduced to $\omega(z)$ as we worked before for the Hermitian matrix models. The details can be discussed based on the Lax pair. The explicit formula for $\omega(z)$ is

$$z\omega(z) = \frac{1}{2} \sum_{j=1}^{m} j g_j \left(1 - (-vz)^{-j}\right)\left(z^j - (-v)^j\right), \qquad (5.75)$$

which can be further simplified to

$$z\omega(z) = \frac{1}{2} \sum_{j=1}^{m} j g_j \left(z^j + z^{-j}\right) + \frac{1}{2}, \qquad (5.76)$$

if the parameters satisfy the conditions

$$\sum_{j=1}^{m} j g_j \left((-v)^j + (-v)^{-j}\right) + 1 = 0. \qquad (5.77)$$

Then, using the equivalent relation, $\rho(\theta) d\theta = \sigma(z) dz$, where $\sigma(z) = \omega(z)/(\pi i)$, the density formula can be obtained either in the z space or θ space.

5.4.2 Weak Couplings: One-Cut Cases

The weak coupling densities on one interval can be derived by assuming u_n approach to 1 and v_n approach to v as $n \to \infty$ by referring the Gross-Witten weak coupling density model. The square root factor in the density can be formulated by the matrix

$$\hat{L}^{(1)} = \begin{pmatrix} z - 1 & zv \\ 1 & 0 \end{pmatrix}, \qquad (5.78)$$

which gives $(\operatorname{tr} \hat{L}^{(1)})^2 - 4 \det \hat{L}^{(1)} = (z - 1)^2 + 4vz$. In this case, $\hat{L}_{n \pm k}$ will be replaced by the $\hat{L}^{(1)}$, and $T_{n \pm j}$ will be replaced by T_n in the reduction to get $\omega(z)$. Here, let us just talk about some special cases.

When $U(z) = g_1(z + z^{-1})$, there is

$$z\omega(z) = \frac{1}{2} g_1 \left(1 + z^{-1}\right) \sqrt{(z - 1)^2 + 4vz}, \qquad (5.79)$$

subject to the condition $2g_1v = 1$ with $g_1 \geq 1/2$, which is corresponding to the Gross-Witten weak coupling density discussed before.

When $U(z) = \sum_{j=1}^{m} g_j(z^j + z^{-j})$ for $m = 2$ or 3, there is

$$z\omega(z) = \frac{1}{2}f(z)(1 + z^{-1})\sqrt{(z-1)^2 + 4vz},$$ (5.80)

where

$$f(z) = g_1 - 4g_2v + 3g_3(1 - 4v + 6v^2) + [2g_2 - 6g_3v](z + z^{-1}) + 3g_3(z^2 + z^{-2}),$$ (5.81)

subject to the condition

$$g_1v + 2g_2(2v - 3v^2) + 3g_3(3v(1 - v)(1 - 3v) + v^3) = \frac{1}{2}.$$ (5.82)

The $\omega(z)$ function so defined has the following asymptotics

$$z\omega(z) = \frac{1}{2}z(U_0'(z) - U_0'(z^{-1})) + \frac{1}{2} + O((z^{1/2} + z^{-1/2})^{-2}),$$ (5.83)

as $|z^{1/2} + z^{-1/2}| \to \infty$, where $U_0(z) = \sum_{j=1}^{3} g_j z^j$. These results can be applied to study the transition problems to be discussed in the next chapter. The density function also needs to be non-negative when the variable is transformed into the θ space, while in the z space the "density" could be complex function. The z space is used for the mathematical calculation such as the asymptotics for an easier computation using the complex integrals.

The $\omega(z)$ functions obtained above are defined in the outside of the cut $\hat{\Omega} = \{z = e^{i\theta} | \cos(\theta/2) \leq \sqrt{1-v}, \theta \in [-\pi, \pi]\}$ in the complex plane. The eigenvalue density $\rho(\theta)$ is defined by $\rho(\theta)d\theta = \sigma(z)dz$, where $\sigma(z) = \omega(z)/\pi i$ is defined on the $\Omega = \{|z| = 1\} \backslash \hat{\Omega}$ on the unit circle, where Ω is the complement of $\hat{\Omega}$ in the unit circle, which will be discussed in detail in next chapter. As a remark, the function $y(z) = z\omega(z) - \frac{1}{2}z(U_0'(z) - U_0'(z^{-1})) - \frac{1}{2}$, satisfies the following relations, $y(z)$ is analytic when $z \in \mathbb{C} \backslash \{\hat{\Omega} \cup \{0\} \cup \{\infty\}\}$; $y(z)|_{\hat{\Omega}^+} + y(z)|_{\hat{\Omega}^-} = -z(U_0'(z) - U_0'(z^{-1})) - 1$; $y(z) \to 0$ as $z \to 0$ or ∞. These properties are needed in discussing the density models.

5.4.3   Weak Coupling: Two-Cut Case

Consider the potential

$$U(z) = g_1(z + z^{-1}) + g_2(z^2 + z^{-2}),$$ (5.84)

or $U(z) = U_0(z) + U_0(z^{-1})$ where $U_0(z) = g_1 z + g_2 z^2$. The density formula can be obtained by using the matrix

$$\hat{L}^{(2)} = \begin{pmatrix} z - u^{(1)} & zv^{(1)} \\ 1 & 0 \end{pmatrix} \begin{pmatrix} z - u^{(2)} & zv^{(2)} \\ 1 & 0 \end{pmatrix}. \tag{5.85}$$

Let us first consider the case when $u^{(1)} = u^{(2)} = 1$, $g_1 = 0$ and $v^{(1)} + v^{(2)} = 2$. Also there is a relation $v^{(1)}v^{(2)} = 1 - \frac{1}{4g_2}$ coming from the expansion of ω to meet with the potential, which is different from the $m = 1$ or 3 cases because when the degree of the potential polynomial is different, the number of the terms in the expansion changes. In this case, the ω function is given by $z\omega(z) = g_2(1 + z^{-2})\sqrt{\Lambda^2 - 4\det \hat{L}^{(2)}}$, where $\Lambda = \mathrm{tr}\,\hat{L}^{(2)} = z^2 + 1$ and $\det \hat{L}^{(2)} = z^2 v^{(1)}v^{(2)}$, or

$$z\omega(z) = g_2(z + z^{-1})\sqrt{(z - z^{-1})^2 + g_2^{-1}}, \quad g_2 \geq \frac{1}{4}. \tag{5.86}$$

The $\omega(z)$ satisfies

$$z\omega(z) = \frac{1}{2}z(U_0'(z) - U_0'(z^{-1})) + \frac{1}{2} + O((z + z^{-1})^{-2}), \tag{5.87}$$

as $|z + z^{-1}| \to \infty$ including $z \to \infty$ and $z \to 0$, where $U_0(z) = g_2 z^2$. Note that in the expansion, we need first to change the term $z - z^{-1}$ in the square root to $z + z^{-1}$ before the expansion calculations since we consider large $z + z^{-1}$. This is important for all of the weak coupling models considered in Chaps. 5 and 6.

The critical point $g_2 = 1/4$ is not corresponding to the parameter bifurcation like we did before, because when $g_2 = 1/4$ there is $v^{(1)}v^{(2)} = 0$, but $v^{(1)} + v^{(2)} = 2$. That means at the critical point we can not have $v^{(1)} = v^{(2)}$ as we worked before if we want to reduce $L^{(2)}$ to $(L^{(1)})^2$. If we consider $u^{(1)} = u^{(2)} = i$ and $v^{(1)} + v^{(2)} = 2i$. The relation between $v^{(1)}v^{(2)}$ and g_2 becomes $v^{(1)}v^{(2)} = -1 - \frac{1}{4g_2}$. The critical point becomes $g_2 = -1/4$ which still leads to $v^{(1)}v^{(2)} = 0$. If we consider the general case with $g_1 \neq 0$, the condition $v^{(1)}v^{(2)} = 0$ at the critical point is still there. So the transition model for the second degree potential is different from the Gross-Witten transition model. The physical background for this phenomenon is not clear yet. Mathematically, the odd terms and even terms in the potential function usually lead to different properties. We will see in Sect. 6.4 that everything works fine except that it is not a parameter bifurcation case at the critical point.

For the $\omega(z)$ defined by (5.86), the analytic function $y(z) = z\omega(z) - \frac{1}{2}z(U_0'(z) - U_0'(z^{-1})) - \frac{1}{2}$, defined in the outside of the cut $\hat{\Omega} = \{z = e^{i\theta} \,||\cos\theta| \leq \sqrt{1 - \frac{1}{4g_2}}\}$, has the properties that $y(z)$ is analytic when $z \in \mathbb{C}\backslash\{\hat{\Omega} \cup \{0\} \cup \{\infty\}\}$, $y(z)|_{\hat{\Omega}^+} + y(z)|_{\hat{\Omega}^-} = -z(U_0'(z) - U_0'(z^{-1})) - 1$, and $y(z) \to 0$ as $z \to 0$ or ∞.

5.4.4 Weak Coupling: Three-Cut Case

Consider the potential

$$U(z) = g_3\big(z^3 + z^{-3}\big),\tag{5.88}$$

or $U(z) = U_0(z) + U_0(z^{-1})$ where $U_0(z) = g_3 z^3$. The interesting result in this model is the critical point $T_c = 1/g_3^c = 6$, which is discussed in [22] for the four-dimensional quantum chromodynamics to create the string model.

Denote

$$\hat{L}^{(3)} = \begin{pmatrix} z-1 & zv^{(1)} \\ 1 & 0 \end{pmatrix} \begin{pmatrix} z-1 & zv^{(2)} \\ 1 & 0 \end{pmatrix} \begin{pmatrix} z-1 & zv^{(3)} \\ 1 & 0 \end{pmatrix},\tag{5.89}$$

and define

$$z\omega(z) = \frac{3}{2}g_3\big(1 + z^{-3}\big)\sqrt{\Lambda^2 - 4\det \hat{L}^{(3)}},\tag{5.90}$$

with $v^{(1)} + v^{(2)} + v^{(3)} = 3$. Then $\Lambda = \operatorname{tr}\hat{L}^{(3)} = z^3 - 1$, and $\det \hat{L}^{(3)} = -z^3 v^{(1)} v^{(2)} v^{(3)}$. If $6g_3 v^{(1)} v^{(2)} v^{(3)} = 1$, then

$$z\omega(z) = \frac{3}{2}g_3\big(z^{3/2} + z^{-3/2}\big)\sqrt{\big(z^{3/2} - z^{-3/2}\big)^2 + \frac{2}{3}g_3^{-1}}, \quad g_3 \geq \frac{1}{6},\tag{5.91}$$

which implies the asymptotics

$$z\omega(z) = \frac{1}{2}z\big(U_0'(z) - U_0'(z^{-1})\big) + \frac{1}{2} + O\big((z^{3/2} + z^{-3/2})^{-2}\big),\tag{5.92}$$

as $|z^{3/2} + z^{-3/2}| \to \infty$.

If $g_3 = 1/6$, then $z\omega(z) = \frac{1}{4}(z^{3/2} + z^{-3/2})^2$. In this case, we can choose $v^{(1)} = v^{(2)} = v^{(3)} = 1$, and then $L^{(3)} = (L^{(1)})^3$, that means this is a parameter bifurcation case. For the strong density with this potential discussed before in this section, we can choose $v = 1$ at the critical point to have $g_3 = 1/6$, and the $z\omega(z)$ meets with the critical case obtained for the weak coupling. It is seen that the transition model in this case is caused by the parameter bifurcation in the sense that v becomes $v^{(1)}$, $v^{(2)}$ and $v^{(3)}$, which extends the Gross-Witten transition model.

Also, the function $y(z) = z\omega(z) - \frac{1}{2}z(U_0'(z) - U_0'(z^{-1})) - \frac{1}{2}$, satisfies the following properties. The $y(z)$ is analytic when $z \in \mathbb{C}\backslash\{\hat{\Omega} \cup \{0\} \cup \{\infty\}\}$; $y(z)|_{\hat{\Omega}+} + y(z)|_{\hat{\Omega}-} = -z(U_0'(z) - U_0'(z^{-1})) - 1$; $y(z) \to 0$ as $z \to 0$ or ∞, where $\hat{\Omega}$ is the union of the three cuts $|\cos(3\theta/2)| \leq \sqrt{1 - \frac{1}{6g_3}}$ for $z = e^{i\theta}$ on the unit circle at the points $z = -1$, $z = e^{i\pi/3}$ and $z = e^{-i\pi/3}$.

References

1. Cresswell, C., Joshi, N.: The discrete first, second and thirty-fourth Painlevé hierarchies. J. Phys. A **32**, 655–669 (1999)
2. Grammaticos, B., Nijhoff, F.W., Papageorgiou, V., Ramani, A., Satsuma, J.: Linearization and solutions of the discrete Painlevé III equation. Phys. Lett. A **185**, 446–452 (1994)
3. Grammaticos, B., Ohta, Y., Ramani, A., Sakai, H.: Degenerate through coalescence of the q-Painlevé VI equation. J. Phys. A **31**, 3545–3558 (1998)
4. Gross, D.J., Witten, E.: Possible third-order phase transition in the large-N lattice gauge theory. Phys. Rev. D **21**, 446–453 (1980)
5. McLeod, J.B., Wang, C.B.: Discrete integrable systems associated with the unitary matrix model. Anal. Appl. **2**, 101–127 (2004)
6. Periwal, V., Shevitz, D.: Unitary-matrix models as exactly solvable string theories. Phys. Rev. Lett. **64**, 1326–1329 (1990)
7. Periwal, V., Shevitz, D.: Exactly solvable unitary matrix models: multicritical potentials and correlations. Nucl. Phys. B **344**, 731–746 (1990)
8. Rossi, P., Campostrini, M., Vicari, E.: The large-N expansion of unitary-matrix models. Phys. Rep. **302**, 143–209 (1998)
9. Simon, B.: Orthogonal Polynomials on the Unit Circle, Vol. 1: Classical Theory. AMS Colloquium Series. AMS, Providence (2005)
10. Simon, B.: Orthogonal Polynomials on the Unit Circle, Vol. 2: Spectral Theory. AMS Colloquium Series. AMS, Providence (2005)
11. Szegö, G.: Orthogonal Polynomials, 4th edn. American Mathematical Society Colloquium Publications, vol. 23. AMS, Providence (1975)
12. Wang, C.B.: Orthonormal polynomials on the unit circle and spatially discrete Painlevé II equation. J. Phys. A **32**, 7207–7217 (1999)

Chapter 6
Transitions in the Unitary Matrix Models

The first-, second- and third-order phase transitions, or discontinuities, in the unitary matrix models will be discussed in this chapter. The Gross-Witten third-order phase transition is described in association with the string equation in the unitary matrix model, and it will be generalized by considering the higher degree potentials. The critical phenomena (second-order divergences) and third-order divergences are discussed similarly to the critical phenomenon in the planar diagram model, but a different Toda lattice and string equation will be applied here by using the double scaling method. The discontinuous property in the first-order transition model of the Hermitian matrix model discussed before will recur in the first-order transition model of the unitary matrix model, indicating a common mathematical background behind the first-order discontinuities. The purpose of this chapter is to further confirm that the string equation method can be widely applied to study phase transition problems in matrix models, and that the expansion method based on the string equations can work efficiently to find the power-law divergences considered in the transition problems.

6.1 Large-N Models and Partition Function

For the potential $V(z) = s(z + z^{-1})$, we have discussed the linear equations satisfied by the orthogonal polynomials $p_n = p_n(z, s)$ on the unit circle in the n and z directions in last chapter. In the z direction, we have obtained in Sect. 5.3 that $\frac{\partial}{\partial z}\Phi_n = A_n(z)\Phi_n$, where

$$A_n(z) = \begin{pmatrix} \frac{s}{2} + \frac{s}{2z^2} + \frac{n-2sx_nx_{n+1}}{2z} & s(x_{n+1} - \frac{x_n}{z})z^{-1} \\ s(x_n - \frac{x_{n+1}}{z}) & -\frac{s}{2} - \frac{s}{2z^2} - \frac{n-2sx_nx_{n+1}}{2z} \end{pmatrix}, \tag{6.1}$$

and x_n satisfies the string equation

$$\frac{n}{s}x_n = -\left(1 - x_n^2\right)(x_{n+1} + x_{n-1}), \tag{6.2}$$

C.B. Wang, *Application of Integrable Systems to Phase Transitions*,
DOI 10.1007/978-3-642-38565-0_6, © Springer-Verlag Berlin Heidelberg 2013

Fig. 6.1 Three X curves in
the unitary model

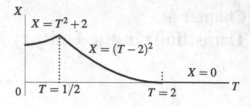

with $x_n \in [-1, 1]$. Then, we have that

$$-\frac{z^2}{s^2} \det A_n = \frac{1}{4}\left(z + z^{-1} + T\right)^2 - \frac{X}{4},\tag{6.3}$$

where $T = n/s$, and

$$X = \frac{(x_n')^2 - T^2 x_n^2}{1 - x_n^2} + 4x_n^2,\tag{6.4}$$

with $' = d/ds$.

In Sect. D.2, it is discussed that the following values of X (see Fig. 6.1) are the singular points of the hypergeometric-type differential equation satisfied by the elliptic integrals,

$$X = \begin{cases} T^2 + 2, & 0 < T \le 1/2, \\ (T - 2)^2, & 1/2 \le T \le 2, \\ 0, & T \ge 2. \end{cases}\tag{6.5}$$

Here, the T values are divided in the three regions for studying the transition problems. Based on the reduction formulation $\frac{1}{n\pi}\sqrt{\det A_n}\, dz = \sigma(z)dz$, we will consider the transition problems for the following phases in the next two sections,

$$\sigma(z) = \begin{cases} \frac{1}{2T\pi iz}\sqrt{(z + z^{-1} + T)^2 - T^2 - 2}, & 0 < T \le 1/2, \\ \frac{1}{2T\pi iz}\sqrt{(z + z^{-1} + 2)(z + z^{-1} + 2T - 2)}, & 1/2 \le T \le 2, \\ \frac{1}{2T\pi iz}(z + z^{-1} + T), & T \ge 2. \end{cases}\tag{6.6}$$

The derivatives of the partition function $Z_n = n! h_0 h_1 \cdots h_{n-1}$ [13, 14] in the s direction are important for discussing the phase transitions, which are similar to the results obtained in the Hermitian matrix models. For the h_n's, by the discussions in Sect. 5.3, there are

$$h_0' = -2x_1 h_0, \qquad h_n' = -2x_n x_{n+1} h_n, \quad n \ge 1,\tag{6.7}$$

where $' = d/ds$. Then, there is

$$\frac{d}{ds} \ln Z_n = -2x_1 - 2\sum_{j=1}^{n-1} x_j x_{j+1}.\tag{6.8}$$

We will discuss later that the x_n's satisfy the following Toda lattice,

$$x_1' = (1 - x_1^2)(x_2 - 1), \qquad x_n' = (1 - x_n^2)(x_{n+1} - x_{n-1}), \quad n \geq 2. \qquad (6.9)$$

Therefore, we have

$$\frac{d^2}{ds^2} \ln Z_n = 2(1 - x_n^2)(1 - x_{n-1}x_{n+1}), \qquad (6.10)$$

for $n \geq 2$. Equation (6.10) implies that the second-order derivative of the free energy will be continuous if we study the reduced models using the bifurcation transition method as discussed in the Hermitian matrix models. But the term $x_{n-1}x_{n+1}$ in (6.10) will cause a second-order divergence in the s direction to be explained in Sect. 6.3.

Since the s direction is so important for the transition problems, we need to go through the details of the system in this direction in order to show how the second-order divergence is caused. We will see that this divergence is because the continuum Painlevé III equation is involved. The discussions for the equations satisfied by the x_n or the u_n and v_n in the s direction can be found, for example, in [4, 5, 10]. As we have experienced in Chap. 4, the Lax pairs are the basic models for investigating the divergent behaviors. Here, we want to setup the Lax pairs for the Toda lattice and the continuum Painlevé III or V equation. Let us consider the linear equation in the s direction in order to get the Lax pair for the Toda lattice or continuum Painlevé equation. It has been discussed in Sect. 5.3 that for the potential $V(z) = s(z + z^{-1})$, the orthogonal polynomials $p_n(z)$ satisfy the following equation in the s equation,

$$z^{-1} \frac{\partial}{\partial s} \frac{p_n}{x_n} = \sum_{k=n-2}^{n-1} b_k p_k, \qquad (6.11)$$

where

$$b_k h_k = \frac{\partial}{\partial s} \left(\frac{h_n}{x_n} \right) \delta_{k,n-1} - \frac{1}{x_n} \oint z^2 p_k \bar{p}_n e^{s(z+1/z)} \frac{dz}{2\pi i z}, \qquad (6.12)$$

for $k = n - 2$ and $n - 1$. Therefore, we have obtained

$$\frac{\partial}{\partial s} \frac{p_n}{x_n} = b_{n-1} z p_{n-1} + b_{n-2} z p_{n-2}, \qquad (6.13)$$

where

$$b_{n-1} = \frac{1}{h_{n-1}} \frac{\partial}{\partial s} \left(\frac{h_n}{x_n} \right) - \frac{h_n}{h_{n-1} x_n} (u_n - v_n + u_{n-1} - v_{n-1}), \qquad (6.14)$$

$$b_{n-2} = -\frac{h_n}{h_{n-2} x_n}. \qquad (6.15)$$

One of the consistency conditions for (6.13) and the recursion formula can be found by comparing the leading coefficients on both sides of (6.13). Since the formula for b_{n-1} above involves $\partial h_n / \partial s$, we also need the formula of $\partial h_n / \partial s$ given by (6.7). Therefore we have

$$\frac{\partial}{\partial s}\left(\frac{1}{x_n}\right) = b_{n-1},$$ (6.16)

$$\frac{1}{h_n}\frac{\partial h_n}{\partial s} = 2(u_n - v_n).$$ (6.17)

It can be proved by direct calculations that (6.13) and the recursion formula are consistent if these two equations hold. Now, let us simplify the equations above. By using the relations

$$\frac{h_n}{h_{n-1}} = \frac{v_n}{u_n} = 1 - x_n^2,$$ (6.18)

$$u_n - v_n = -x_n x_{n+1},$$ (6.19)

the b_{n-1} and b_{n-2} can be simplified to

$$b_{n-1} = -\frac{1}{x_n^2}(1 - x_n^2)(x_{n+1} - x_{n-1}),$$ (6.20)

$$b_{n-2} = -\frac{1}{x_n}(1 - x_{n-1}^2)(1 - x_n^2),$$ (6.21)

and the equations above for the consistency become the following equation

$$x_n' = (1 - x_n^2)(x_{n+1} - x_{n-1}),$$ (6.22)

where $x_n' = dx_n/ds$, which is the Toda lattice (6.9) needed above.

So far, we have obtained the linear equations

$$z(p_n + v_n p_{n-1}) = p_{n+1} + u_n p_n,$$ (6.23)

$$\frac{\partial p_n}{\partial z} = n p_{n-1} + s\frac{v_n v_{n-1}}{u_n u_{n-1}} p_{n-2},$$ (6.24)

$$\frac{\partial p_n}{\partial s} = \frac{(1 - x_n^2)(x_{n+1} - x_{n-1})}{x_n}(p_n - z p_{n-1}) - (1 - x_n^2)p_{n-1},$$ (6.25)

for the orthogonal polynomials $p_n(z, s)$, where

$$u_n = -\frac{x_{n+1}}{x_n},$$ (6.26)

$$v_n = -\frac{x_{n+1}}{x_n}(1 - x_n^2),$$ (6.27)

and $x_n = p_n(0, s)$. The consistency condition for the Lax pair (6.23) and (6.24) is the string equation, the consistency condition for the Lax pair (6.23) and (6.25) is the Toda lattice in terms of u_n and v_n or x_n, and the consistency condition for the Lax pair (6.24) and (6.25) is the continuum Painlevé III or V with the details explained in Sect. C.3. Briefly, the function $1/u_{n-1} = -x_{n-1}/x_n$ satisfies the continuum Painlevé III equation, and $v_n/(v_n - u_n) = 1 - x_n^{-2}$ satisfies the continuum Painlevé V equation. In addition, the u_n and v_n satisfy an alternate discrete Painlevé II equation explained in Sect. 5.3.1.

Similarly to the z equation discussed in last chapter, the s equation (6.25) can be changed to the following matrix form,

$$\frac{\partial}{\partial s}\begin{pmatrix} p_n \\ \tilde{p}_n \end{pmatrix} = \begin{pmatrix} -z^{-1} - x_n x_{n+1} & x_{n+1} + x_n z^{-1} \\ x_{n+1} + x_n z & -z - x_n x_{n+1} \end{pmatrix}\begin{pmatrix} p_n \\ \tilde{p}_n \end{pmatrix}, \qquad (6.28)$$

where $\tilde{p}_n$ is defined by $z p_n = p_{n+1} - x_{n+1}\tilde{p}_n$. Furthermore, let

$$\begin{pmatrix} \psi_n \\ \hat{\psi}_n \end{pmatrix} = e^{\frac{s}{2}(\eta^2 + \eta^{-2})}\begin{pmatrix} \eta^{-n+1/2} & 0 \\ 0 & \eta^{-n-1/2} \end{pmatrix}\begin{pmatrix} p_n \\ \tilde{p}_n \end{pmatrix}, \qquad (6.29)$$

where $\eta = z^{1/2}$, then we have

$$\frac{\partial}{\partial s}\begin{pmatrix} \psi_n \\ \hat{\psi}_n \end{pmatrix} = \begin{pmatrix} \frac{1}{2}(\eta^2 - \eta^{-2}) - x_n x_{n+1} & x_{n+1}\eta + x_n \eta^{-1} \\ x_n \eta + x_{n+1}\eta^{-1} & -\frac{1}{2}(\eta^2 - \eta^{-2}) - x_n x_{n+1} \end{pmatrix}\begin{pmatrix} \psi_n \\ \hat{\psi}_n \end{pmatrix}. \qquad (6.30)$$

The z variable sometimes needs to be changed to $\eta = \sqrt{z}$ for the technical convenience in the discussions as seen above. There might be some interesting properties in physics related to this simple transformation. In the following sections, we are going to apply the results obtained from the integrable systems to study the phase transition or discontinuity problems in the unitary matrix models.

6.2 First-Order Discontinuity with Two Cuts

Now, let us consider the $\sigma(z)$ defined by (6.6) for $0 < T < 2$. If we make a change of variable $z = e^{i\theta}$, then $\sigma(z)$ can be changed to a form of the eigenvalue density $\rho(\theta)$ considered in physics. By the relation $\sigma(z)dz = \rho(\theta)d\theta$, we can get

$$\rho(\theta) = \begin{cases} \frac{2}{T\pi}\sqrt{[\frac{T+2+\sqrt{T^2+2}}{4} - \sin^2\frac{\theta}{2}][\frac{T+2-\sqrt{T^2+2}}{4} - \sin^2\frac{\theta}{2}]}, & 0 < T \le 1/2, \\[2mm] \frac{2}{T\pi}\cos\frac{\theta}{2}\sqrt{\frac{T}{2} - \sin^2\frac{\theta}{2}}, & 1/2 \le T \le 2, \end{cases} \qquad (6.31)$$

where $\theta \in \Omega_\theta$ and

$$\Omega_\theta = \begin{cases} \{\theta \,|\, (\cos\theta + \frac{T+\sqrt{T^2+2}}{2})(\cos\theta + \frac{T-\sqrt{T^2+2}}{2}) \ge 0\}, & 0 < T \le 1/2, \\[2mm] \{\theta \,|\, \cos\theta \ge 1 - T\}, & 1/2 \le T \le 2. \end{cases} \qquad (6.32)$$

When $0 < T \le 1/2$, Ω_θ is corresponding to the arcs Ω_1 and Ω_2 in Fig. 6.2. When $T \ge 1/2$, Ω_2 disappears and Ω_θ is corresponding to Ω_1 only. The union of

Fig. 6.2 Cuts (*thick arcs*) and eigenvalue range (*thin arcs*)

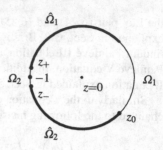

$\Omega = \Omega_1 \cup \Omega_2$ and $\hat{\Omega} = \hat{\Omega}_1 \cup \hat{\Omega}_2$ is the unit circle, so Ω is the compliment of $\hat{\Omega}$ in the unit circle. When $0 < T < 1/2$, the ρ function does not satisfy the non-negativity condition since the density function on the arc across $z = -1$ (see Fig. 6.2) is negative. This is a case like the model discussed in Sect. 4.1.2. This negative-positive model can be changed to a non-negative model by constructing the density on two coupled arcs passing the point $z = 1$, similar to the discussion in Sect. 4.1.2. We are not going to discuss the details here since we want to keep the discussions in terms of ρ for consistency. As T is increased to $1/2$, the two cuts on the unit circle merge together at the point $z = -1$, that will cause a first-order discontinuity to be discussed in the following.

Define the ω function as

$$\omega(z) = \begin{cases} \frac{1}{2Tz}\sqrt{(z + z^{-1} + T)^2 - T^2 - 2}, & 0 < T \le 1/2, \\ \frac{1}{2Tz}\sqrt{(z + z^{-1} + 2)(z + z^{-1} + 2T - 2)}, & 1/2 \le T \le 2, \end{cases} \qquad (6.33)$$

for z in the complex plane outside the cut(s) $\hat{\Omega}$ and the points $z = 0$ and ∞, with the following asymptotics

$$z\omega(z) = \begin{cases} \frac{1}{2T}(z + z^{-1}) + \frac{1}{2} + O((z + z^{-1} + T)^{-1}), & 0 < T \le 1/2, \\ \frac{1}{2T}(z + z^{-1}) + \frac{1}{2} + O((z + z^{-1} + 2)^{-1}), & 1/2 \le T \le 2, \end{cases} \qquad (6.34)$$

as $z \to 0$ or $z \to \infty$, where $U(z) = T^{-1}(z + z^{-1})$. Note that for this special potential, there is $U(z) = z(U_0'(z) - U_0'(1/z))$ where $U_0(z) = T^{-1}z$. This could cause a confusion that one may still search the analytic function $z\omega(z)$ with the asymptotics $\frac{1}{2}U(z)$ for the higher degree potentials. But the analytic function $z\omega(z)$ generally need to have an asymptotics with the leading term $\frac{1}{2}z(U_0'(z) - U_0'(1/z))$ in order to satisfy the variational equation discussed in Sect. 5.1 because of the two essential singular points $z = 0$ and ∞. Also, the two O terms in the above expansions are different.

Lemma 6.1 *The $\sigma(z)$ defined by (6.6) for $0 < T < 2$ satisfies*

$$(P) \int_\Omega \frac{\sigma(\zeta)}{z - \zeta} d\zeta = -\frac{1}{2}U'(z) + \frac{1}{2z}, \qquad (6.35)$$

for an inner point z in Ω, where $\int_\Omega \sigma(z)dz = 1$.

Proof Let C^+ and C^- be the closed counterclockwise outside and inside edges of the unit circle respectively, and $C^* = C^+ \cup C^-$. By Cauchy theorem, there is

$$\left\{ \int_{C^*} - \int_{|z|=\delta} - \int_{|z|=R} \right\} \left(\omega(z) - \frac{1}{2z} U(z) \right) dz = 0,$$

which implies the following by the asymptotics at 0 and ∞,

$$\int_{C^*} \left(\omega(z) - \frac{1}{2z} U(z) \right) dz \to \int_{|z|=\hat{R}} \frac{dz}{2z} + \int_{|z|=\delta} \frac{dz}{2z} = 2\pi i,$$

as $\delta \to 0$ and $R \to \infty$. Since $\int_{C^*} U(z) dz/z = 0$ and $\omega(z)|_{\hat{\Omega}+} + \omega(z)|_{\hat{\Omega}-} = 0$, we have $\int_{\Omega} \frac{1}{\pi i} \omega(z)|_{\Omega} dz = 1$. Therefore $\int_{\Omega} \sigma(z) d\sigma = \int_{\Omega} \frac{1}{\pi i} \omega(z)|_{\Omega} dz = 1$.

Consider a point z in Ω. Choose a small $\varepsilon > 0$ and make a small circle of radius ε with center z. Let the "semicircle" inside of the unit circle $|z| = 1$ be γ_ε^-, and the "semicircle" outside the unit circle be γ_ε^+. Remove the small arc around z ($|z| = 1$) enclosed by the ε circle from the unit circle. The remaining arc is denoted as $C(\varepsilon)$. As $\varepsilon \to 0$, $C(\varepsilon)$ becomes the unit circle. All the closed contours are oriented counterclockwise. Then by Cauchy theorem, there is

$$\left(\int_{C^+(\varepsilon) \cup \gamma_\varepsilon^+} + \int_{C^-(\varepsilon) \cup \gamma_\varepsilon^-} \right) \left(\zeta \omega(\zeta) - \frac{1}{2T} (\zeta + \zeta^{-1}) - \frac{1}{2} \right) \frac{d\zeta}{\zeta - z} = 0,$$

where $\gamma_\varepsilon^\pm$ are given in the above discussion. Since the small semicircles γ_ε^+ and γ_ε^- have opposite directions around the point z, there is

$$\left(\int_{\gamma_\varepsilon^+} + \int_{\gamma_\varepsilon^-} \right) \left(\zeta \omega(\zeta) - \frac{1}{2T} (\zeta + \zeta^{-1}) - \frac{1}{2} \right) \frac{d\zeta}{\zeta - z} \to 0,$$

as $\varepsilon \to 0$. Because the integrals of $\zeta \omega(\zeta)$ along the inside and outside edges of the cuts are canceled, there is

$$\left(\int_{C^+(\varepsilon)} + \int_{C^-(\varepsilon)} \right) \zeta \omega(\zeta) \frac{d\zeta}{\zeta - z} = 2 \int_{\Omega(\varepsilon)} \zeta \omega(\zeta) \frac{d\zeta}{\zeta - z} \to 2(P) \int_{\Omega} \zeta \omega(\zeta) \frac{d\zeta}{\zeta - z},$$

as $\varepsilon \to 0$, where $\Omega(\varepsilon)$ is the Ω without the small arc at the point z. Also, we have

$$\frac{1}{\pi i} \int_{C(\varepsilon)} \left(\frac{1}{2T} (\zeta + \zeta^{-1}) + \frac{1}{2} \right) \frac{d\zeta}{\zeta - z}$$

$$= \frac{1}{2\pi i} \int_{C(\varepsilon) \cup \gamma_\varepsilon^-} \frac{1}{\zeta - z} d\zeta + \frac{1}{2T\pi i} \int_{C(\varepsilon) \cup \gamma_\varepsilon^-} \frac{\zeta}{\zeta - z} d\zeta$$

$$+ \frac{1}{2T\pi i} \int_{C(\varepsilon) \cup \gamma_\varepsilon^+} \frac{\zeta^{-1}}{\zeta - z} d\zeta$$

$$- \frac{1}{2\pi i} \int_{\gamma_\varepsilon^-} \frac{1}{\zeta - z} d\zeta - \frac{1}{2T\pi i} \int_{\gamma_\varepsilon^-} \frac{\zeta}{\zeta - z} d\zeta - \frac{1}{2T\pi i} \int_{\gamma_\varepsilon^+} \frac{\zeta^{-1}}{\zeta - z} d\zeta$$

$$\to \frac{1}{2} + \frac{1}{2T} (z - z^{-1}) = \frac{1}{2} z U'(z) + \frac{1}{2},$$

as $\varepsilon \to 0$, where the three integrals along $C(\varepsilon) \cup \gamma_\varepsilon^-$ or $C(\varepsilon) \cup \gamma_\varepsilon^+$ vanish. Therefore, we get

$$(P) \int_\Omega \frac{\zeta \sigma(\zeta)}{\zeta - z} d\zeta = \frac{1}{2} z U'(z) + \frac{1}{2},$$

and then (6.35) is proved. □

Lemma 6.2 *For an inner point z of $\hat{\Omega}$, there is*

$$\int_\Omega \frac{\sigma(\zeta)}{z - \zeta} d\zeta = -\frac{1}{2} U'(z) + \frac{1}{2z} + \omega(z), \tag{6.36}$$

for $0 < T < 2$.

Proof If we keep using the notations in the proof of last lemma, then for z in $\hat{\Omega}$ there is

$$\left(\int_{C^+(\varepsilon) \cup \gamma_\varepsilon^+} + \int_{C^-(\varepsilon) \cup \gamma_\varepsilon^-} \right) \frac{\zeta \omega(\zeta) - \frac{1}{2} U(\zeta) - \frac{1}{2}}{\zeta - z} d\zeta = 0,$$

based on the asymptotics at $\zeta = 0$ and $\zeta = \infty$. Further, since ω has opposite signs on the two edges of the cut(s), there are

$$\int_{C^+(\varepsilon) \cup C^-(\varepsilon)} \frac{\zeta \omega(\zeta)}{\zeta - z} d\zeta + \int_{\gamma_\varepsilon^+ \cup \gamma_\varepsilon^-} \frac{\zeta \omega(\zeta)}{\zeta - z} d\zeta \to 2 \int_\Omega \frac{\zeta \omega(\zeta)}{\zeta - z} d\zeta + 2\pi i z \omega(z),$$

and

$$\left\{ 2 \int_{C(\varepsilon)} + \int_{\gamma_\varepsilon^+ \cup \gamma_\varepsilon^-} \right\} \frac{\frac{1}{2} U(\zeta) + \frac{1}{2}}{\zeta - z} d\zeta$$

$$= 2 \int_{C(\varepsilon)} \frac{\frac{1}{2T}(\zeta + \zeta^{-1}) + \frac{1}{2}}{\zeta - z} d\zeta$$

$$= -2 \int_{\gamma_\varepsilon^-} \frac{\frac{1}{2T}\zeta + \frac{1}{2}}{\zeta - z} d\zeta - 2 \int_{\gamma_\varepsilon^+} \frac{\frac{1}{2T}\zeta^{-1}}{\zeta - z} d\zeta \to 2\pi i \left(\frac{1}{2T}(z - z^{-1}) + \frac{1}{2} \right),$$

as $\varepsilon \to 0$. Then, we get the following

$$\int_\Omega \frac{\zeta \sigma(\zeta)}{\zeta - z} d\zeta = \frac{1}{2} z U'(z) - z \omega(z) + \frac{1}{2}, \tag{6.37}$$

and this lemma is proved. □

Now, consider

$$E = -\int_\Omega U(z) \sigma(z) dz - \int_\Omega \int_\Omega \ln|z - \zeta| \sigma(z) \sigma(\zeta) dz d\zeta. \tag{6.38}$$

We will see a first-order discontinuity at the critical point $T = 1/2$.

Lemma 6.3

$$\frac{dE}{dT} = \begin{cases} \frac{1}{T^2}\int_\Omega (z+z^{-1})\sigma(z)dz - 2\int_{\hat{\Omega}_1}\omega(\eta)d\eta\frac{d}{dT}\int_{\Omega_2}\sigma(z)dz, & 0 < T \le 1/2, \\ \frac{1}{T^2}\int_\Omega (z+z^{-1})\sigma(z)dz, & 1/2 \le T < 2, \end{cases}$$

(6.39)

where $\Omega = \Omega_1 \cup \Omega_2$ *with* $1 \in \Omega_1$ *and* $-1 \in \Omega_2$, *and* $\hat{\Omega}_1$ *is the cut between* Ω_1 *and* Ω_2, *where all arcs are oriented counterclockwise.*

Proof When $0 < T \le 1/2$, we have

$$\frac{dE}{dT} = -\int_\Omega \frac{dU(z)}{dT}\sigma(z)dz$$

$$-2\left(\int_{\Omega_1}+\int_{\Omega_2}\right)\left(\int_\Omega \ln|z-\zeta|\sigma(\zeta)d\zeta + \frac{1}{2}U(z)\right)\frac{d\sigma(z)}{dT}dz.$$

By the lemmas above, for $z \in \Omega_1$ there is

$$\int_\Omega \ln|z-\zeta|\sigma(\zeta)d\zeta + \frac{1}{2}U(z) = \int_\Omega \ln|z_0-\zeta|\sigma(\zeta)d\zeta + \frac{1}{2}U(z_0),$$

where z_0 is the start point of Ω_1 (Fig. 6.2), and for $z \in \Omega_2$ there is

$$\int_\Omega \ln|z-\zeta|\sigma(\zeta)d\zeta + \frac{1}{2}U(z) = \int_\Omega \ln|z_0-\zeta|\sigma(\zeta)d\zeta + \frac{1}{2}U(z_0) + \int_{\hat{\Omega}_1}\omega(z)dz.$$

Then the result is true by using $\frac{d}{dT}\int_{\Omega_1\cup\Omega_2}\sigma(z)dz = \frac{d}{dT}1 = 0$. When $1/2 \le T < 2$, the $(\cdots)$ part in the second term on the right hand side of the formula $\frac{dE}{dT}$ above is constant because Ω is now a connected arc, which can be moved to the outside of the integral leading the entire integral to vanish. $\square$

Lemma 6.4 *As* $T \to 1/2 - 0$, *there are the following expansions,*

$$T = \frac{1}{2} - \frac{3}{4}\varepsilon^2,$$

(6.40)

$$z_\pm = -1 \pm i\varepsilon + \cdots,$$

(6.41)

where $z_\pm$ *are the end points of* Ω_2 *defined by* $z_\pm + z_\pm^{-1} + T + \sqrt{T^2+2} = 0$.

Proof Substituting the expansions $z = -1 + c_1\varepsilon + c_2\varepsilon^2 + \cdots$ and $T = \frac{1}{2} + T_1\varepsilon + T_2\varepsilon^2$ into the equation $z + z^{-1} + T + \sqrt{T^2+2} = 0$, we can get the expansion results. $\square$

Theorem 6.1 *The E function given by* (6.38) *has discontinuous first-order derivative at* $T = 1/2$,

$$E'(1/2+0) - E'(1/2-0) = 2 + \frac{1}{\sqrt{3}}\ln(2-\sqrt{3}),$$

(6.42)

where $' = d/dT$.

Proof According to (6.38), we only need to consider

$$I_0 \equiv 2 \int_{\hat{\Omega}_1} \omega(z)dz \frac{d}{dT} \int_{\Omega_2} \sigma(z)dz. \qquad (6.43)$$

By Lemma 6.4, if we make a transformation $z = -e^{i\varepsilon\phi}$ around $z = -1$, then as T approaches to $1/2$, there is

$$\sigma(z)dz \sim \frac{\sqrt{3}}{\pi} \varepsilon^2 \sqrt{1 - \phi^2} d\phi,$$

where ϕ is from -1 to 1, that implies

$$\frac{d}{dT} \int_{\Omega_2} \sigma(z)dz \to -\frac{2}{\sqrt{3}}.$$

Also, as $T \to 1/2 - 0$, we have

$$\int_{\hat{\Omega}_1} \omega(z)dz \to 2i \int_{\pi/3}^{\pi} \cos\frac{\theta}{2} \sqrt{1 - \frac{1}{4}\sin^2\frac{\theta}{2}} d\theta = -\frac{\sqrt{3}}{2} - \frac{1}{4}\ln(2 - \sqrt{3}),$$

which implies $E'(1/2 + 0) - E'(1/2 - 0) = \lim_{\varepsilon \to 0} I_0 = 2 + \frac{1}{\sqrt{3}}\ln(2 - \sqrt{3})$, since $\int_{\Omega}(z + z^{-1})\sigma(z)dz$ is continuous.   □

We have seen the exact same term $\frac{1}{\sqrt{3}}\ln(2 - \sqrt{3})$ in the model discussed in Sect. 4.1.2, which could be related to an interesting physical problem.

6.3 Double Scaling Associated with Gross-Witten Transition

Consider the $\sigma(z)$ defined by (6.6) for $T > 1/2$. By the transformation $\sigma(z)dz = \rho(\theta)d\theta$ with $z = e^{i\theta}$, there are the following weak and strong coupling densities [3]

$$\rho(\theta) = \begin{cases} \frac{2}{T\pi}\cos\frac{\theta}{2}\sqrt{\frac{T}{2} - \sin^2\frac{\theta}{2}}, & 1/2 \le T \le 2, \\ \frac{1}{2\pi}(1 + \frac{2}{T}\cos\theta), & T \ge 2, \end{cases} \qquad (6.44)$$

where $\theta \in \Omega_\theta$ and

$$\Omega_\theta = \begin{cases} \{\theta \,|\, |\sin\frac{\theta}{2}| \le \sqrt{T/2}, |\theta| \le \pi\}, & 1/2 \le T \le 2, \\ \{\theta \,|\, |\theta| \le \pi\}, & T \ge 2. \end{cases} \qquad (6.45)$$

Then the free energy

$$E = -\frac{2}{T} \int_{\Omega_\theta} \cos(\theta)\rho(\theta)d\theta - \int_{\Omega_\theta} \int_{\Omega_\theta} \ln\left|\sin\frac{\theta - \theta'}{2}\right| \rho(\theta)\rho(\theta')d\theta d\theta'$$

$$+ \frac{1}{2\pi} \int_{-\pi}^{\pi} \ln\left|\sin\frac{\theta}{2}\right| d\theta \qquad (6.46)$$

has the following explicit formulas [3],

$$E = \begin{cases} -\frac{2}{T} - \frac{1}{2}\ln\frac{T}{2} + \frac{3}{4}, & T \le 2, \\ -\frac{1}{T^2}, & T \ge 2. \end{cases} \tag{6.47}$$

It can be checked that the free energy has continuous first- and second-order derivatives, and its third-order derivative is discontinuous at $T = 2$. This is a brief description of the well known Gross-Witten third-order phase transition found in 1980, see [3].

In the following, we are going to discuss the reduction of the eigenvalue densities ρ from the coefficient matrix A_n given by (6.1) in the Lax pair. The determinant of A_n can be written as [11]

$$\sqrt{\det A_n(z)}$$
$$= \frac{1}{i}\sqrt{\left(\frac{s}{2} + \frac{s}{2z^2} + \frac{n - 2sx_nx_{n+1}}{2z}\right)^2 + \frac{s^2}{z}\left(x_n - \frac{x_{n+1}}{z}\right)\left(x_{n+1} - \frac{x_n}{z}\right)}. \tag{6.48}$$

Consider $n/s = T$ and $u_n = -x_{n+1}/x_n$. Then the string equation (6.2) becomes $T/(1 - x_n^2) = u_n + 1/u_{n-1}$, or asymptotically as n and $s \to \infty$,

$$u_n \sim \left[\frac{T}{2(1 - x_n^2)} + \sqrt{\left(\frac{T}{2(1 - x_n^2)}\right)^2 - 1}\right]^{-1}.$$

If $T = 2(1 - x_n^2)(\le 2)$, then $u_n \sim 1$, or $x_{n+1} \sim -x_n \sim x_{n-1}$, that implies

$$\frac{1}{n\pi}\sqrt{\det A_n(z)}dz \sim \frac{2}{\pi T}\cos\frac{\theta}{2}\sqrt{\frac{T}{2} - \sin^2\frac{\theta}{2}}d\theta, \tag{6.49}$$

where $z = e^{i\theta}$. If $T > 2$, then $u_n < 1$, or $x_n \to 0$, that implies

$$\frac{1}{n\pi}\sqrt{\det A_n(z)}dz \sim \frac{1}{2\pi}\left(1 + \frac{2}{T}\cos\theta\right)d\theta. \tag{6.50}$$

These results are corresponding to the weak and strong coupling densities (6.44) obtained in [3].

Now, let us consider a critical phenomenon, a second-order discontinuity related but differing from the third-order discontinuity talked above. The string equation (6.2) can be reduced to

$$gv = \frac{1}{2}, \tag{6.51}$$

by taking $x_{n-1} \sim -x_n \sim x_{n+1}$ according to the above reduction, where $g = s/n = T^{-1}$ and $1 - x_n^2 \sim v$. Consider the ε expansions

$$g = \frac{1}{2} + \varepsilon, \tag{6.52}$$

and

$$v = \frac{1}{1+2\varepsilon} = 1 - 2\varepsilon + O(\varepsilon^2). \tag{6.53}$$

Then, there is $x_n^2 \sim 2\varepsilon$. In the s direction, as $s \to n/2+0$ we have $x_n = O(\sqrt{s - \frac{n}{2}})$, and then

$$dx_n/ds = O(|s - n/2|^{-1/2}). \tag{6.54}$$

Come back to the formula $\frac{d^2}{ds^2}\log Z_n = 2(1 - x_n^2)(1 - x_{n-1}x_{n+1})$ obtained in Sect. 6.3. By the discussion in Sect. C.3, we have

$$x_{n-1}x_{n+1} = \frac{(nx_n)^2/s^2 - (x_n')^2}{4(1 - x_n^2)^2}, \tag{6.55}$$

where $' = d/ds$, that implies

$$\frac{d^2}{ds^2}\log Z_n = 2(1 - x_n^2) - \frac{(nx_n)^2/s^2 - (x_n')^2}{2(1 - x_n^2)}. \tag{6.56}$$

According to the discussion above, we then have the critical phenomenon

$$\frac{d^2 F_n}{ds^2} = O((dx_n/ds)^2) = O(|s - n/2|^{-1}), \tag{6.57}$$

with the critical point $s = n/2$, where $F_n = -\log Z_n$ which is different from the free energy $E = -\lim \frac{1}{n^2}\log Z_n$ discussed above. The formula of F_n is close to the free energy in statistical mechanics [18]. Also, the original physical model [3] in the unitary matrix model involves the parameter n in the discussion of the free energy, which means the F_n considered here has a physical background. Since $x_n = O(\sqrt{\varepsilon})$ and $g = 1/2 + \varepsilon > 1/2$, the transition in the above critical phenomenon is between the strong coupling phase in the Gross-Witten transition model and the phase $\frac{1}{n\pi}\sqrt{\det A_n(z)}dz$ which is on the same side $T = g^{-1} < 2$ of the weak coupling phase in the Gross-Witten transition model, with a weaker coupling structure since there are four roots for $\det A_n(z) = 0$ in the complex plane. One can use the method discussed in Sect. 4.3.1 to get (6.56) by the free energy formula in terms of the density based on the linear equations in the s direction discussed in Sect. 6.1. The details are omitted here since the discussion in Sect. 6.1 to derive (6.10) is sufficient.

The above discussions can be summarized as follows. If we consider the local variable s, then the second-order derivative of $\log Z_n$ is divergent, $\frac{d^2 F_n}{ds^2} = O(|s - n/2|^{-1})$ as $s \to n/2 + 0$. If we consider the temperature variable $T = n/s$ with the reduction $x_{n-1} \sim -x_n \sim x_{n+1}$ then we get a third-order discontinuity with

$$\frac{d^3 E}{dg^3} = -4v_1 \frac{d\varepsilon}{dg}(1 + O(\varepsilon)) \to 8, \tag{6.58}$$

as $g \to 1/2 + 0$ and $d^3 E/dg^3 = 0$ as $g \to 1/2 - 0$. The method can be extended to higher degree potentials to get other critical phenomena to be discussed in the later sections. These results are all based on the string equations to create the relations between the parameters in the model so that the critical phenomena can be derived by using the ε-expansions.

The model for this critical phenomenon has a closer structure to the original discrete integrable system since we only supply the condition (6.51) to the integrable model (6.56). Note that this critical phenomenon model is not included in the phases discussed in Sect. 6.1 where the singular values of X are from the hypergeometric-type differential equation. It is interesting to investigate what has happened at the critical point $T = 2$, such as where the consistency or the integrability has been transited to. In the following, we use the double scaling method to discuss that the discrete integrable system can be reduced to the continuum integrable system to explain the transition of the integrability.

Let us first consider the string equation in terms of u_n and v_n

$$\frac{n}{s} = v_n + \frac{v_n}{u_n u_{n-1}}, \tag{6.59}$$

$$\frac{n}{s} = v_n \frac{u_{n-1} - v_{n-1}}{u_n - v_n} + \frac{v_n}{u_n u_{n+1}} \frac{u_{n+1} - v_{n+1}}{u_n - v_n}. \tag{6.60}$$

discussed in Sect. 5.3. It is given in last chapter that (6.59) and (6.60) are the consistency conditions for the Lax pair

$$z(p_n + v_n p_{n-1}) = p_{n+1} + u_n p_n, \tag{6.61}$$

$$\frac{\partial p_n}{\partial z} = n p_{n-1} + s \frac{v_n v_{n-1}}{u_n u_{n-1}} p_{n-2}. \tag{6.62}$$

We are going to apply the double scalings to the u_n and v_n functions to show the scaling properties around $n/s = 2$ based on the discussions in [10, 13, 14, 17]. The double scaling is also used in other literatures for different problems. Here, we apply the double scaling method to discuss the phase transition problem.

Denote $d_n = v_n - u_n$. Equations (6.59) and (6.60) then become

$$\frac{n}{s} = (u_n + d_n)\left(1 + \frac{1}{u_n u_{n-1}}\right), \tag{6.63}$$

$$\frac{n}{s} = (u_n + d_n)\left(\frac{d_{n-1}}{d_n} + \frac{1}{u_n u_{n+1}} \frac{d_{n+1}}{d_n}\right). \tag{6.64}$$

In this section, we show that as n and $s \to \infty$, if we make the following double scaling in large-N asymptotics,

$$\frac{2s}{n} = 1 + c_0 x/n^\beta, \tag{6.65}$$

$$u_n = 1 + \frac{c_1}{n^\alpha} \frac{u_x}{u} + \cdots, \tag{6.66}$$

$$d_n = \frac{c_2}{n^\beta} u^2 + \cdots, \tag{6.67}$$

with proper constants c_0, c_1, c_2 and $0 < \alpha < \beta$, then (6.63) and (6.64) will tend to the continuum Painlevé II equation

$$u'' = xu + 2u^3, \tag{6.68}$$

where x is defined by (6.65), and the Lax pair (6.61) and (6.62) will become to the Lax pair for the continuum Painlevé II equation.

Theorem 6.2 *As n and $s \to \infty$, under the assumptions (6.65), (6.66) and (6.67) with*

$$\alpha = 1/3, \qquad \beta = 2/3, \tag{6.69}$$

$$c_0 = -1/2^{1/3}, \qquad c_1 = 2^{1/3}, \qquad c_2 = -2^{2/3}, \tag{6.70}$$

the discrete equations (6.63) and (6.64) are asymptotic to the continuum Painlevé II equation (6.68).

Proof As shown in [17], $\Delta x \equiv x(n+1, s) - x(n, s)$ has the following property,

$$x(n+1, s) - x(n, s) = -\frac{\beta}{c_2 n^{1-\beta}} + \frac{c_3 s}{n} \frac{-1+\beta}{c_2 n^{1-\beta}} + O\left(\frac{1}{n^{2-\beta}}\right)$$

$$= -\frac{1}{c_0 n^{1-\beta}} + O\left(\frac{1}{n}\right),$$

that implies $u_n - u_{n-1} = \frac{c_1}{n^\alpha}\left(\frac{u_x}{u}\right)_x \frac{-1}{c_0 n^{1-\beta}} + \cdots$. Write (6.63) as

$$\frac{n}{s} - 2 = \frac{u_n + d_n}{u_n u_{n-1}}\left(u_n - u_{n-1} + (1 - u_n)(1 - u_{n-1})\right) + 2\frac{d_n}{u_n}.$$

After substituting (6.65), (6.66) and (6.67) into the above equation, we get

$$-\frac{2c_0 x}{n^\beta} = \frac{c_1}{n^\alpha}\left(\frac{u_x}{u}\right)_x \frac{-1}{c_0 n^{1-\beta}} + \frac{c_1^2}{n^{2\alpha}} \frac{u_x^2}{u^2} + 2\frac{c_2}{n^\beta} u^2 + \cdots.$$

For the coefficients above, we choose $\beta = 1 + \alpha - \beta$, $\beta = 2\alpha$, $\frac{c_1}{2c_0^2} = 1$, $-\frac{c_1^2}{2c_0} = 1$, and $\frac{c_2}{c_0} = 2$. It can be seen that (6.69) and (6.70) provide the unique solution for these relations. Then the asymptotic equation discussed above becomes

$$x = \frac{u_{xx}}{u} - 2u^2 + \cdots.$$

As $n, s \to \infty$, this is the continuum Painlevé II equation.

Also, (6.64) can be written as

$$\frac{n}{s} - 2 = \frac{u_n + d_n}{u_n u_{n-1}} \frac{d_{n-1}}{d_n} \left(u_n - u_{n-1} + (1 - u_n)(1 - u_{n-1}) \right)$$
$$+ 2\frac{d_n}{u_n} + \frac{u_n + d_n}{u_n u_{n-1}} \left(\frac{d_{n+1}}{d_n} - \frac{d_{n-1}}{d_n} \right),$$

which also approaches the continuum Painlevé II equation by noting that $d_{n+1} - d_{n-1} = O(\frac{1}{n^{1+\beta}})$ given by (6.67). $\qquad\qquad\Box$

The above discussions imply that

$$x_n^2 = 1 - (1 - x_n^2) = \frac{u_n - v_n}{u_n} = O\left(\frac{2s}{n} - 1\right), \qquad (6.71)$$

as n and s are large. This asymptotics is consistent with the result discussed before. To complete the discussion for the cause of the transition at the critical point in terms of double scaling, we also need to consider the double scaling limit of the Lax pair. It is known that the continuum Painlevé II equation $u'' = xu + 2u^3$ has the Lax pair [6]

$$\Psi_x = \begin{pmatrix} -i\lambda & iu \\ -iu & i\lambda \end{pmatrix} \Psi,$$

$$\Psi_\lambda = \begin{pmatrix} -4i\lambda^2 - 2iu^2 - ix & 4i\lambda u - 2u_x \\ -4i\lambda u - 2u_x & 4i\lambda^2 + 2iu^2 + ix \end{pmatrix} \Psi,$$

where $\Psi = (\psi_1, \psi_2)^T$. Eliminating ψ_2 we get

$$\psi_{xx} = \frac{u_x}{u}\psi_x + \left(\frac{i\lambda u_x}{u} - \lambda^2 \right)\psi + u^2\psi,$$

$$\psi_\lambda = \left(-ix - 2iu^2 - \frac{2\lambda u_x}{u} \right)\psi + \left(4\lambda + 2i\frac{u_x}{u} \right)\psi_x,$$

where $\psi = \psi_1$. Let $\psi = e^{ix\lambda + \frac{4i}{3}\lambda^3} \phi$. We then have

$$\phi_{xx} = \left(u^2 + \frac{2i\lambda u_x}{u} \right)\phi + \left(-2i\lambda + \frac{u_x}{u} \right)\phi_x, \qquad (6.72)$$

$$\phi_\lambda = 2i\left(-x - u^2 + \frac{2i\lambda u_x}{u} \right)\phi + 2i\left(-2i\lambda + \frac{u_x}{u} \right)\phi_x, \qquad (6.73)$$

where the first equation is a different version of the Schrödinger equation. This is the Lax pair in scalar form for the continuum Painlevé II equation. We are going to show in the following that the Lax pair of the coupled discrete equation can be asymptotically reduced to these two equations.

To discuss the asymptotics of the Lax pair (6.61) and (6.62), let

$$p_n(z) = (-1)^n \phi_n(z). \tag{6.74}$$

This trivial transformation is important. It will be seen that the function that tends to ϕ is ϕ_n, not p_n. In fact, $\phi_n e^{sz} \to \phi$. Therefore (6.61), in terms of ϕ_n, becomes

$$-z\big(\phi_n - (u_n + d_n)\phi_{n-1}\big) = \phi_{n+1} - u_n \phi_n. \tag{6.75}$$

Also it is not hard to see that (6.62) becomes

$$\phi_{n,z} = -n\phi_{n-1} + s\left(1 + \frac{d_n}{u_n}\right)\left(1 + \frac{d_{n-1}}{u_{n-1}}\right)\phi_{n-2}.$$

To derive (6.73), we need to express $\phi_{n,z}$ in terms of ϕ_n and ϕ_{n+1}. Notice that $v_n = u_n + d_n$, and

$$v_{n-1}\phi_{n-2} = \frac{1}{z}\phi_n + \phi_{n-1} - \frac{u_{n-1}}{z}\phi_{n-1},$$

$$n = s\left(v_n + \frac{v_n}{u_n u_{n-1}}\right),$$

so that from the equation for $\phi_{n,z}$ above we have

$$\phi_{n,z} = -n\phi_{n-1} + s\frac{v_n}{u_n u_{n-1}}\left(\frac{1}{z}\phi_n + \phi_{n-1} - \frac{u_{n-1}}{z}\phi_{n-1}\right)$$

$$= -sv_n\phi_{n-1} + \frac{sv_n}{u_n u_{n-1}z}\phi_n - \frac{sv_n}{u_n z}\phi_{n-1}.$$

Apply the recursion formula

$$v_n\phi_{n-1} = \phi_n + \frac{1}{z}\phi_{n+1} - \frac{u_n}{z}\phi_n$$

to the above equation. After some simplifications, we obtain

$$\frac{1}{s}\phi_{n,z} + \phi_n = -\left(z + \frac{1}{u_n}\right)\frac{1}{z^2}(\phi_{n+1} - \phi_n) + (u_n - 1)\left(z + \frac{1}{u_n}\right)\frac{1}{z^2}\phi_n$$

$$+ \frac{v_n - u_n}{u_n u_{n-1}z}\phi_n + \frac{u_n - u_{n-1}}{u_n u_{n-1}z}\phi_n. \tag{6.76}$$

Now we are ready to prove the following theorem.

Theorem 6.3 *With the assumptions (6.65), (6.66) and (6.67) as $n, s \to \infty$, and*

$$z = -1 + \frac{2^{4/3}i}{n^{1/3}}\lambda, \tag{6.77}$$

$$\phi_n e^{sz} = \phi(\lambda, x) + o(1), \tag{6.78}$$

the equations (6.75) and (6.76) are asymptotic to (6.72) and (6.73) respectively.

Proof In order to get (6.72), write (6.75) as

$$\phi_{n+1} + \phi_{n-1} - 2\phi_n = (u_n + d_n - 1)(\phi_n - \phi_{n-1}) - (1+z)(\phi_n - (u_n + d_n)\phi_{n-1})$$
$$- d_n \phi_n.$$

Since

$$(\phi_n - \phi_{n-1})e^{sz} = \phi_x \Delta x + \cdots,$$
$$(\phi_{n+1} + \phi_{n-1} - 2\phi_n)e^{sz} = \phi_{xx} \Delta x^2 + \cdots,$$

the equation above becomes

$$\phi_{xx} \frac{1}{c_0^2 n^{2/3}} = \frac{c_1}{n^{1/3}} \frac{u_x}{u} \phi_x \frac{-1}{c_0 n^{1/3}} + \frac{2i\lambda}{c_0 n^{1/3}} \left(\phi_x \frac{-1}{c_0 n^{1/3}} - \frac{c_1}{n^{1/3}} \frac{u_x}{u} \phi \right) - \frac{c_2}{n^{2/3}} u^2 \phi$$
$$+ o\left(n^{-2/3}\right),$$

or

$$\phi_{xx} = -c_0 c_1 \frac{u_x}{u} \phi_x - 2i\lambda \phi_x - c_0 c_1 2i\lambda \frac{u_x}{u} \phi - c_0^2 c_2 u^2 \phi + o(1),$$

where $o(1)$ represents the remaining terms which tend to zero as $n, s \to \infty$. Therefore, if we take $n, s \to \infty$, (6.72) is obtained.

Substituting the asymptotic formulas into (6.76), we have

$$\frac{2}{n^{2/3}} \frac{1}{2^{4/3}i} \phi_\lambda = -\frac{2^{1/3}}{n^{2/3}} \left(2^{4/3} i\lambda - 2^{1/3} \frac{u_x}{u} \right) \phi_x + \frac{2^{1/3}}{n^{2/3}} \frac{u_x}{u} \left(2^{4/3} i\lambda - 2^{1/3} \frac{u_x}{u} \right) \phi$$
$$+ \frac{2^{2/3}}{n^{2/3}} u^2 \phi - \frac{2^{2/3}}{n^{2/3}} \left(\frac{u_x}{u} \right)_x \phi.$$

After some simplifications, we obtain

$$\phi_\lambda = 2i \left[u^2 - \left(\frac{u_x}{u} \right)_x + 2i\lambda \frac{u_x}{u} - \left(\frac{u_x}{u} \right)^2 \right] \phi - 2i \left(2i\lambda - \frac{u_x}{u} \right) \phi_x + o(1).$$

As $n, s \to \infty$, it is reduced to (6.73).   □

If we refer Fig. 6.2 to study this scaling deformation, the point z in (6.77) is corresponding to the point z_0 in Fig. 6.2, but not the points $z_\pm$, and there is no Ω_2 in this case. This means the integrability is transited at $z = -1$ as the arc $\Omega = \Omega_1$ is becoming large until to the unit circle. The vertical direction in $z = -1 + \frac{2^{4/3}i}{n^{1/3}} \lambda$ is consistent with the transition process shown in the Gross-Witten model.

One can experience that the s equation can not join with the double scaling above to become an equation in the new integrable system. This property indicates that the parameters or variable n, s and z in the system perform different roles in the reduction of the transition models, and the models can be distinguished according to which parameter direction(s) the integrable system is deformed along to form a transition model.

If the integrable system is reduced in the n direction, then at the critical point $T = 2$, there is a third-order transition, which is the Gross-Witten transition model. If the integrable system is reduced in the s direction, then the parameter n is kept in the density and free energy, and the transition is of second-order, in which case there is

$$\frac{dg}{d \ln v} = 0, \tag{6.79}$$

at the critical point $g = 1/2$, according to the strong coupling condition given in Sect. 5.4.2. If the integrable system is reduced in both n and s directions with the n parameter controlling the free energy and the s parameter controlling the X variable in the density model, then we get a first-order discontinuity. Note that the weak and strong coupling densities in the Gross-Witten transition model can be obtained separately from the string equation and the Toda lattice. The first case is associated with the n and z equations, and the second case is related to the z and s equations as shown by (6.4) and (6.5) in association with the discussions in Sect. D.2. The reduction conditions in these two cases can not be applied together. For example, if we apply $1 - x_n^2 = T/2$ to the equation $X = (T - 2)^2$, then we get $x_n' = 0$, which is a contradiction. In the first case, the x_n's are more like "particles" as the reduction $x_{n+1} \sim -x_n \sim x_{n-1}$ has shown. In the second case, the x_n's perform as "waves" since the x_n satisfies a differential equation for the elliptic function in the s direction. These two performances indicate a duality of the x_n in the Gross-Witten model, which is a new property derived from the integrable system and deserves further investigations in physics.

6.4 Third-Order Transitions for the Multi-cut Cases

In this section, we discuss the transition model with the potential

$$U(z) = \frac{1}{T}\left(z^m + z^{-m}\right), \tag{6.80}$$

where m is an positive integer. Define the ω function by

$$z\omega(z) = \frac{m}{2T}\left(z^m + z^{-m}\right) + \frac{1}{2}, \tag{6.81}$$

for $T \geq 2m$ where z is in the complex plane. And define the ω function by

$$z\omega(z) = \frac{m}{2T}\left(z^{m/2} + z^{-m/2}\right)\sqrt{\left(z^{m/2} + z^{-m/2}\right)^2 + \frac{2}{m}(T - 2m)}, \tag{6.82}$$

Fig. 6.3 Cuts (*thick arcs*) and eigenvalue range (*thin arcs*) for $m = 3$

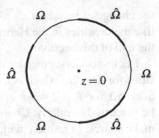

for $T \leq 2m$ with $z \in \mathbb{C} \setminus \hat{\Omega}$ where $\hat{\Omega}$ is the union of cuts on the unit circle,

$$\hat{\Omega} = \left\{ z = e^{i\theta} \mid |\cos(m\theta/2)| \leq \sqrt{1 - (T/2m)}, |\theta| \leq \pi \right\}. \tag{6.83}$$

Figure 6.3 shows the three cuts $\hat{\Omega}$ and the complement of $\hat{\Omega}$ in the unit circle (three arcs Ω) for $m = 3$. In this case, we will see that critical point is $T = 6$ which is possibly another expected transition remarked by Gross and Witten in [3] (Sect. IV) when they discuss an approximation for the four-dimensional QCD $U(N)$ gauge theory. According to [3], the physical model is expected to be constructed based on the analysis of the critical value [2, 15] and the bounded properties of the one-plaquette Wilson loop. As a remark, when $m = 1$ there is only one cut (arc) passing through the point $z = -1$. When $m = 2$, there are two cuts (arcs) passing through $z = i$ and $z = -i$ respectively.

In both cases discussed above ($T \leq 2m$ and $T \geq 2m$), we have

$$z\omega(z) = \frac{m}{2T}\left(z^m + z^{-m}\right) + \frac{1}{2} + O\left(\left(z^{m/2} + z^{-m/2}\right)^{-2}\right), \tag{6.84}$$

as $|z^{m/2} + z^{-m/2}| \to \infty$. Define $\sigma(z)$ by

$$\sigma(z) = \frac{1}{\pi i} \omega(z), \tag{6.85}$$

on $\Omega = \{|z| = 1\}$ for $T \geq 2m$ and on $\Omega = \{z \mid |z| = 1\} \setminus \hat{\Omega}$ for $T \leq 2m$. It is not hard to get $\int_{\Omega} \sigma(z)dz = 1$. Since $\sigma(ze^{i\frac{2\pi}{m}})d(ze^{i\frac{2\pi}{m}}) = \sigma(z)dz$, we have

$$\int_{\Omega_j} \sigma(z)dz = \frac{1}{m}, \tag{6.86}$$

where Ω_j's are the disjoint arcs of Ω such that $\Omega = \bigcup_{j=1}^{m} \Omega_j$. This property will be applied to simplify the calculation of the free energy. As before, $\sigma(z)$ can be changed to $\rho(\theta)$ by

$$\rho(\theta)d\theta = \sigma(z)dz, \tag{6.87}$$

where $z = e^{i\theta}$. The square root in (6.82) takes alternative positive and negative signs on the cuts such that $\rho(\theta)$ is non-negative for $z \in \Omega$. The polynomial outside the square root can balance the negative sign from the square root. In fact, if we make the change of variables $z^{m/2} + z^{-m/2} = 2\cos\phi = 2x$, then these density models are like the densities in the Hermitian matrix models. More discussions will be given at the end of this section.

Let us now first consider the variational equation satisfied by the eigenvalue densities for $T \geq 2m$ which is a case similar to the strong coupling model discussed in Sect. 5.1 for $T \geq 2$ when $m = 1$. We will see that these unitary matrix models can be discussed similarly. Choose a point z on the unit circle and a small $\varepsilon > 0$, then make a circle of radius ε with center z. Denote the "semicircle" inside the unit circle $|z| = 1$ as γ_ε^-, and the "semicircle" outside the unit circle as γ_ε^+. Remove the small arc around z enclosed by the ε circle from the unit circle. The remaining big arc is denoted as $\Omega(\varepsilon)$. As $\varepsilon \to 0$, $\Omega(\varepsilon)$ becomes the unit circle, and $\int_{\Omega(\varepsilon)} \to (P)\int_\Omega$ to be used in the following discussions. Denote $\Omega^- = \Omega(\varepsilon) \cup \gamma_\varepsilon^-$ and $\Omega^+ = \Omega(\varepsilon) \cup \gamma_\varepsilon^+$. All the closed contours are oriented counterclockwise. Then, we have for $T \geq 2m$,

$$\frac{1}{\pi i} \int_{\Omega(\varepsilon)} \frac{\zeta \omega(\zeta)}{\zeta - z} d\zeta$$

$$= \frac{1}{2\pi i} \int_{\Omega^-} \frac{1}{\zeta - z} d\zeta + \frac{m}{2T\pi i} \int_{\Omega^-} \frac{\zeta^m}{\zeta - z} d\zeta + \frac{m}{2T\pi i} \int_{\Omega^+} \frac{\zeta^{-m}}{\zeta - z} d\zeta$$

$$- \frac{1}{2\pi i} \int_{\gamma_\varepsilon^-} \frac{1}{\zeta - z} d\zeta - \frac{m}{2T\pi i} \int_{\gamma_\varepsilon^-} \frac{\zeta^m}{\zeta - z} d\zeta - \frac{m}{2T\pi i} \int_{\gamma_\varepsilon^+} \frac{\zeta^{-m}}{\zeta - z} d\zeta$$

$$\to \frac{1}{2} + \frac{m}{2T} \left(z^m - z^{-m} \right) = \frac{1}{2} z U'(z) + \frac{1}{2},$$

as $\varepsilon \to 0$, where the three integrals along Ω^+ or Ω^- vanish. Based on this result, we have the following lemma.

Lemma 6.5 *The $\sigma(z)$ defined by (6.85) satisfies*

$$(P) \int_\Omega \frac{\zeta \sigma(\zeta)}{\zeta - z} d\zeta = \frac{1}{2} z U'(z) + \frac{1}{2}, \tag{6.88}$$

where $|z| = 1$ for $T \geq 2m$, and z is an inner point of Ω for $T \leq 2m$.

Proof We have proved the $T \geq 2m$ case above. When $T \leq 2m$, since z is an inner point of Ω, we can choose a small $\varepsilon > 0$ and make a small circle of radius ε with center z to remove a small arc from Ω. Let $\Omega_0^\pm(\varepsilon)$ be reduced from the unit circle $|\zeta| = 1$ by removing the small arc at z and changing the portions at the cuts to the inside and outside edges respectively, both oriented counterclockwise. Then by Cauchy theorem, there is

$$\left(\int_{\Omega_0^+(\varepsilon) \cup \gamma_\varepsilon^+} + \int_{\Omega_0^-(\varepsilon) \cup \gamma_\varepsilon^-} \right) \left(\zeta \omega(\zeta) - \frac{m}{2T} \left(\zeta^m + \zeta^{-m} \right) - \frac{1}{2} \right) \frac{d\zeta}{\zeta - z} = 0,$$

where $\gamma_\varepsilon^\pm$ are given in the above discussion. Since the small semicircles γ_ε^+ and γ_ε^- have opposite directions around the point z, there is

$$\left(\int_{\gamma_\varepsilon^+} + \int_{\gamma_\varepsilon^-} \right) \left(\zeta\omega(\zeta) - \frac{m}{2T}(\zeta^m + \zeta^{-m}) - \frac{1}{2} \right) \frac{d\zeta}{\zeta - z} \to 0,$$

as $\varepsilon \to 0$. Because the integrals of $\zeta\omega(\zeta)$ along the inside and outside edges of the cuts $\hat{\Omega}$ are canceled, there is

$$\left(\int_{\Omega_0^+(\varepsilon)} + \int_{\Omega_0^-(\varepsilon)} \right) \zeta\omega(\zeta) \frac{d\zeta}{\zeta - z} = 2 \int_{\Omega(\varepsilon)} \zeta\omega(\zeta) \frac{d\zeta}{\zeta - z} \to 2(\text{P}) \int_{\Omega} \zeta\omega(\zeta) \frac{d\zeta}{\zeta - z},$$

as $\varepsilon \to 0$, where $\Omega(\varepsilon)$ is the Ω without the small arc at the point z. Also, we have

$$\frac{1}{\pi i} \left(\int_{\Omega_0^+(\varepsilon)} + \int_{\Omega_0^-(\varepsilon)} \right) \left(\frac{m}{2T}(\zeta^m + \zeta^{-m}) + \frac{1}{2} \right) \frac{d\zeta}{\zeta - z} \to zU'(z) + 1,$$

as $\varepsilon \to 0$. Then the lemma is proved.   □

The $\gamma_\varepsilon^\pm(\varepsilon)$ play different roles in the two cases, $T \geq 2m$ and $T \leq 2m$ discussed above. When $T \leq 2m$, we need to use $\Omega_0^\pm(\varepsilon)$ so that the properties of the $\omega(z)$ at the cuts can be applied. In the following, we first convert the formula in the above lemma into a new form so that it can be applied to calculate the free energy function.

Theorem 6.4 *The $\sigma(z)$ defined by (6.85) satisfies*

$$(\text{P}) \int_{\Omega} \frac{\sigma(\zeta)}{z - \zeta} d\zeta = -\frac{1}{2}U'(z) + \frac{1}{2z}, \tag{6.89}$$

for both $T \leq 2m$ and $T \geq 2m$ where z is an inner point of Ω.

At the critical point $T = 2m$, both the weak and strong densities become $\rho^c(\theta) = \frac{1}{\pi} \cos^2 \frac{m\theta}{2}$, for $\theta \in [-\pi, \pi]$. To discuss the transition at this critical point, consider the free energy function

$$E = \int_{\Omega} (-U(z))\sigma(z)dz - \int_{\Omega} \int_{\Omega} \ln|z - \zeta|\sigma(z)\sigma(\zeta)dzd\zeta. \tag{6.90}$$

When $T \geq 2m$, the theorem above implies that

$$E = -\frac{1}{2\pi T} \int_{-\pi}^{\pi} \cos(m\theta) \left(1 + \frac{2m}{T} \cos(m\theta) \right) d\theta$$
$$- \frac{1}{2\pi} \int_{-\pi}^{\pi} \ln|2\sin\theta/2| \left(1 + \frac{2m}{T} \cos(m\theta) \right) d\theta - \frac{1}{T},$$

that gives

$$E = -\frac{m}{T^2}, \quad T \geq 2m. \tag{6.91}$$

When $T \leq 2m$, the eigenvalues z are distributed on m arcs, and the integral over each arc is equal to $1/m$ according to (6.86), so that $\frac{d}{dT} \int_{\Omega_j} \sigma(z)dz = 0$ for each arc Ω_j. Then, we have $\frac{dE}{dT} = -\int_{\Omega} \frac{dU(z)}{dT} \sigma(z)dz$. In fact, if we denote the z_j as the start point of each arc Ω_j, and take integral from z_j to z for an inner point z in Ω_j on both sides of (6.89), then there is

$$\int_{\Omega} \ln|\zeta - z|\sigma(\zeta)d\zeta - \int_{\Omega} \ln|\zeta - z_j|\sigma(\zeta)d\zeta = -\frac{1}{2}\big(U(z) - U(z_j)\big), \quad z \in \Omega_j,$$

for $j = 1, \ldots, m$. Multiplying $\frac{d\sigma(z)}{dz}$ on both sides and taking integral from z_j to z, and using the result $\int_{\Omega_j} \frac{d}{dT}\sigma(z)dz = 0$ obtained above, we get

$$\int_{\Omega_j} \left(\int_{\Omega} \ln|\zeta - z|\sigma(\zeta)d\zeta \right) \frac{d\sigma(z)}{dT}dz = -\frac{1}{2} \int_{\Omega_j} U(z) \frac{d\sigma(z)}{dT}dz,$$

which implies

$$\int_{\Omega} \left(\int_{\Omega} \ln|\zeta - z|\sigma(\zeta)d\zeta \right) \frac{d\sigma(z)}{dT}dz = -\frac{1}{2} \int_{\Omega} U(z) \frac{d\sigma(z)}{dT}dz. \qquad (6.92)$$

Because

$$\frac{dE}{dT} = -\int_{\Omega} \frac{dU(z)}{dT}\sigma(z)dz - \int_{\Omega} U(z)\frac{d\sigma(z)}{dT}dz$$

$$- 2\int_{\Omega}\int_{\Omega} \ln|\zeta - z|\sigma(\zeta)\frac{d\sigma(z)}{dT}d\zeta dz,$$

we then get $\frac{dE}{dT} = -\int_{\Omega} \frac{dU(z)}{dT}\sigma(z)dz$ for $T < 2m$, or

$$\frac{dE}{dT} = \frac{m}{2T^3} \int_{\Omega_0^+ \cup \Omega_0^-} \left(z^m + z^{-m}\right)\left(z^{m/2} + z^{-m/2}\right)$$

$$\times \sqrt{\left(z^{m/2} + z^{-m/2}\right)^2 + \frac{2}{m}(T - 2m)}\frac{dz}{2\pi i z}, \qquad (6.93)$$

where $\Omega_0^{\pm}$ are the outside and inside edges respectively of the unit circle according to the cuts $\hat{\Omega}$. Both Ω_0^- and Ω_0^- are oriented counterclockwise so that the integrals $\int_{\Omega_0^+}$ and $\int_{\Omega_0^-}$ are canceled at the cuts resulting $\frac{1}{2}(\int_{\Omega_0^+} + \int_{\Omega_0^-}) = \int_{\Omega}$, that is why $\int_{\Omega}$ is changed to $\frac{1}{2}\int_{\Omega_0^+ \cup \Omega_0^-}$ in the above. The integrand above is asymptotic to

$$z^{2m} + z^{-2m} + \frac{T}{m}\left(z^m + z^{-m}\right) - \frac{T}{2m^2}(T - 4m) + o(1),$$

as $z \to \infty$ or $z \to 0$. Then there is

$$\frac{dE}{dT} = \frac{2}{T^2} - \frac{1}{2mT}, \quad T \leq 2m, \qquad (6.94)$$

by changing $\int_{\Omega_0^+(\varepsilon)}$ to $\int_{|z|=R}$, and $\int_{\Omega_0^-(\varepsilon)}$ to $\int_{|z|=\delta}$ with $R \to \infty$ and $\delta \to 0$. Since $E(2m) = -\frac{1}{4m}$, we have

$$E = -\frac{2}{T} - \frac{1}{2m}\ln\frac{T}{2m} + \frac{3}{4m}, \quad T \leq 2m. \tag{6.95}$$

Theorem 6.5 *The free energy*

$$E = \begin{cases} -\frac{2}{T} - \frac{1}{2m}\ln\frac{T}{2m} + \frac{3}{4m}, & T \leq 2m, \\ -\frac{m}{T^2}, & T \geq 2m, \end{cases} \tag{6.96}$$

has continuous first- and second-order derivatives, and discontinuous third-order derivative,

$$\frac{d^3}{dT^3}E(2m-) = \frac{10}{(2m)^4} < \frac{12}{(2m)^4} = \frac{d^3}{dT^3}E(2m+). \tag{6.97}$$

The above result shows that the coefficient of $\ln\frac{T}{2m}$ changes as m changes. If this property can be related to other transition problems, it would be interesting. As studied in the transition models in statistical mechanics, such as the Kosterlitz-Thouless transition for the two dimensional XY model, the coefficient of the logarithmic term in the free energy is always important.

The density models discussed above are created on the unit circle. We can change these models to the real line, for example, for the $m = 2$ case. If we make a bilinear transformation

$$z = \frac{1+i\eta}{\eta+i}, \tag{6.98}$$

then the upper half plane $\text{Im}\,\eta > 0$ is transformed to the unit disk $|z| < 1$, and the lower half plane $\text{Im}\,\eta < 0$ is transformed to $|z| > 1$. Consequently, we have

$$\omega(z)dz = \frac{8\eta}{T(\eta^2+1)^2}\sqrt{4\left(\frac{\eta^2-1}{\eta^2+1}\right)^2 - T}\,d\eta \equiv \hat{\omega}(\eta)d\eta, \tag{6.99}$$

by substituting the bilinear transformation into the $\omega(z)$, where $\hat{\omega}(\eta)$ is defined in the η-plane except the cuts $\Omega_\eta = [-\eta_2, -\eta_1] \cup [\eta_1, \eta_2]$ with

$$\eta_1 = \sqrt{\frac{2-\sqrt{T}}{2+\sqrt{T}}}, \quad \eta_2 = \sqrt{\frac{2+\sqrt{T}}{2-\sqrt{T}}}, \quad T \leq 4. \tag{6.100}$$

The $\rho(\theta)$ is transferred to a model in η space by the following relations,

$$\rho(\theta)d\theta = \sigma(z)dz = \frac{1}{\pi i}\omega(z)dz = \frac{1}{\pi i}\hat{\omega}(\eta)\Big|_{\Omega_\eta^-}d\eta \equiv \hat{\rho}(\eta)d\eta. \tag{6.101}$$

The potential is also transferred as follows,

$$U(z) = \frac{2}{T} \frac{4\eta^2 - (\eta^2 - 1)^2}{(\eta^2 + 1)^2} \equiv \hat{U}(\eta). \tag{6.102}$$

It can be seen that as $\eta \to \pm i$,

$$\hat{\omega}(\eta) = -\frac{1}{2}\hat{U}'(\eta) - \frac{2\eta}{\eta^4 - 1} + O(\eta^2 + 1), \tag{6.103}$$

where $-2\eta/(\eta^4 - 1) \sim \eta/(\eta^2 + 1)$ as $\eta \to \pm i$. Then, the lower edge Ω_η^- of Ω_η is where the density is defined such that the integral of the density $\hat{\rho}(\eta)$ is equal to 1, that means $\hat{\omega}(\eta) = \pi i \rho(\eta)$ on Ω_η^- as given above. One can also discuss the phase transition between the following models in the η-plane. The strong coupling model can be obtained by $\frac{1}{\pi i}\omega(z)dz = \frac{1}{\pi i}\hat{\omega}(\eta)d\eta = \hat{\rho}(\eta)d\eta$, where

$$\hat{\rho}(\eta) = \frac{1}{\pi(\eta^2 + 1)}\left(1 - \frac{4}{T}\frac{\eta^4 - 6\eta^2 + 1}{(\eta^2 + 1)^2}\right), \quad T \geq 4, \tag{6.104}$$

with $-\infty < \eta < \infty$. The weak coupling model can be obtained by $\frac{1}{\pi i}\hat{\omega}(\eta)|_{\Omega_\eta^-}d\eta = \hat{\rho}(\eta)d\eta$, which gives

$$\hat{\rho}(\eta) = \frac{8\eta}{\pi T(\eta^2 + 1)^2}\sqrt{e^{-\pi i}\left(4\left(\frac{\eta^2 - 1}{\eta^2 + 1}\right)^2 - T\right)}, \quad T \leq 4, \tag{6.105}$$

for $\eta \in \Omega_\eta$. We see that $\hat{\rho}(\eta)$ meets with the non-negativity condition similarly as in the Hermitian matrix models since there is the factor η outside the square root to eliminate the negative sign of the square root on the left interval.

The Gross-Witten third-order phase transition model can be generalized by considering higher degree potentials as seen above. For the readers who are interested in the related problems, it should be noted that the Gross-Witten transition is also associated with other important problems, such as the Chern-Simons matrix models, two-dimensional Yang-Mills theory, and XY model as discussed in the literatures, for example, see [8, 12, 16]. The potential $s_1(z + z^{-1}) + s_2(z^2 + z^{-2}) = 2s_1\cos\theta + 2s_2\cos 2\theta$ has been considered in [7] for the Villain approximation of the XY model. Here, we focus on the discontinuities of the free energy based on the eigenvalue densities and string equations in the unitary matrix models.

6.5 Divergences for the One-Cut Cases

Let us first look at the parameter relation

$$g_1(v + v^{-1}) - 2g_2(v^2 + v^{-2}) + 3g_3(v^3 + v^{-3}) = 1, \tag{6.106}$$

for the strong coupling phase, given by (5.77) for the potential $U(z) = \sum_{j=1}^{m} g_j(z^j + z^{-j})$, including the cases $m = 1, 2$, and 3. We are going to discuss the transitions to the one-cut weak coupling cases given in Sect. 5.4.2.

When $m = 1$, the condition $g_1(v + v^{-1}) = 1$ implies that

$$\frac{dg_1}{dv} = \frac{g_1(1 - v^2)}{v(1 + v^2)}, \tag{6.107}$$

so that $\frac{dg_1}{dv} = 0$ at the critical point $v = 1$ and $g_1 = 1/2$. This is the case of strong coupling phase with $g_1 < 1/2$. The weak coupling phase with $g_1 v = 1/2$ and $g_1 > 1/2$ given in Sect. 5.4.2, however, does not have the property $\frac{dg_1}{dv} = 0$ at the critical point $g_1 = 1/2$ because $g_1 - g_1^c = O(v - 1)$. At this critical point, the divergence for $d^2 \log Z_n/ds^2$ has been discussed before. In the following, when we discuss the divergences of the derivatives of the free energy we will also pay attention to the property of $\frac{dg_1}{dv}$ since it is interesting in physics. We will show that for $m = 2$ or 3 there is always $\frac{dg_1}{dv} = 0$ at the critical point, no matter g_1 approaches to the critical point from left or right of the critical point. When g_1 approaches to g_1^c from right (weak coupling case), there is $g_1 - g_1^c = O(|v - 1|^m)$. For the strong coupling cases, there is $g_1 - g_1^c = O(|v - 1|^{2m})$.

When $m = 2$, we have obtained in Sect. 5.4.2 for the weak coupling case that there is the following relation

$$g_1 v + 2g_2(2v - 3v^2) - \frac{1}{2} = 0, \tag{6.108}$$

based on the high order string equation by taking $u_n \sim 1$ and $v_n \sim v$, where $g_1 = s_1/n$ and $g_2 = s_2/n$. As a remark, at the critical point both the weak and strong density functions become $\rho_2^c(\theta) = \frac{c_2}{\pi} \cos^4 \frac{\theta}{2}$, for a constant c_2, that means this transition is different from the models discussed in last section.

For simplicity, let us consider the expansion in the g_1 direction, and other g_j's are chosen as constants $g_j = g_j^c$. Consider the ε expansion of v,

$$v = 1 + v_1 \varepsilon + v_2 \varepsilon^2 + v_3 \varepsilon^3 + \cdots. \tag{6.109}$$

Then, the relation (6.108) becomes

$$\left(g_1 - g_1^c\right)v + g_1^c - 2g_2^c - 1/2 + \left(g_1^c - 8g_2^c\right)v_1 \varepsilon$$
$$+ \left(g_1^c - 8g_2^c\right)v_2 \varepsilon^2 - 6g_2^c v_1^2 \varepsilon^2 + \left(g_1^c - 8g_2^c\right)v_3 \varepsilon^3 - 12g_2^c v_1 v_2 \varepsilon^3 + \cdots = 0,$$

where the first term $(g_1 - g_1^c)v$ is to decide the order of $g_1 - g_1^c$. If we choose $g_1 = g_1^c + \varepsilon$, the coefficients can not be determined. If we choose

$$g_1 = g_1^c + \varepsilon^2, \tag{6.110}$$

and

$$g_1^c = 2/3, \qquad g_2^c = 1/12, \tag{6.111}$$

then all the coefficients can be determined, and specially $v_1 = \sqrt{2}$.

If we restrict $g_2 = 1/12$, then (6.108) gives

$$g_1 = \frac{1}{2}(v + v^{-1}) - \frac{1}{3},\tag{6.112}$$

which implies

$$g_1 \geq \frac{2}{3}.\tag{6.113}$$

And the critical case is given by the vanishing point of

$$\frac{dg_1}{dv} = \frac{1}{2v}(v - v^{-1}),\tag{6.114}$$

which is $v = 1$. Also, if we substitute $g_2 = 1/12$ and $g_3 = 0$ into (6.106), then

$$g_1(v + v^{-1}) = \frac{1}{6}(v^2 + v^{-2}) + 1,\tag{6.115}$$

which implies $\frac{dg_1}{dv} = 0$ at the critical point $v = 1$ and $g_1 = 2/3$.

The formula $\frac{\partial^2}{\partial s_1^2} \log Z_n = 2(1 - x_n^2)(1 - x_{n-1}x_{n+1})$ is still true since the formula for $\partial h_n / \partial s_1$ and the Toda lattice are the same as the $m = 1$ case, but the string equation becomes (5.62) (see [1]). By the Toda lattice and (5.62), we can get

$$x_{n+1} - x_{n-1} = \frac{x_n'}{1 - x_n^2},$$

$$\frac{3s_2}{2}x_n(x_{n+1} + x_{n-1})^2 - s_1(x_{n+1} + x_{n-1}) - \frac{s_2 x_n'' + n x_n}{1 - x_n^2} - \frac{3s_2 x_n (x_n')^2}{2(1 - x_n^2)^2} = 0,$$

similarly as the discussion in Sect. C.3, where $' = \partial/\partial s_1$, that implies

$$\frac{\partial^2}{\partial s_1^2} \log Z_n = O\left((x_n')^2\right) + O\left(x_n''/x_n\right).\tag{6.116}$$

Since $v \sim 1 - x_n^2$, we have $x_n^2 \sim -\sqrt{2}\varepsilon$ where we choose $\varepsilon < 0$. Then in the s_1 direction, there is $x_n = O(|s_1 - \frac{2n}{3}|^{1/4})$, and we get the following critical phenomenon,

$$\frac{\partial^2 F_n}{\partial s_1^2} = O(x_n''/x_n) = O\left(\left|s_1 - \frac{2n}{3}\right|^{-2}\right),\tag{6.117}$$

as $s_1 \to \frac{2n}{3} + 0$, with the critical point $s_1 = \frac{2n}{3}$, where $F_n = -\ln Z_n$ different from $E = -\lim_{n\to\infty} \frac{1}{n^2} \ln Z_n$ to be discussed next. As a remark, the g parameters in (6.108) need to be changed to the s parameters by using $g_j = s_j/n$ when we discuss the problem in the s_1 direction. Since we are interested in how to get the critical exponent in the critical phenomenon, the details are omitted.

According to the formula (6.10), the asymptotic assumption $u_n \sim 1$ and $v_n \sim v$ as $n \to \infty$ gives $\frac{\partial^2}{\partial s_1^2} \log Z_n \sim 2(1 - x_n^2)^2 \sim 2v^2$. Then we can get the second-order derivative of the free energy,

$$\frac{\partial^2}{\partial g_1^2} E = -2v^2. \qquad (6.118)$$

The assumptions $u_n \sim 1$ and $v_n \sim v$ imply that the above result is for the weak coupling density as we have studied in last chapter. The third-order derivative of E then has the following power-law divergence behavior,

$$\frac{\partial^3}{\partial g_1^3} E = -4v_1 \frac{d\varepsilon}{dg_1}(1 + O(\varepsilon)) = O(|g_1 - g_1^c|^{-1/2}), \qquad (6.119)$$

as $g_1 \to g_1^c + 0$.

Note that if $\varepsilon > 0$, then the x_n in (6.116) is imaginary, which is different from the case for the potential $U(z) = g(z + z^{-1})$ discussed in Sect. 6.3. The imaginary x_n brings a problem similar to the problem with the cuts in the complex plane considered in Sect. 4.2.2 for the Hermitian matrix model with potential $g_2\eta^2 + g_4\eta^4$. We have discussed there that the double scaling method involves complex coefficient. If one is interested in the double scaling for the current model to get a higher order integrable system, the double scaling method could also involve complex coefficients. It would be interesting to investigate whether this property is related to the renormalization theory, and whether the parameter relations in terms of s_j and n are related to the β function in quantum chromodynamics.

When $m = 3$, we have obtained in Sect. 5.4.2 for the weak coupling case that there is the following relation,

$$g_1v + 2g_2(2v - 3v^2) + 3g_3(3v(v - 1)(3v - 1) + v^3) - \frac{1}{2} = 0, \qquad (6.120)$$

for the parameters g_j and v, where $g_j = s_j/n$. This relation was obtained from the high order string equation by taking the reduction $u_n \sim 1$ and $v_n \sim v$. As a remark, at the critical point both weak and strong density functions become $\rho_3^c(\theta) = \frac{c_3}{\pi} \cos^6 \frac{\theta}{2}$ for a constant c_3.

As before, let us consider the expansion in the g_1 direction, and choose other g_j as constants $g_j = g_j^c$. Consider the ε expansion of v,

$$v = 1 + v_1\varepsilon + v_2\varepsilon^2 + v_3\varepsilon^3 + \cdots . \qquad (6.121)$$

Then, (6.120) becomes

$$(g_1 - g_1^c)v$$
$$+ g_1^c - 2g_2^c + 3g_3^c - 1/2$$
$$+ (g_1^c - 8g_2^c + 27g_3^c)v_1\varepsilon$$

$$+ \left(g_1^c - 8g_2^c + 27g_3^c\right)v_2\varepsilon^2 + \left(-6g_2^c + 54g_3^c\right)v_1^2\varepsilon^2$$
$$+ \left(g_1^c - 8g_2^c + 27g_3^c\right)v_3\varepsilon^3 + \left(-12g_2^c + 108g_3^c\right)v_1v_2\varepsilon^3 + 30g_3^c v_1^3\varepsilon^3 + \cdots = 0.$$

As in the $m = 2$ case, we need to choose

$$g_1 = g_1^c + \varepsilon^3, \tag{6.122}$$

in order to determine all the coefficients. By comparing the coefficients of the ε powers in the above equation, we get

$$g_1^c = 3/4, \qquad g_2^c = 3/20, \qquad g_3^c = 1/60, \tag{6.123}$$

and $v_1 = -2^{1/3}$. The third-order derivative of E has the following power-law divergence in the g_1 direction,

$$\frac{\partial^3}{\partial g_1^3}E = -4v_1\frac{d\varepsilon}{dg_1}\left(1 + O(\varepsilon)\right) = O\left(|g_1 - g_1^c|^{-2/3}\right), \tag{6.124}$$

as $g_1 \to g_1^c + 0$. The critical phenomenon is left to interested readers.

If we consider the s_2 or a general direction s_j and calculate the second-order derivative of $\ln Z_n$, it can be experienced that the computations are very complicated. Better methods are expected to find the formulation of the correlation function, and the critical exponents in the second- and third-order divergences are possibly related. The correlation function theory is an important subject, which is associated to the quantum inverse scattering and matrix Riemann-Hilbert problems [9]. Here, we just introduce some basic properties to motivate interested researchers to develop new methods to improve this investigation.

In the previous chapters, we have explained the basic techniques for creating the transition models by using the string equations in the Hermitian and unitary matrix models. In the next chapter, we will introduce the density models derived from the Laguerre or Jacobi polynomials.

References

1. Cresswell, C., Joshi, N.: The discrete first, second and thirty-fourth Painlevé hierarchies. J. Phys. A **32**, 655–669 (1999)
2. Foerster, D.: On condensation of extended structures. Phys. Lett. B **77**, 211–213 (1978)
3. Gross, D.J., Witten, E.: Possible third-order phase transition in the large-N lattice gauge theory. Phys. Rev. D **21**, 446–453 (1980)
4. Hisakado, M.: Unitary matrix model and the Painlevé III. Mod. Phys. Lett. A **11**, 3001–3010 (1996)
5. Hisakado, M., Wadati, M.: Matrix models of two-dimensional gravity and discrete Toda theory. Mod. Phys. Lett. A **11**, 1797–1806 (1996)
6. Its, A.R., Novokshenov, Yu.: The Isomonodromy Deformation Method in the Theory of Painlevé Equations. Lecture Notes in Mathematics, vol. 1191. Springer, Berlin (1986)

7. Janke, W., Kleinert, H.: How good is the Villain approximation? Nucl. Phys. B **270**, 135–153 (1986)
8. Klebanov, I.R., Maldacena, J., Seiberg, N.: Unitary and complex matrix models as 1D type 0 strings. Commun. Math. Phys. **252**, 275–323 (2004)
9. Korepin, V.E., Bogoliubov, N.M., Izergin, A.G.: Quantum Inverse Scattering Method and Correlation Functions. Cambridge University Press, Cambridge (1993)
10. McLeod, J.B., Wang, C.B.: Discrete integrable systems associated with the unitary matrix model. Anal. Appl. **2**, 101–127 (2004)
11. McLeod, J.B., Wang, C.B.: Eigenvalue density in Hermitian matrix models by the Lax pair method. J. Phys. A, Math. Theor. **42**, 205205 (2009)
12. Morozov, A.Yu.: Unitary integrals and related matrix models. Theor. Math. Phys. **162**, 1–33 (2010)
13. Periwal, V., Shevitz, D.: Unitary-matrix models as exactly solvable string theories. Phys. Rev. Lett. **64**, 1326–1329 (1990)
14. Periwal, V., Shevitz, D.: Exactly solvable unitary matrix models: multicritical potentials and correlations. Nucl. Phys. B **344**, 731–746 (1990)
15. Polyakov, A.M.: String representations and hidden symmetries for gauge fields. Phys. Lett. B **82**, 247–250 (1979)
16. Szabo, R.J., Tierz, M.: Chern-Simons matrix models, two-dimensional Yang-Mills theory and the Sutherland model. J. Phys. A **43**, 265401 (2010)
17. Wang, C.B.: Orthonormal polynomials on the unit circle and spatially discrete Painlevé II equation. J. Phys. A **32**, 7207–7217 (1999)
18. Yeomans, J.M.: Statistical Mechanics of Phase Transitions. Oxford University Press, London (1994)

Chapter 7
Marcenko-Pastur Distribution and McKay's Law

When the Lax pair method for the eigenvalue density is applied to the Laguerre and Jacobi polynomials, the Marcenko-Pastur distribution, McKay's law and their generalizations can be obtained. The logarithmic divergences based on these density models will be discussed in this chapter by using the elliptic integrals and expansion method discussed before. Even though the associated matrix models and transitions have not been widely studied so far, the density models provide an alternative interpretation for the state changes and two-phase models of the random systems that are now important in applications typically in complexity subjects such as behavioral sciences and econophysics. The power-law distribution is also a very interesting research direction that has drawn the attention of many researchers to find the formulation of the power-law distributions. The Laplace transform is applied to discuss this problem in this chapter, showing that the power exponent in the power-law distribution can be obtained from the power behavior of the density model at the end point(s) of the interval where the eigenvalues are distributed.

7.1 Laguerre Polynomials and Densities

Consider the Laguerre polynomials $L_n^{(\alpha)}(x)$ [35],

$$\int_0^\infty L_m^{(\alpha)}(x)L_n^{(\alpha)}(x)x^\alpha e^{-x}dx = \Gamma(\alpha+1)\binom{n+\alpha}{n}\delta_{m,n}, \qquad (7.1)$$

where $\alpha > -1$, and $\Gamma(\cdot)$ is the Gamma function. The Laguerre polynomials satisfy the recursion formula

$$xL_n^{(\alpha)}(x) = -(n+1)L_{n+1}^{(\alpha)}(x) + (2n+1+\alpha)L_n^{(\alpha)}(x) - (n+\alpha)L_{n-1}^{(\alpha)}(x), \quad (7.2)$$

and the differential equation

$$x\frac{d}{dx}L_n^{(\alpha)}(x) = nL_n^{(\alpha)}(x) - (n+\alpha)L_{n-1}^{(\alpha)}(x). \qquad (7.3)$$

C.B. Wang, *Application of Integrable Systems to Phase Transitions*, DOI 10.1007/978-3-642-38565-0_7, © Springer-Verlag Berlin Heidelberg 2013

Explicitly, there are

$$L_0^{(\alpha)}(x) = 1, \qquad L_1^{(\alpha)}(x) = -x + 1 + \alpha,$$

and

$$L_n^{(\alpha)}(x) = \sum_{k=0}^{n} \binom{n+k}{n-k} \frac{(-x)^k}{k!}, \qquad n \geq 2.$$

Now, choose $\Phi_n(x) = x^{\alpha/2} e^{-x/2} (L_n^{(\alpha)}(x), L_{n-1}^{(\alpha)}(x))^T$. It can be verified that $\Phi_n(x)$ satisfies the following equation [35]

$$\frac{\partial}{\partial x} \Phi_n = A_n(x) \Phi_n, \tag{7.4}$$

where

$$A_n(x) = \frac{1}{x} \begin{pmatrix} -\frac{x-\alpha}{2} + n & -n - \alpha \\ n & \frac{x-\alpha}{2} - n \end{pmatrix}, \tag{7.5}$$

with $\operatorname{tr} A_n(x) = 0$. It can be calculated that

$$\sqrt{\det A_n(x)} = \frac{n}{2x} \sqrt{\left(\left(1 + \sqrt{\frac{n+\alpha}{n}}\right)^2 - \frac{x}{n}\right)\left(\frac{x}{n} - \left(1 - \sqrt{\frac{n+\alpha}{n}}\right)^2\right)}. \tag{7.6}$$

Rescale the variable $x = n\eta$, and parameters $q = \frac{n}{n+\alpha}$, $\eta_+ = (1 + \frac{1}{\sqrt{q}})^2$ and $\eta_- = (1 - \frac{1}{\sqrt{q}})^2$. Then we get the density model associated with the Laguerre polynomials [25]

$$\frac{1}{n\pi} \sqrt{\det A_n(x)} dx = \frac{1}{2\pi\eta} \sqrt{(\eta_+ - \eta)(\eta - \eta_-)} d\eta, \tag{7.7}$$

as discussed in Sect. 1.1 with the unified model. The density function on the right hand side is the Marcenko-Pastur distribution [23]. Also see [30]. Note that the original Marcenko-Pastur distribution $\frac{Q}{2\pi\lambda} \sqrt{(\lambda_+ - \lambda)(\lambda - \lambda_-)}$ has a factor $Q > 1$. If we consider $q < 1$ in (7.7), then by the changes of the parameter $q = Q^{-1}$ and the variable $\eta = Q\lambda$, there is

$$\frac{1}{2\pi\eta} \sqrt{(\eta_+ - \eta)(\eta - \eta_-)} d\eta = \frac{Q}{2\pi\lambda} \sqrt{(\lambda_+ - \lambda)(\lambda - \lambda_-)} d\lambda,$$

where $\lambda_+ = (1 + \frac{1}{\sqrt{Q}})^2$ and $\lambda_- = (1 - \frac{1}{\sqrt{Q}})^2$. To be consistent with the previous discussions for the variational equation, we consider the density in (7.7) with $q < 1$. The density is still the Marcenko-Pastur distribution.

The Lax pair method can generalize this density formula to provide more models for researching the random phenomena associated with the densities. Let us first consider the variational equation that the Marcenko-Pastur density satisfies. Consider the potential

$$W(\eta) = \eta + (1 - q^{-1}) \ln \eta. \tag{7.8}$$

Define

$$\omega_0(\eta) = \frac{1}{2\eta}\sqrt{(\eta - 1 - q^{-1})^2 - 4q^{-1}}, \qquad (7.9)$$

which satisfies

$$\omega_0(\eta) = \frac{1}{2}(1 - (1 + q^{-1})\eta^{-1}) + O(\eta^{-2}), \qquad (7.10)$$

as $\eta \to \infty$, and

$$\operatorname*{res}_{\eta=0} \omega_0(\eta) = \frac{1}{2}(1 - q^{-1}), \qquad (7.11)$$

by choosing the branch of $\omega_0(\eta)$ such that $\sqrt{(1 - q^{-1})^2} = 1 - q^{-1}$, which stands for the square root of the product of $0 - (1 - q^{-1/2})^2$ and $0 - (1 + q^{-1/2})^2$, where $q < 1$. Note that the leading term on the right hand side of (7.10) is not $\frac{1}{2}W'(\eta)$, but $\frac{1}{2}(1 - (1 + q^{-1})\eta^{-1})$. This is a fundamental difference from the asymptotics in the Hermitian matrix models in which the ω function is always asymptotic to $\frac{1}{2}W'(\eta)$. This difference comes from the singular point $\eta = 0$ in the density model above. Also, the normalization of the density is not controlled by the asymptotics at the infinity, but by both the properties at ∞ and 0 that we are going to explain in the following.

Define

$$\rho_0(\eta) = \frac{1}{\pi i}\omega_0(\eta)\Big|_{\Omega^+}, \qquad \eta \in \Omega = [(1 - q^{-1/2})^2, (1 + q^{-1/2})^2], q < 1. \qquad (7.12)$$

Theorem 7.1 *The $\rho_0(\eta)$ defined by (7.12) satisfies*

$$\int_\Omega \rho_0(\eta)d\eta = 1, \qquad (7.13)$$

and

$$(P)\int_\Omega \frac{\rho_0(\eta)}{\lambda - \eta}d\eta = \frac{1}{2}W'(\lambda). \qquad (7.14)$$

Proof Let Γ_R and Γ_ε be counterclockwise circles of radii R and ε respectively with center 0, and Ω^* be the closed counterclockwise contour around the upper and lower edges of Ω. See Fig. 7.1. Then according to Cauchy theorem, there is

$$\left\{\int_{\Omega^*} + \int_{\Gamma_\varepsilon} - \int_{\Gamma_R}\right\}\omega_0(\eta)d\eta = 0,$$

which implies

$$-\int_\Omega \rho_0(\eta)d\eta + \frac{1}{2}(1 - q^{-1}) + \frac{1}{2}(1 + q^{-1}) = 0,$$

and then (7.13) is proved.

Fig. 7.1 Contours for
Marcenko-Pastur distribution

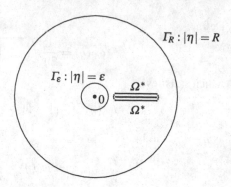

To prove the second formula, we need to change the $\Omega^* = (-\Omega^-) \cup \Omega^+$ at the singular point $\lambda \in \Omega$, where the "$-$" sign before Ω^- stands for the opposite orientation. The Ω^- and Ω^+ around λ can be changed to semicircles of ε radius. If we still use the previous notations, then

$$\left\{ \int_{\Omega^*} + \int_{\Gamma_\varepsilon} - \int_{\Gamma_R} \right\} \frac{\omega_0(\eta)}{\eta - \lambda} d\eta = 0,$$

which implies

$$-(\mathrm{P}) \int_\Omega \frac{\rho_0(\eta)}{\eta - \lambda} d\eta + \frac{\frac{1}{2}(1 - q^{-1})}{-\lambda} - \frac{1}{2} = 0,$$

where we have used $\eta \omega_0(\eta)|_{\eta=0} = \frac{1}{2}\sqrt{(1 - q^{-1})^2} = \frac{1}{2}(1 - q^{-1})$, and $\int_{-\pi}^0 \omega_0(\lambda + \varepsilon e^{i\theta})d\theta + \int_0^\pi \omega_0(\lambda + \varepsilon e^{i\theta})d\theta \to 0$ as $\varepsilon \to 0$ since the integral path of the first integral is below the Ω and the integral path of the second integral is above Ω. So the theorem is proved. $\qquad \square$

The variational equation for the Marcenko-Pastur distribution is also discussed in the Wishart ensembles, for example, see [5, 39]. The Marcenko-Pastur distribution has been widely applied in econophysics and relevant random researches for studying the distribution of the positive eigenvalues, for example, see [22, 28], typically for analyzing the financial data. More literatures can be found in the journals such as Physica A and Physical Review E. In the following, we are going to apply the method discussed in the previous chapters to give a generalized model with the potential

$$W(\eta) = g_1\eta + g_2\eta^2 + g_3\eta^3 + \left(1 - q^{-1}\right)\ln \eta. \tag{7.15}$$

For the convenience in discussion, denote

$$W_0(\eta) = g_1\eta + g_2\eta^2 + g_3\eta^3. \tag{7.16}$$

We discuss this potential instead of the second degree potential $g_1\eta + g_2\eta^2$ because we want the density model to satisfy the non-negativity condition for the two-cut case based on the experiences in the previous chapters. Similar to the ω function in the Hermitian matrix model for $m = 2$ discussed in Chap. 3, the ω function for the one-cut model is now defined as

$$\omega_1(\eta) = \frac{1}{2\eta}\left(g_1 + 2g_2(\eta + a) + 3g_3\left(\eta^2 + a\eta + a^2 + 2b^2\right)\right)\sqrt{(\eta - a)^2 - 4b^2}, \quad (7.17)$$

for $\eta \in \mathbb{C}\backslash\Omega_1$, where $\Omega_1 = [a - 2b, a + 2b]$, and it satisfies

$$\omega_1(\eta) = \frac{1}{2}W_0'(\eta) - \frac{1}{2}\left(1 + q^{-1}\right)\eta^{-1} + O\left(\eta^{-2}\right), \quad (7.18)$$

as $\eta \to \infty$, if the parameters satisfy the condition

$$g_1 a + 2g_2\left(a^2 + 2b^2\right) + 3g_3\left(a^3 + 6ab^2\right) = 1 + q^{-1}. \quad (7.19)$$

By the asymptotics of (7.17) as $\eta \to 0$, it is not hard to derive that

$$\operatorname*{res}_{\eta = 0} \omega_1(\eta) = \frac{1}{2}\left(1 - q^{-1}\right), \quad (7.20)$$

if the parameters satisfy another condition

$$\left(g_1 + 2g_2 a + 3g_3\left(a^2 + 2b^2\right)\right)\sqrt{a^2 - 4b^2} = 1 - q^{-1}, \quad a^2 > 4b^2. \quad (7.21)$$

Note that $\arg(0 - (a - 2b)) = \pi$ and $\arg(0 - (a + 2b)) = \pi$, that imply $\sqrt{a^2 - 4b^2}$ which stands for $\sqrt{(0 - (a - 2b))(0 - (a + 2b))}$ is negative. Since $g_1 + 2g_2 a + 3g_3(a^2 + 2b^2)$ is positive to keep the density function non-negative, we see that $1 - q^{-1}$ needs to be negative. If we choose

$$a = 1 + b^2, \quad (7.22)$$

then $\sqrt{a^2 - 4b^2} = 1 - b^2$ if we choose $b > 1$. Equations (7.19) and (7.21) now become

$$2g_2 + 6g_3 a = \frac{q^{-1} - b^2}{b^2(1 - b^2)}, \quad (7.23)$$

$$g_1 - 3g_3\left(1 + b^4\right) = \frac{2b^2 + b^4 - q^{-1}(1 + 2b^2)}{b^2(1 - b^2)}, \quad (7.24)$$

where $b > 1$ and $q < 1$. According to [37], the Laguerre polynomials are corresponding to the discrete Painlevé IV equation. Then the generalized models should be linked to the high order discrete Painlevé IV equations.

Define

$$\rho_1(\eta) = \frac{1}{\pi i}\omega_1(\eta)\bigg|_{\Omega_1^+}, \quad \eta \in \Omega_1 = [a - 2b, a + 2b]. \quad (7.25)$$

Theorem 7.2 *If the parameters* a, b *and* g_j $(j = 1, 2, 3)$ *satisfy the conditions* *(7.19) and (7.21), then the* $\rho_1(\eta)$ *defined by (7.25) satisfies*

$$\int_{\Omega_1} \rho_1(\eta) d\eta = 1, \tag{7.26}$$

and

$$(P) \int_{\Omega_1} \frac{\rho_1(\eta)}{\lambda - \eta} d\eta = \frac{1}{2} W'(\lambda). \tag{7.27}$$

Proof Let Γ_R be a large counterclockwise circle of radius R, $|\eta| = R$, Γ_ε be a counterclockwise circle of radius ε with center 0, and Ω_1^* be the closed counterclockwise contour around the upper and lower edges of Ω_1. The Cauchy theorem implies

$$\left\{ \int_{\Omega_1^*} + \int_{\Gamma_\varepsilon} - \int_{\Gamma_R} \right\} \omega_1(\eta) d\eta = 0.$$

Then, we have

$$- \int_{\Omega_1} \rho_1(\eta) d\eta + \frac{1}{2}(1 - q^{-1}) + \frac{1}{2}(1 + q^{-1}) = 0,$$

and then (7.26) is proved.

To prove the second formula, we need to change the small portions of the contour $\Omega_1^* = (-\Omega_1^-) \cup \Omega_1^+$ at the singular point $\lambda \in \Omega_1$ to be semicircles of ε radius. If we still use the previous notations, then

$$\left\{ \int_{\Omega_1^*} + \int_{\Gamma_\varepsilon} - \int_{\Gamma_R} \right\} \frac{\omega_1(\eta) - \frac{1}{2} W_0'(\eta)}{\eta - \lambda} d\eta = 0,$$

where as $\varepsilon \to 0$, there are

$$\int_{\Omega_1^*} \frac{\omega_1(\eta)}{\eta - \lambda} d\eta \to -2\pi i (P) \int_{\Omega_1} \frac{\rho_1(\eta)}{\eta - \lambda} d\eta, \qquad \int_{\Omega_1^*} \frac{W_0'(\eta)}{\eta - \lambda} d\eta = 2\pi i W_0'(\lambda),$$

$$\int_{\Gamma_\varepsilon} \frac{\omega_1(\eta)}{\eta - \lambda} d\eta \to -\frac{2\pi i}{\lambda} \operatorname*{res}_{\eta=0} \omega_1(\eta), \qquad \int_{\Gamma_\varepsilon} \frac{W_0'(\eta)}{\eta - \lambda} d\eta = 0,$$

and

$$\int_{\Gamma_R} \frac{\omega_1(\eta) - \frac{1}{2} W_0'(\eta)}{\eta - \lambda} d\eta = 0.$$

For the first formula above, we have used $\int_{-\pi}^{0} \omega_1(\lambda + \varepsilon e^{i\theta}) d\theta + \int_{0}^{\pi} \omega_1(\lambda + \varepsilon e^{i\theta}) d\theta \to 0$ as $\varepsilon \to 0$ for the calculations around the point λ, where the integral path of the first integral is below the Ω and the integral path of the second integral is above Ω_1. Then

$$(\text{P})\int_{\Omega_1}\frac{\rho_1(\eta)}{\lambda-\eta}d\eta = \frac{1}{2}W_0'(\lambda) + \left(1-q^{-1}\right)\frac{1}{2\lambda}.$$

So the theorem is proved. $\square$

As a remark, $y(\eta) = \omega_1(\eta) - \frac{1}{2}W_0'(\eta)$ satisfies the following relations: $y(\eta)$ is analytic for $\eta \in \mathbb{C}\backslash(\{0\} \cup \Omega)$ with a first-order pole $\eta = 0$; $y(\eta)|_{\Omega+} + y(\eta)|_{\Omega-} = -W_0'(\eta)$; $y(\eta) \to 0$ as $\eta \to \infty$. If the cut Ω_1 is split to two parts, there should be a phase transition or discontinuity according to the discussions in the previous chapters, that is what we will discuss in the next section for the generalized model. The phase transitions or high-order phase transitions for the complex systems have become one of the important research directions in recent years, specially for the random problems in econophysics and behavioral finance, for example, see [11, 17, 41, 42]. There are also other methods to study the complex systems in applications such as the quantum methods, for example, see [8, 12, 15]. The phase transition formulations in the matrix models introduce a new idea to study the dynamic states of the system and the criticality that are important for investigating the anomalies and bound of rationality in the behavioral finance.

7.2 Divergences Related to Marcenko-Pastur Distribution

Referring the free energy functions in the Hermitian and unitary matrix models, we consider the following quantity,

$$E = \int_{\Omega}\left(\eta + \left(1-q^{-1}\right)\ln(\eta)\right)\rho_0(\eta)d\eta - \int_{\Omega}\int_{\Omega}\ln|\eta - \lambda|\rho_0(\eta)\rho_0(\lambda)d\eta d\lambda, \quad (7.28)$$

for the Marcenko-Pastur distribution with $0 < q < 1$. When $q = 1$, the Marcenko-Pastur distribution becomes

$$\rho^c(\eta) = \frac{1}{2\pi}\sqrt{\frac{4-\eta}{\eta}}, \quad \eta \in [0, 4]. \quad (7.29)$$

We have the following ε-expansions around $q = 1$,

$$q^{-1} = 1 + \varepsilon, \quad (7.30)$$

$$\eta_- = \left(1 - q^{-1/2}\right)^2 = \frac{1}{4}\varepsilon^2 + \cdots, \quad (7.31)$$

$$\eta_+ = \left(1 + q^{-1/2}\right)^2 = 4 + 2\varepsilon + \cdots, \quad (7.32)$$

Fig. 7.2 Contour for
logarithmic singularity

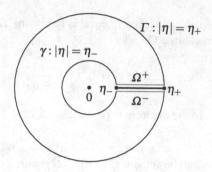

where $\varepsilon > 0$. Since $\rho_0(\eta)$ is defined on a single interval $\Omega = [\eta_-, \eta_+]$, Theorem 7.1 implies

$$\frac{d}{dq} E = q^{-2} \int_\Omega \ln(\eta) \rho_0(\eta) d\eta. \tag{7.33}$$

If we take the branch cut for $\ln \eta$ as $\arg(\eta) = 2\pi$, then there is

$$\left\{ \int_{\Omega^+} + \int_\Gamma - \int_{\Omega^-} + \int_\gamma \right\} \ln(\eta) \omega_0(\eta) d\eta = 0,$$

for the closed contour $\Omega^+ \cup \Gamma \cup (-\Omega^-) \cup \gamma$ with positive orientation, where the "$-$" sign before Ω^- stands for the opposite orientation, Ω^+ and Ω^- are the upper and lower edges of Ω, Γ is the circle $|\eta| = \eta_+$, and γ is the circle $|\eta| = \eta_-$. See Fig. 7.2.

Then, we have

$$\int_{\eta_-}^{\eta_+} \ln(\eta) \rho_0(\eta) d\eta = - \operatorname{Re} \frac{1}{2\pi i} \left(\int_\Gamma + \int_\gamma \right) \ln(\eta) \omega_0(\eta) d\eta.$$

It can be calculated that

$$\int_\Gamma \ln(\eta) \omega_0(\eta) d\eta = 2\pi i + O(\varepsilon),$$

and

$$\int_\gamma \ln(\eta) \omega_0(\eta) d\eta = -2\pi i \varepsilon \ln \varepsilon + O(\varepsilon),$$

as $\varepsilon \to 0$, that imply

$$\frac{d}{dq} E = -1 + |q - 1| \ln |q - 1| + O(|q - 1|), \tag{7.34}$$

as $q \to 1$ with $q < 1$. Therefore, as $q \to 1$ we have

$$\frac{d^2}{dq^2} E = O(\ln |q - 1|), \tag{7.35}$$

that gives a logarithmic divergence. This property indicates that as the parameter $q(< 1)$ in the Marcenko-Pastur distribution approaches to the critical point $q = 1$, the quantity E has a second-order logarithmic divergence.

When $q > 1$, the potential is changed to $W(\eta) = q\eta + (1 - q)\ln\eta$. Correspondingly, for the original Marcenko-Pastur distribution

$$\rho_0 = \frac{q}{2\pi\eta}\sqrt{(\eta_+ - \eta)(\eta - \eta_-)}d\eta, \quad q > 1, \tag{7.36}$$

where $\eta_\pm = (1 \pm \frac{1}{\sqrt{q}})^2$, there is

$$\frac{d}{dq}E = \int_{\eta_-}^{\eta_+} (\eta - \ln\eta)\rho_0(\eta)d\eta = 2 - \varepsilon\ln\varepsilon + O(\varepsilon), \tag{7.37}$$

as $\varepsilon \to 0$, where $q = 1 + \varepsilon$. Since the potential function is different in the $q < 1$ and $q > 1$ cases, the application background is unusual. But the second-order derivatives of E are of order $O(\ln|q - 1|)$ as $q \to 1$. If we consider large q, then

$$\frac{d}{d\tau}E = -\frac{1}{\tau^2}\int_{\eta_-}^{\eta_+} (\eta - \ln\eta)\rho_0(\eta)d\eta = O(\tau^{-2}), \tag{7.38}$$

as $\tau \to 0$, where $q = 1/\tau$, and ρ_0 becomes a delta function. One can experience that it is hard to integrate the function $\ln(\eta)\rho_0(\eta)$ for the Marcenko-Pastur distribution. But such quantities are important in physics. Interested readers may find some relations to the mathematical discussions for the entropy problems such as von Neumann entropy studied in the entanglement theories, for example, see [20, 38].

For the potential $W(\eta) = g_1\eta + g_2\eta^2 + g_3\eta^3 + (1 - q^{-1})\ln\eta$ discussed in last section, let us consider the following quantities,

$$E(\rho_j) = \int_{\Omega_j} W(\eta)\rho_j(\eta)d\eta - \int_{\Omega_j}\int_{\Omega_j} \ln|\eta - \lambda|\rho_j(\eta)\rho_j(\lambda)d\eta d\lambda, \tag{7.39}$$

for $j = 1$ and 2, where ρ_1 is defined by (7.25) with $\Omega_1 = [a - 2b, a + 2b]$, and ρ_2 is defined by

$$\rho_2(\eta) = \frac{1}{2\pi\eta}(3g_3\eta - 1)\,\mathrm{Re}\sqrt{e^{-\pi i}[(\eta - u)^2 - x_1^2][(\eta - u)^2 - x_2^2]}, \tag{7.40}$$

for $\eta \in \Omega_2 = [u - x_2, u - x_1] \cup [u + x_1, u + x_2] \equiv [\eta_-^{(1)}, \eta_+^{(1)}] \cup [\eta_-^{(2)}, \eta_+^{(2)}]$, where

$$x_1^2 = u^2 - w - 2v, \tag{7.41}$$

$$x_2^2 = u^2 - w + 2v, \tag{7.42}$$

and

$$u = (a_1 + a_2)/2 > x_2 > x_1 > 0, \quad v = b_1 b_2, \quad w = a_1 a_2 - b_1^2 - b_2^2, \tag{7.43}$$

as in the split density model discussed in Sect. 3.4, with the parameter conditions given in the following.

We choose $g_3 > 0$, and we need $3g_3\eta - 1 \leq 0$ when $\eta \in [\eta_-^{(1)}, \eta_+^{(1)}]$ and $3g_3\eta - 1 \geq 0$ when $\eta \in [\eta_-^{(2)}, \eta_+^{(2)}]$ such that ρ_2 is non-negative since the square root is negative on the left interval. Define the analytic function

$$\omega_2(\eta) = \frac{1}{2\eta}(3g_3\eta - 1)\sqrt{[(\eta - u)^2 - x_1^2][(\eta - u)^2 - x_2^2]}, \qquad (7.44)$$

for η in the complex plane outside the point $\eta = 0$ and the cuts Ω_2. If the parameters satisfy

$$2g_2 + 6g_3 u = -1, \qquad (7.45)$$

$$g_1 - 3g_3 w = 2u, \qquad (7.46)$$

$$w = 1 + q^{-1}, \qquad (7.47)$$

$$v^2 = q^{-1} > 1, \qquad (7.48)$$

then

$$\omega_2(\eta) = \frac{1}{2} W_0'(\eta) - \frac{1}{2}(1 + q^{-1})\eta^{-1} + O(\eta^{-2}), \qquad (7.49)$$

as $\eta \to \infty$, and

$$\operatorname*{res}_{\eta=0} \omega_2(\eta) = \frac{1}{2}(1 - q^{-1}), \qquad (7.50)$$

where $W_0(\eta) = g_1\eta + g_2\eta^2 + g_3\eta^3$. Note that the residue at $\eta = 0$ is equal to $-\frac{1}{2}\sqrt{w^2 - 4v^2} = -\frac{1}{2}\sqrt{(1 - q^{-1})^2} = -\frac{1}{2}(q^{-1} - 1)$, which is negative because the point $\eta = 0$ is at the left of two cuts Ω_2 and the square root is positive at $\eta = 0$ while it is negative between the two cuts, that is different from the one-cut case. The residue and the asymptotics at infinity need to have the above results so that the density is normalized and satisfies the variational equation as explained in last section. The split density ρ_2 has the following properties,

$$\begin{cases} (P) \int_{\Omega_2} \frac{\rho_2(\lambda)}{\eta - \lambda} d\lambda = \frac{1}{2} W'(\eta), & \eta \in (\eta_-^{(1)}, \eta_+^{(1)}) \cup (\eta_-^{(2)}, \eta_+^{(2)}), \\ \int_{\Omega_2} \frac{\rho_2(\lambda)}{\eta - \lambda} d\lambda = \frac{1}{2} W'(\eta) - \omega_2(\eta), & \eta \in (\eta_+^{(1)}, \eta_-^{(2)}), \end{cases} \qquad (7.51)$$

where $W(\eta) = W_0(\eta) + (1 - q^{-1}) \ln \eta$ and $' = \partial/\partial\eta$. The first formula above is the variational equation, and the proof is same as the ρ_1 discussed in last section. To prove the second formula, consider a small circle γ_η^* around a point $\eta \in (\eta_+^{(1)}, \eta_-^{(2)})$ of ε radius with another small circle γ_0^* of ε radius with center 0, and the counter-clockwise contours Ω_2^* around the edges of the two cuts Ω_2. By Cauchy theorem, there is

$$\frac{1}{2\pi i} \int_{\Omega_2^* \cup \gamma_\eta^* \cup \gamma_0^*} \frac{\omega_2(\lambda)}{\lambda - \eta} d\lambda = \frac{1}{2\pi i} \int_{|\lambda|=R} \frac{\omega_2(\lambda)}{\lambda - \eta} d\lambda,$$

where R is a large number. As $R \to \infty$ and $\varepsilon \to 0$, the above equation becomes

$$\text{(P)} \int_{\Omega_2} \frac{\rho_2(\lambda)}{\eta - \lambda} d\lambda + \omega_2(\eta) - \frac{1}{2\eta}(1 - q^{-1}) = \frac{1}{2} W_0'(\lambda),$$

that shows the second formula above since $W'(\eta) = W_0'(\eta) + (1 - q^{-1})\eta^{-1}$.

If we consider the derivative in the q direction, there is

$$\frac{d}{dq} E(\rho_2) = \int_{\Omega_2} \frac{dW(\eta)}{dq} \rho_2(\eta) d\eta$$

$$- 2\left(\int_{\eta_-^{(1)}}^{\eta_+^{(1)}} + \int_{\eta_-^{(2)}}^{\eta_+^{(2)}} \right) \left(\int_{\Omega_2} \ln|\eta - \lambda|\rho_2(\lambda)d\lambda - \frac{1}{2}W(\eta) \right) \frac{d\rho_2(\eta)}{dq} d\eta. \tag{7.52}$$

All other parameters will be considered as functions of q. For $\eta \in (\eta_-^{(1)}, \eta_+^{(1)})$, there is

$$\int_{\Omega_2} \ln|\eta - \lambda|\rho_2(\lambda)d\lambda - \frac{1}{2}W(\eta) = \int_{\Omega_2} \ln|\eta_-^{(1)} - \lambda|\rho_2(\lambda)d\lambda - \frac{1}{2}W(\eta_-^{(1)}), \tag{7.53}$$

and for $\eta \in (\eta_-^{(2)}, \eta_+^{(2)})$, there is

$$\int_{\Omega_2} \ln|\eta - \lambda|\rho_2(\lambda)d\lambda - \frac{1}{2}W(\eta)$$

$$= \int_{\Omega_2} \ln|\eta_-^{(1)} - \lambda|\rho_2(\lambda)d\lambda - \frac{1}{2}W(\eta_-^{(1)}) - \int_{\eta_+^{(1)}}^{\eta_-^{(2)}} \omega_2(\eta)d\eta.$$

Consequently, there is the following result by using $\int_{\Omega_2} \frac{d}{dq}\rho_2(\eta)d\eta = \frac{d}{dq}\int_{\Omega_2} \rho_2(\eta)d\eta = 0$ as before,

$$\frac{d}{dq} E(\rho) = \begin{cases} \int_{\Omega_1} \frac{dW(\eta)}{dq}\rho_1(\eta)d\eta, & \rho = \rho_1, \\ \int_{\Omega_2} \frac{dW(\eta)}{dq}\rho_2(\eta)d\eta + 2\int_{\eta_+^{(1)}}^{\eta_-^{(2)}} \omega_2(\eta)d\eta\frac{d}{dq}\int_{\eta_-^{(2)}}^{\eta_+^{(2)}} \rho_2(\eta)d\eta, & \rho = \rho_2. \end{cases} \tag{7.54}$$

The critical point can be obtained when both ρ_1 and ρ_2 become

$$\rho_c(\eta) = \frac{1}{4\pi}(\eta - 2)^2 \sqrt{\frac{4 - \eta}{\eta}}, \quad \eta \in [0, 4], \tag{7.55}$$

with the following parameter values,

$$a_c = 2, \quad b_c = 1, \quad g_1^c = 5, \quad 2g_2^c = -3, \quad 3g_3^c = \frac{1}{2}. \tag{7.56}$$

In the following discussions, we choose $g_3 = g_3^c = 1/6$ as a constant, and take g_1, g_2, u, v and w as functions of q. Then, there are the following expansions,

$$q^{-1} = 1 + \varepsilon, \varepsilon > 0, \tag{7.57}$$

$$g_1 = 5 + c_2\varepsilon, \qquad g_2 = -\frac{3}{2} - \frac{1}{4}\left(c_2 - \frac{1}{2}\right)\varepsilon, \tag{7.58}$$

$$u = 2 + \frac{1}{2}\left(c_2 - \frac{1}{2}\right)\varepsilon, \qquad v^2 = 1 + \varepsilon, \qquad w = 2 + \varepsilon, \tag{7.59}$$

$$x_1^2 = (2c_2 - 3)\varepsilon + \frac{1}{4}\left(\left(c_2 - \frac{1}{2}\right)^2 + 1\right)\varepsilon^2 + \cdots, \tag{7.60}$$

$$x_2^2 = 4 + (2c_2 - 1)\varepsilon + \frac{1}{4}\left(\left(c_2 - \frac{1}{2}\right)^2 - 1\right)\varepsilon^2 + \cdots, \tag{7.61}$$

$$\eta_-^{(1)} = u - x_2 = \frac{1}{16}\varepsilon^2 + \cdots, \qquad \eta_-^{(2)} = u + x_2 = 4 + \left(c_2 - \frac{1}{2}\right)\varepsilon + \cdots, \tag{7.62}$$

with a constant c_2, obtained from the parameter relations (7.45) to (7.48). We need to discuss the asymptotic expansion of the term

$$I_0 \equiv 2\int_{\eta_+^{(1)}}^{\eta_-^{(2)}} \omega_2(\eta)d\eta \frac{d}{dq}\int_{\eta_-^{(2)}}^{\eta_+^{(2)}} p_2(\eta)d\eta, \tag{7.63}$$

in the formula (7.54) to check whether it affects the discontinuity. The expression

$$\int_{\eta_-^{(2)}}^{\eta_+^{(2)}} p_2(\eta)d\eta = \frac{1}{2\pi}\int_{\eta_-^{(2)}}^{\eta_+^{(2)}} \frac{3g_3\eta - 1}{\eta}\sqrt{[x_2^2 - (\eta - u)^2][(\eta - u)^2 - x_1^2]}d\eta$$

can be changed to

$$\frac{3g_3}{2\pi}\int_{x_1}^{x_2} \frac{x}{x + u}\sqrt{(x_2^2 - x^2)(x^2 - x_1^2)}dx + \frac{3g_3u - 1}{2\pi}\int_{x_1}^{x_2} \frac{\sqrt{(x_2^2 - x^2)(x^2 - x_1^2)}}{x + u}dx,$$

where $x = \eta - u$. We have the asymptotics

$$\int_{x_1}^{x_2} \frac{x}{x + u}\sqrt{(x_2^2 - x^2)(x^2 - x_1^2)}dx = 4\pi\left(\frac{1}{2} - \frac{4}{3\pi}\right) + O(\varepsilon), \tag{7.64}$$

as $x_1 \to 0+$, obtained according to the result given in Sect. B.1 by noting that $u^2 - x_2^2 = O(\varepsilon^2)$ and $x_2^2 \sim 4$. The above asymptotics also implies

$$\int_{x_1}^{x_2} \frac{\sqrt{(x_2^2 - x^2)(x^2 - x_1^2)}}{x + u}dx = 4 - \pi + O(x_1^2\ln x_1),$$

by using the asymptotics $\int_{x_1}^{x_2} \sqrt{(x_2^2 - x^2)(x^2 - x_1^2)}dx = 8/3 + O(x_1^2 \ln x_1)$ given in Sect. B.1. Then, there is

$$\int_{\eta_-^{(2)}}^{\eta_+^{(2)}} \rho_2(\eta)d\eta = \frac{1}{2} - \frac{4}{3\pi} + O(\varepsilon), \tag{7.65}$$

as $\varepsilon \to 0$. For the factor $\int_{\eta_+^{(1)}}^{\eta_-^{(2)}} \omega_2(\eta)d\eta$ in I_0, the substitution $x = \eta - u$ changes the expression

$$\int_{\eta_+^{(1)}}^{\eta_-^{(2)}} \omega_2(\eta)d\eta = \frac{1}{2} \int_{\eta_+^{(1)}}^{\eta_-^{(2)}} \frac{3g_3\eta - 1}{\eta} \sqrt{[(\eta - u)^2 - x_2^2][(\eta - u)^2 - x_1^2]}d\eta,$$

to

$$\frac{3g_3}{2\pi} \int_{-x_1}^{x_1} \frac{x}{x+u} \sqrt{(x_2^2 - x^2)(x_1^2 - x^2)}dx$$

$$+ \frac{3g_3u - 1}{2\pi} \int_{-x_1}^{x_1} \frac{\sqrt{(x_2^2 - x^2)(x_1^2 - x^2)}}{x+u}dx,$$

where the two integrals above have the following asymptotics,

$$\int_{-x_1}^{x_1} \frac{\sqrt{(x_2^2 - x^2)(x_1^2 - x^2)}}{x+u}dx = \frac{x_2}{u} \int_{-x_1}^{x_1} \sqrt{x_1^2 - x^2}dx + O(\varepsilon^2) = \frac{\pi}{2}x_1^2 + O(\varepsilon^2),$$

and

$$\int_{-x_1}^{x_1} \frac{x}{x+u} \sqrt{(x_2^2 - x^2)(x_1^2 - x^2)}dx$$

$$= \frac{x_2}{u} \int_{-x_1}^{x_1} x \left(1 - \frac{x}{u}\right) \sqrt{(x_2^2 - x^2)(x_1^2 - x^2)}dx + O(\varepsilon^3)$$

$$= -\frac{1}{u} \int_{-x_1}^{x_1} x^2 \sqrt{(x_2^2 - x^2)(x_1^2 - x^2)}dx + O(\varepsilon^3)$$

$$= x_1^4 \frac{(\Gamma(3/2))^2}{\Gamma(3)} + O(\varepsilon^3).$$

Because $x_1^2 = O(\varepsilon)$, the above calculations imply $\int_{\eta_+^{(1)}}^{\eta_-^{(2)}} \omega_2(\eta)d\eta = O(\varepsilon^2)$. Finally, we get

$$I_0 = O(\varepsilon^2), \tag{7.66}$$

as $\varepsilon \to 0$. This property shows that I_0 does not affect the discontinuity to be discussed in the following, because the divergence will occur at the second-order derivative due to the singular point $\eta = 0$ in the potential. This is then different from the results in the Hermitian matrix models.

Next, let us consider the asymptotics of $\int_{\Omega_2} \eta^j \rho_2(\eta) d\eta$ for $j = 1$ and 2 in the formula

$$\frac{d}{dq} E(\rho_2) = \int_{\Omega_2} \left(\frac{dg_1}{dq} \eta + \frac{dg_2}{dq} \eta^2 + q^{-2} \ln \eta \right) \rho_2(\eta) d\eta + I_0, \quad q < 1. \quad (7.67)$$

By the asymptotic expansion of $\omega_2(\eta) = \frac{3g_3\eta - 1}{2\eta} \sqrt{((\eta - u)^2 - x_1^2)((\eta^2 - u)^2 - x_2^2)}$ as $\eta \to \infty$, we can get

$$\int_{\Omega_2} \eta \rho_2(\eta) d\eta = \frac{-1}{2\pi i} \int_{|\eta| = R} \eta \omega_2(\eta) d\eta = 1/2 + O(\varepsilon). \quad (7.68)$$

and

$$\int_{\Omega_2} \eta^2 \rho_2(\eta) d\eta = \frac{-1}{2\pi i} \int_{|\eta| = R} \eta^2 \omega_2(\eta) d\eta = 1 + O(\varepsilon). \quad (7.69)$$

For the term $\int_{\Omega_2} \ln(\eta) \rho_2(\eta) d\eta$ in (7.67), we can consider the contour integral $\int_l \ln(\eta) \omega_2(\eta) d\eta = 0$, where the integral is along $l = \overline{\Omega} \cup \Gamma \cup (-\underline{\Omega}) \cup \gamma$ with

$$\overline{\Omega} = \left[\eta_-^{(1)}, \eta_+^{(1)} \right]^+ \cup \left[\eta_+^{(1)}, \eta_-^{(2)} \right] \cup \left[\eta_-^{(2)}, \eta_+^{(2)} \right]^+,$$

$$\underline{\Omega} = \left[\eta_-^{(1)}, \eta_+^{(1)} \right]^- \cup \left[\eta_+^{(1)}, \eta_-^{(2)} \right] \cup \left[\eta_-^{(2)}, \eta_+^{(2)} \right]^-,$$

$\Gamma = \{ \eta_+^{(2)} e^{i\theta} | 0 \le \theta \le 2\pi \}$ and $\gamma = \{ \eta_-^{(1)} e^{i\theta} | 2\pi \ge \theta \ge 0 \}$ with positive orientation. Then it can be calculated that

$$\int_{\Omega_2} \ln(\eta) \rho_2(\eta) d\eta = \frac{\eta_-^{(1)}}{2\pi} \text{Re} \int_0^{2\pi} \left(\ln \eta_-^{(1)} + i\theta \right) \omega_2 \left(\eta_-^{(1)} e^{i\theta} \right) e^{i\theta} d\theta$$

$$- \frac{\eta_+^{(2)}}{2\pi} \text{Re} \int_0^{2\pi} \left(\ln \eta_+^{(2)} + i\theta \right) \omega_2 \left(\eta_+^{(2)} e^{i\theta} \right) e^{i\theta} d\theta. \quad (7.70)$$

Based on the ε-expansions given above, we get

$$\int_{\Omega_2} \ln(\eta) \rho_2(\eta) d\eta = -13/6 - \varepsilon \ln \varepsilon + O(\varepsilon), \quad (7.71)$$

where the leading term is obtained from the integral on Γ, the second term is from the integral on γ, and the contributions of other integrals are of higher orders. The above discussions then imply

$$\frac{d}{dq} E(\rho_2) = -\frac{1}{4} \left(c_2 + \frac{1}{2} \right) - \frac{13}{6} - \varepsilon \ln \varepsilon + O(\varepsilon), \quad (7.72)$$

as $\varepsilon \to 0$, where $\varepsilon = q^{-1} - 1 > 0$.

For ρ_1, there are similar results, but the discussions are relatively easier. We first have

$$\frac{d}{dq}E(\rho_1) = \int_{\Omega_1}\left(\frac{dg_1}{dq}\eta + \frac{dg_2}{dq}\eta^2 + q^{-2}\ln\eta\right)\rho_1(\eta)d\eta, \quad q < 1. \tag{7.73}$$

By the parameter relations (7.23) and (7.24), there are the following expansions,

$$q^{-1} = 1 + \varepsilon, \quad \varepsilon > 0, \quad b^2 = 1 + \frac{1}{2}\varepsilon - \frac{c_1}{12}\varepsilon^2 + \cdots, \tag{7.74}$$

$$g_1 = 5 + c_1\varepsilon + \cdots, \quad g_2 = -\frac{3}{2} - \frac{c_1}{6}\varepsilon + \cdots, \tag{7.75}$$

$$\eta_- = a - 2b = \frac{1}{16}\varepsilon^2 + \cdots, \quad \eta_+ = a + 2b = 4 + \frac{3}{2}\varepsilon + \cdots, \tag{7.76}$$

as $\varepsilon \to 0$, with a constant c_1. Then, we have $\int_{\Omega_1}\eta\rho_1(\eta)d\eta = 1/2 + O(\varepsilon)$ and $\int_{\Omega_1}\eta^2\rho_1(\eta)d\eta = 1 + O(\varepsilon)$ similar to (7.68) and (7.69). And similar to (7.70), it can be calculated that $\int_{\Omega_1}\ln(\eta)\rho_1(\eta)d\eta$ is equal to

$$\frac{\eta_-}{2\pi}\,\mathrm{Re}\int_0^{2\pi}(\ln\eta_- + i\theta)\omega_1(\eta_- e^{i\theta})e^{i\theta}d\theta - \frac{\eta_+}{2\pi}\,\mathrm{Re}\int_0^{2\pi}(\ln\eta_+ + i\theta)\omega_1(\eta_+ e^{i\theta})e^{i\theta}d\theta, \tag{7.77}$$

which gives

$$\int_{\Omega_1}\ln(\eta)\rho_1(\eta)d\eta = -13/6 - 2\varepsilon\ln\varepsilon + O(\varepsilon). \tag{7.78}$$

It should be noted that the coefficient of $\varepsilon\ln\varepsilon$ is changed to -2 if comparing with the coefficient -1 in (7.71). So we have the following ε-expansion,

$$\frac{d}{dq}E(\rho_1) = -\frac{c_1}{3} - \frac{13}{6} - 2\varepsilon\ln\varepsilon + O(\varepsilon), \tag{7.79}$$

as $\varepsilon \to 0$, where $\varepsilon = q^{-1} - 1 > 0$.

By comparing (7.72) with (7.79), it can be seen that if $4c_1 - 3c_2 \neq 3/2$, then the first-order derivative dE/dq is discontinuous. If $c_1 = c_2 = 3/2$, then $4c_1 - 3c_2 = 3/2$, and we have

$$\frac{d}{dq}E(\rho) = \begin{cases} -\frac{8}{3} - 2|q - 1|\ln|q - 1| + O(|q - 1|), & \rho = \rho_1, 0 < q < 1, \\ -\frac{8}{3} - |q - 1|\ln|q - 1| + O(|q - 1|), & \rho = \rho_2, 0 < q < 1, \end{cases} \tag{7.80}$$

as $q \to 1$. Finally we get the following second-order logarithmic divergence,

$$\frac{d^2}{dq^2}E = O\left(\ln|q - 1|\right), \tag{7.81}$$

as $q \to 1 - 0$. It is interesting to note that in the case $c_1 = c_2 = 3/2$, for both ρ_1 and ρ_2 phases the parameters g_1, g_2 and q stay on the same side of the critical point with the expansions

$$g_1 = 5 + \frac{3}{2}\varepsilon > 5, \qquad g_2 = -\frac{3}{2} - \frac{1}{4}\varepsilon < -\frac{3}{2}, \qquad q < 1. \qquad (7.82)$$

This property further indicates that the one-side transition would be specific for the models reduced from the Laguerre polynomials or the generalizations since the polynomials are defined on one side of the 0 point. The discontinuities of first or second order discussed above are to extend the Marcenko-Pastur distribution case for studying a wider class of random behavioral problems. One can investigate more complicated models, but the divergences would always occur at the second-order derivatives because of the singular point 0.

7.3 Jacobi Polynomials and Logarithmic Divergences

The Jacobi polynomials $P_n^{(\alpha,\beta)}(x)$ are defined on the interval $[-1, 1]$, and orthogonal with the weight $w(x) = (1 - x)^{\alpha}(1 + x)^{\beta}$,

$$\int_{-1}^{1} P_n^{(\alpha,\beta)}(x) P_{n'}^{(\alpha,\beta)}(x) w(x) dx$$

$$= \frac{2^{\alpha+\beta+1}}{2n + \alpha + \beta + 1} \frac{\Gamma(n + \alpha + 1)\Gamma(n + \beta + 1)}{\Gamma(n + +1)\Gamma(n + \alpha + \beta + 1)} \delta_{nn'}, \qquad (7.83)$$

where $\alpha > -1$, $\beta > -1$. Specially

$$P_0^{(\alpha,\beta)}(x) = 1, \qquad P_1^{(\alpha,\beta)}(x) = \frac{1}{2}(\alpha + \beta + 2)x + \frac{1}{2}(\alpha - \beta).$$

The Jacobi polynomials satisfy the following recursion formula and the differential equation [35]:

$$2n(n + \alpha + \beta)(2n + \alpha + \beta - 2)P_n^{(\alpha,\beta)}(x)$$

$$= (2n + \alpha + \beta - 1)\left[(2n + \alpha + \beta)(2n + \alpha + \beta - 2)x + \alpha^2 - \beta^2\right]P_{n-1}^{(\alpha,\beta)}(x)$$

$$\quad - 2(n + \alpha - 1)(n + \beta - 1)(2n + \alpha + \beta)P_{n-2}^{(\alpha,\beta)}(x), \quad n = 2, 3, \ldots, \quad (7.84)$$

and

$$(2n + \alpha + \beta)(1 - x^2)\frac{d}{dx}P_n^{(\alpha,\beta)}(x)$$

$$= -n\left[(2n + \alpha + \beta)x + \beta - \alpha\right]P_n^{(\alpha,\beta)}(x) + 2(n + \alpha)(n + \beta)P_{n-1}^{(\alpha,\beta)}(x). \quad (7.85)$$

Now, choose $\Phi_n(x) = w(x)^{1/2}(P_n^{(\alpha,\beta)}(x), P_{n-1}^{(\alpha,\beta)}(x))^T$. It can be verified that $\Phi_n(x)$ satisfies the following equation [35]

$$\frac{\partial}{\partial x}\Phi_n = A_n(x)\Phi_n, \tag{7.86}$$

where

$$A_n(x) = \frac{1}{1-x^2}\begin{pmatrix} -\frac{\alpha^2-\beta^2}{2(2n+\alpha+\beta)} - (n+\frac{\alpha+\beta}{2})x & \frac{2(n+\alpha)(n+\beta)}{2n+\alpha+\beta} \\ -\frac{2n(n+\alpha+\beta)}{2n+\alpha+\beta} & \frac{\alpha^2-\beta^2}{2(2n+\alpha+\beta)} + (n+\frac{\alpha+\beta}{2})x \end{pmatrix}. \tag{7.87}$$

As a remark, the coefficients in the matrix A_n here do not have square roots for the parameters, that are different from the coefficients in the differential equations for the Jacobi polynomials given in other literatures. Let $p = \alpha/n$ and $q = \beta/n$. The above formula implies

$$\sqrt{\det A_n(x)}$$

$$= \frac{n}{1-x^2}\sqrt{\frac{4(1+p)(1+q)(1+p+q)}{(2+p+q)^2} - \left(\frac{p^2-q^2}{2(2+p+q)} + \left(1+\frac{p+q}{2}\right)x\right)^2}.$$

Specially, if $p = q$, there is the following result,

$$\frac{1}{n\pi}\sqrt{\det A_n(x)}dx = \frac{1}{\pi(1-x^2)}\sqrt{1+2q-(q+1)^2x^2}dx. \tag{7.88}$$

If we make change of the variable and parameter, $x = \eta/c$ and $q = c/2 - 1$, then there is

$$\frac{1}{\pi(1-x^2)}\sqrt{1+2q-(q+1)^2x^2}dx = \frac{c}{2\pi(c^2-\eta^2)}\sqrt{4(c-1)-\eta^2}d\eta. \tag{7.89}$$

The density function on the right hand side above is the McKay's law [24]. If $p = -q$ with $-1 < q < 1$, then

$$\frac{1}{n\pi}\sqrt{\det A_n(x)}dx = \frac{1}{\pi(1-x^2)}\sqrt{1-q^2-x^2}dx. \tag{7.90}$$

It will discussed later that this density model does not have transition, which is then different from the McKay's law. A general eigenvalue density based on the Jacobi polynomials is discussed in the following.

When $1 + p > 0$, $1 + q > 0$, and $1 + p + q > 0$, denote

$$\omega(\eta) = \frac{1}{1-\eta^2}\sqrt{\left(\frac{p^2-q^2}{2(2+p+q)} + \left(1+\frac{p+q}{2}\right)\eta\right)^2 - \frac{4(1+p)(1+q)(1+p+q)}{(2+p+q)^2}}, \tag{7.91}$$

which can be written as

$$\omega(\eta) = \frac{2 + p + q}{2(1 - \eta^2)} \sqrt{(\eta - \eta_-)(\eta - \eta_+)}, \tag{7.92}$$

where

$$\eta_- = \frac{q^2 - p^2 - 4\sqrt{(1 + p)(1 + q)(1 + p + q)}}{(2 + p + q)^2}, \tag{7.93}$$

$$\eta_+ = \frac{q^2 - p^2 + 4\sqrt{(1 + p)(1 + q)(1 + p + q)}}{(2 + p + q)^2}. \tag{7.94}$$

We first have the following lemma by direct calculations, which will be applied when computing the residue at $\eta = -1$ or $\eta = 1$.

Lemma 7.1

$$\frac{1}{4}\left[p^2 - q^2 - (2 + p + q)^2\right]^2 - 4(1 + p)(1 + q)(1 + p + q) = q^2(2 + p + q)^2. \tag{7.95}$$

Lemma 7.2 *For* $1 + p > 0$, $1 + q > 0$, *and* $1 + p + q > 0$, *there is*

$$-1 \leq \eta_- \leq \eta_+ \leq 1, \tag{7.96}$$

where $\eta_- = -1$ *if* $q = 0$, *and* $\eta_+ = 1$ *if* $p = 0$.

Proof To prove $-1 \leq \eta_-$, consider $0 \leq 4q^2(2 + p + q)^2$. By (7.95), this inequality becomes

$$16(1 + p)(1 + q)(1 + p + q) \leq \left(q^2 - p^2 + (2 + p + q)^2\right)^2.$$

Since $q^2 - p^2 + (2 + p + q)^2 = 2((p + q)(2 + q) + 2) > 2((p + q + 1) - q) > 0$, we get

$$4\sqrt{(1 + p)(1 + q)(1 + p + q)} \leq q^2 - p^2 + (2 + p + q)^2,$$

which implies $-1 \leq \eta_-$, where equality holds only if $q = 0$.

For $\eta_+ \leq 1$, consider $0 \leq 4p^2(2 + p + q)^2$. By (7.95), it becomes

$$16(1 + p)(1 + q)(1 + p + q) \leq \left(p^2 - q^2 + (2 + p + q)^2\right)^2.$$

Since $p^2 - q^2 + (2 + p + q)^2 = 2((p + q)(2 + p) + 2) > 2((p + q + 1) - p) > 0$, we get

$$4\sqrt{(1 + p)(1 + q)(1 + p + q)} \leq p^2 - q^2 + (2 + p + q)^2,$$

which gives $\eta_+ \leq 1$, where equality holds only if $p = 0$. □

The above lemme shows that $\eta = 1$ and $\eta = -1$ are in the domain $\mathbb{C} \backslash \Omega$ where $\Omega = [\eta_-, \eta_+]$, and they are the poles of the analytic function ω. Then the ω function is analytic in $\mathbb{C}\backslash(\{-1\}\cup\Omega\cup\{1\})$. With these basic properties ready, we can now discuss the variational equation.

Corresponding to the weight function $w(x) = (1-x)^{\alpha}(1+x)^{\beta}$, we choose the potential

$$W(\eta) = p\ln(1-\eta) + q\ln(1+\eta), \tag{7.97}$$

for the Jacobi model, and then $W'(\eta) = p/(\eta-1) + q/(\eta+1)$. In the Hermitian models, the ω function is usually asymptotic to $W'(\eta)/2$ as $\eta \to \infty$ as we have discussed before. In the Laguerre models, $W'(\eta)/2$ partially appears in the asymptotics of the ω function. For the Jacobi models, there is no $W'(\eta)/2$ in the asymptotics of the ω function. The following result shows that $W'(\eta)/2$ will come out from the residues of the ω function at the poles ± 1.

Lemma 7.3 *For the $\omega(\eta)$ defined by (7.91), there are*

$$\omega(\eta) = -\left(1 + \frac{p+q}{2}\right)\frac{1}{\eta} + O\left(\frac{1}{\eta^2}\right), \tag{7.98}$$

as $\eta \to \infty$, and

$$\operatorname*{res}_{\eta=1} \omega(\eta) = -\frac{p}{2}, \qquad \operatorname*{res}_{\eta=-1} \omega(\eta) = -\frac{q}{2}. \tag{7.99}$$

Proof The asymptotics (7.98) directly follows from the definition (7.91) of the ω. For the residue at $\eta = 1$, we first note that $(1-\eta^2)\omega(\eta)|_{\eta=1} = \sqrt{p^2} = p$ by using (7.95) with a interchange between p and q, where $\sqrt{p^2} = p$ according to the branch of the analytic function $\omega(\eta)$. Then,

$$\operatorname*{res}_{\eta=1} \omega(\eta) = \lim_{\eta\to 1}(\eta-1)\omega(\eta) = -\frac{(1-\eta^2)\omega(\eta)}{1+\eta}\bigg|_{\eta=1} = -\frac{p}{2}.$$

At $\eta = -1$, there are $(1-\eta^2)\omega(\eta)|_{\eta=-1} = e^{2\pi i/2}q = -q$, and

$$\operatorname*{res}_{\eta=-1} \omega(\eta) = \lim_{\eta\to -1}(\eta+1)\omega(\eta) = \frac{(1-\eta^2)\omega(\eta)}{1-\eta}\bigg|_{\eta=-1} = -\frac{q}{2}.$$

Then the lemma is proved. $\square$

Now, define

$$\rho(\eta) = \frac{1}{\pi i}\omega(\eta)\bigg|_{\Omega^+}, \tag{7.100}$$

for $\eta \in \Omega$.

Theorem 7.3 *The $\rho(\eta)$ defined by (7.100) satisfies the normalization*

$$\int_{\Omega} \rho(\eta)d\eta = 1, \tag{7.101}$$

Fig. 7.3 Contours for
McKay's law

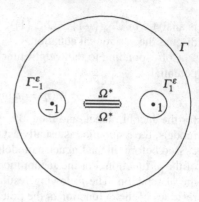

and the variational equation

$$\text{(P)} \int_\Omega \frac{\rho(\eta)}{\eta - \lambda} d\eta = \frac{1}{2} W'(\lambda), \qquad (7.102)$$

where λ is an inner point of Ω.

Proof Let Γ be a large counterclockwise circle of radius R with center 0, Γ^ε_{-1} and Γ^ε_1 be the counterclockwise circles of radius ε with centers -1 and 1 respectively, and Ω^* be the union of the closed counterclockwise contours around the upper and lower edges of Ω, or $\Omega^* = \Omega_- \cup (-\Omega^+)$. See Fig. 7.3. Then Cauchy theorem implies

$$\left\{ \int_{\Omega^*} + \int_{\Gamma^\varepsilon_{-1}} + \int_{\Gamma^\varepsilon_1} - \int_\Gamma \right\} \omega(\eta) d\eta = 0,$$

or

$$-\int_\Omega \rho(\eta) d\eta - \frac{p+q}{2} + \left(1 + \frac{p+q}{2} \right) = 0,$$

and then (7.101) is proved.

To prove (7.102), we need to change Ω^* at the inner point $\lambda \in \Omega$. The Ω^- and Ω^+ around λ can be changed to the semicircles of ε radius. If we still use the previous notations, then

$$\left\{ \int_{\Omega^*} + \int_{\Gamma^\varepsilon_{-1}} + \int_{\Gamma^\varepsilon_1} - \int_\Gamma \right\} \frac{\omega(\eta)}{\eta - \lambda} d\eta = 0,$$

which implies

$$-2\text{(P)} \int_\Omega \frac{\rho(\eta)}{\eta - \lambda} d\eta + \frac{p}{\lambda - 1} + \frac{q}{\lambda + 1} = 0,$$

where the integral $\int_\Gamma$ disappears, and $\int_{\Gamma^\varepsilon_{-1}}$ and $\int_{\Gamma^\varepsilon_1}$ give the last two terms above. Here we have used $\int_{-\pi}^0 \omega(\lambda + \varepsilon e^{i\theta}) d\theta + \int_0^\pi \omega(\lambda + \varepsilon e^{i\theta}) d\theta \to 0$ as $\varepsilon \to 0$ due to the opposite signs of the ω in these two integrals, where the integral path of the first integral is below Ω and the second integral is above Ω.   ☐

If we consider the following quantity

$$E = \int_{\eta_-}^{\eta_+} W(\eta)\rho(\eta)d\eta - \int_{\eta_-}^{\eta_+}\int_{\eta_-}^{\eta_+} \ln|\eta - \lambda|\rho(\eta)\rho(\lambda)d\eta d\lambda, \qquad (7.103)$$

for small p and q by choosing $p = c_1\varepsilon$ and $q = c_2\varepsilon$, then we have

$$\frac{dE}{d\varepsilon} = c_1 \int_{\eta_-}^{\eta_+} \ln(1-\eta)\rho(\eta)d\eta + c_2 \int_{\eta_-}^{\eta_+} \ln(1+\eta)\rho(\eta)d\eta. \qquad (7.104)$$

Then by (7.93) and (7.94) there are

$$\eta_- = -1 + \frac{c_2^2}{2}\varepsilon^2 + \cdots, \qquad \eta_+ = 1 - \frac{c_1^2}{2}\varepsilon^2 + \cdots. \qquad (7.105)$$

To compute $\int_{\eta_-}^{\eta_+} \ln(1-\eta)\rho(\eta)d\eta$, consider

$$\left\{ \int_\Gamma + \int_{\Omega^+} + \int_\gamma - \int_{\Omega^-} \right\} \ln(\eta)\omega(\eta)d\eta = 0,$$

for the closed contour $\Gamma \cup \Omega^+ \cup \gamma \cup (-\Omega^-)$ with positive orientation, where the "−" sign before Ω^- stands for the opposite orientation, Ω^+ and Ω^- are the upper and lower edges of Ω, γ is the small circle $|\eta - 1| = 1 - \eta_+$ (or $\eta - 1 = (\eta_+ - 1)e^{i\theta}$ with $2\pi \geq \theta \geq 0$), and Γ is the circle $|\eta - 1| = 1 - \eta_-$ (or $\eta - 1 = (\eta_- - 1)e^{i\theta}$ with $0 \leq \theta \leq 2\pi$). By the ε-expansions (7.105), we have

$$\int_{\eta_-}^{\eta_+} \ln(1-\eta)\rho(\eta)d\eta$$

$$= \frac{1}{2\pi} \text{Re}\left(\frac{|c_1|}{2}\varepsilon \int_0^{2\pi} \left(\ln(1+\eta_+) + i\theta\right)\sqrt{1 - e^{i\theta}}\,d\theta \right)$$

$$- \frac{1}{2\pi} \text{Re}\left(\int_0^{2\pi} \left(\ln(1-\eta_-) + i\theta\right)\frac{\sqrt{(e^{i\theta}-1)e^{i\theta}}}{1 - e^{i\theta}}\,d\theta \right) + O(\varepsilon), \qquad (7.106)$$

as $\varepsilon \to 0$. By the formulas given in Appendix A, we can get

$$\int_{\eta_-}^{\eta_+} \ln(1-\eta)\rho(\eta)d\eta = -\ln 2 + |c_1|\varepsilon\ln\varepsilon + O(\varepsilon). \qquad (7.107)$$

Similarly, as $\varepsilon \to 0$, there is

$$\int_{\eta_-}^{\eta_+} \ln(1+\eta)\rho(\eta)d\eta = -\ln 2 + |c_2|\varepsilon\ln\varepsilon + O(\varepsilon). \qquad (7.108)$$

Therefore, we have

$$\frac{dE}{d\varepsilon} = -(c_1 + c_2)\ln 2 + (c_1|c_1| + c_2|c_2|)\varepsilon \ln \varepsilon + O(\varepsilon), \tag{7.109}$$

as $\varepsilon \to 0$, that gives a logarithmic divergence for the second-order derivative of E if $c_1|c_1| + c_2|c_2| \neq 0$.

If $p = -q$ for $|q| < 1$ corresponding to $c_1 + c_2 = 0$, then

$$\frac{dE}{dq} = \int_{-\eta_0}^{\eta_0} \ln \frac{1+\eta}{1-\eta} \rho(\eta) d\eta, \tag{7.110}$$

where $\rho(\eta) = \frac{\sqrt{1-q^2-\eta^2}}{\pi(1-\eta^2)}$, given by (7.90), and $\eta_0 = \sqrt{1-q^2} < 1$. It follows that

$$\frac{dE}{dq} = 2\sum_{n=1}^{\infty} \frac{1}{2n-1} \int_{-\eta_0}^{\eta_0} \eta^{2n-1} \frac{\sqrt{\eta_0^2 - \eta^2}}{1-\eta^2} d\eta = 0, \tag{7.111}$$

because of the odd symmetry of the functions, and then E is a constant in this special case.

If p or q is large, we can consider $p = c/\tau$ and $q = c_2/\tau$ with $\tau \to 0$. If $c_2 = 0$ and $c_1 > 0$, then $\eta_- = 1 + O(\tau)$ and $\eta_+ = 1 + O(\tau)$ as $\tau \to 0$, that imply

$$\frac{dE}{d\tau} = O(\ln \tau), \tag{7.112}$$

as $\tau \to 0$. There is similar result if $c_1 = 0$ and $c_2 > 0$. If $c_1 > c_2 > 0$, then $\eta_\pm \to (c_2 - c_1)/(c_1 + c_2)$ as $\tau \to 0$. Then we have

$$\frac{dE}{d\tau} = O(1), \tag{7.113}$$

as $\tau \to 0$. The case $c_2 > c_1 > 0$ is similar. If $c_1 = c_2 > 0$, then $\eta_\pm \to O(\tau^{1/2})$, and

$$\frac{dE}{d\tau} = O\left(\tau^{1/2}\right), \tag{7.114}$$

as $\tau \to 0$.

This type mathematical problems have drawn lot attentions in recent years, specially in the researches for the entanglement problems. Many literatures have been published in the field about the entanglement or entropy with various methods including correlation functions, quantum disorder, quantum phase transition, density matrix and matrix Riemann-Hilbert problems. Interested readers can find the discussions, for example, in [1, 7, 10, 14, 20, 38]. As a remark, the function $\omega(\eta)$ discussed above satisfies the following properties: $\omega(\eta)$ is analytic for $\eta \in \mathbb{C} \setminus (\{-1\} \cup \Omega \cup \{1\})$ with the first-order poles $\eta = \pm 1$; $\omega(\eta)|_{\Omega^+} + \omega(\eta)|_{\Omega^-} = 0$; $\omega(\eta) \to 0$ as $\eta \to \infty$. We know that the scalar Riemann-Hilbert problem: $\omega(\eta)|_{\Omega^+} + \omega(\eta)|_{\Omega^-} = 0$; $\omega(\eta) \to 0$ as $\eta \to \infty$ has the unique trivial solution $\omega(\eta) = 0$. When two poles ± 1 are added in the problem, the solution is not trivial. There would be more interesting properties for the modified problems if more poles are added, that are left for further investigations.

7.4 Integral Transforms for the Density Functions

We have obtained the power-law and logarithmic divergences of the derivatives of the free energy functions based on the integrals of the density functions in the previous discussions. The exponent in the power-law divergence at the critical point is called critical exponent that can be used to classify the diverse transition systems. The systems with the same critical exponent belong to the same universality class according to the renormalization group theory because they have the same scaling behavior and share the same fundamental dynamics. We have seen that the power-law divergence depends on the potential and the expansions of the parameters at the critical point, and the logarithmic divergence is a limit case of the power-law divergence [3]. We have also seen that the power-law divergence is not directly connected to the power formula of the eigenvalue density. In this section, we are going to show that the power formula of the eigenvalue density can be applied to get the power-law distribution by using the integral transforms.

Let us use the Laplace transform to show that the eigenvalue densities in the matrix models can be transformed to the power-law distributions that are important in the complexity researches. Consider the Wigner semicircle

$$\rho(\eta) = \frac{1}{2b^2\pi}\sqrt{4b^2 - (\eta - a)^2}, \qquad (7.115)$$

which satisfies $\int_{a-2b}^{a+2b} \rho(\eta)d\eta = 1$. For the convenience in the discussions below, we consider the case $2b - a > 0$. Let

$$p(x) = c_0 \int_{a-2b}^{a+2b} e^{-x(\eta - a + 2b)^\mu} \rho(\eta)d\eta, \qquad 0 < \mu < \frac{3}{2}, \qquad (7.116)$$

with a constant $c_0 > 0$. We want to show that $p(x)$ is a power-law distribution function for $x \geq 0$.

First, $p(x)$ is well defined for the parameters given above with the corresponding restrictions. It is easy to see that $p(0) = c_0 > 0$, $p(x) > 0$ for $x > 0$, and $p'(x) < 0$ for $x > 0$. To show $\int_0^\infty p(x)dx = 1$, consider

$$I(t) = \int_0^\infty \int_{a-2b}^{a+2b} e^{-x(t\eta - a + 2b)^\mu}\sqrt{4b^2 - (\eta - a)^2}d\eta dx, \qquad (7.117)$$

for $0 \leq t \leq 1$. Taking derivative with respect to t on both sides of the equation above, after simplification by using $\int_0^\infty xe^{-x(t\eta - a + 2b)^\mu}dx = (t\eta - a + 2b)^{-2\mu}$, we have

$$I'(t) = -\mu \int_{a-2b}^{a+2b} \eta\sqrt{4b^2 - (\eta - a)^2}(t\eta - a + 2b)^{-\mu - 1}d\eta. \qquad (7.118)$$

It follows that by taking integral for t from 0 to 1,

$$I(1) - I(0) = \int_{a-2b}^{a+2b} \sqrt{4b^2 - (\eta - a)^2}[(\eta - a + 2b)^{-\mu} - (2b - a)^{-\mu}]d\eta,$$

where $t = 1$ means $t \to 1 - 0$ to avoid the singularity when $\eta = a - 2b$. Since $I(0) = 2b^2\pi/(2b - a)^\mu$, we then get

$$I(1) = \int_{-2b}^{2b} \frac{\sqrt{4b^2 - \zeta^2}}{(\zeta + 2b)^\mu}d\zeta, \tag{7.119}$$

where $\zeta = \eta - a$. The integral is convergent since the integrand is of $O((\zeta + 2b)^{1/2 - \mu})$ as $\zeta \to -2b$ with $1/2 - \mu > -1$. Then, $I(1) = (4b)^{2-\mu}B(3/2, 3/2 - \mu)$, where $B(\cdot, \cdot)$ is the Euler beta function. Therefore, we finally obtain

$$\int_0^\infty p(x)dx = \frac{c_0}{2b^2\pi}I(1) = \frac{c_0}{2b^2\pi}(4b)^{2-\mu}B(3/2, 3/2 - \mu), \tag{7.120}$$

which implies that if

$$c_0 = \frac{(4b)^\mu \pi}{8B(3/2, 3/2 - \mu)}, \tag{7.121}$$

then $\int_0^\infty p(x)dx = 1$ and $p(x)$ is a distribution function.

Next, let us show $p(x)$ has a power-law asymptotics as $x \to \infty$. Consider the following integral

$$\int_0^\delta e^{-x\zeta^\mu}\zeta^{\nu-1}d\eta, \tag{7.122}$$

where $\delta > 0$ is a fixed number. Let $t = x\eta^\mu$. It follows that [26]

$$\int_0^\delta e^{-x\zeta^\mu}\eta^{\nu-1}d\eta = \frac{1}{\mu x^{\frac{\nu}{\mu}}}\int_0^{x\delta^\mu} e^{-t}t^{\frac{\nu}{\mu}-1}dt = \frac{\Gamma(\nu/\mu)}{\mu x^{\nu/\mu}} + o(x^{-\nu/\mu}),$$

as $x \to \infty$. The asymptotics of $p(x)$ as $x \to \infty$ is dominated by the behavior of $\rho(\eta)$ at $\eta = a - 2b$ as known in the Laplace asymptotics method [26]. Since $\rho(\eta) = O((\zeta)^{1/2})$ as $\zeta \to 0$, where $\zeta = \eta - a + 2b$, the asymptotic behavior of $p(x)$ can be obtained by (7.122) for $\nu = 3/2$. Then we get $p(x) = O(x^{-\frac{3}{2\mu}})$ as $x \to \infty$.

Sometimes, the ν can be a different number. For example, for the critical density

$$\rho_c(\eta) = \frac{1}{2b^2\pi}(4b^2 - (\eta - a)^2)^{3/2}, \tag{7.123}$$

we have $\nu = 5/2$, that will give a distribution with the power-law $O(x^{-\frac{5}{2\mu}})$ as $x \to \infty$. As known in the random matrix theory, the eigenvalue density is obtained from the correlation function in large-N asymptotics. When $N = 1$, the correlation function is the Gaussian distribution. If we consider the Laplace transform of the Gaussian distribution

$$p_0(x) = \int_0^\infty e^{-x\eta^2} e^{-\eta^2} d\eta, \tag{7.124}$$

it can by seen that

$$p_0(x) = e^{x^2/4} \int_{x/2}^\infty e^{-t^2} dt = O(x^{-1}), \tag{7.125}$$

as $x \to \infty$. The above discussions can be also applied to other eigenvalue densities, such as the Marcenko-Pastur distribution, McKay's law and Gross-Witten densities.

If we consider the Laplace transform for the Marcenko-Pastur distribution (7.12),

$$\int_0^\infty e^{-\xi\eta} \, \mathrm{Re} \, \frac{1}{\pi i} \omega_0(\eta) d\eta = \int_{\eta_-}^{\eta_+} e^{-\xi\eta} \rho_0(\eta) d\eta, \tag{7.126}$$

then by the asymptotics $\int_0^\delta e^{-\xi\eta} \eta^{\nu-1} d\eta = \frac{\Gamma(\nu)}{\xi^\nu} + o(\xi^{-\nu})$ as $\xi \to \infty$, we can get a power-law $\xi^{-3/2}$, since $\rho_0(\eta) = O((\eta - \eta_-)^{1/2})$ as $\eta \to \eta_-$. If we consider the critical case $q = 1$ with $\eta_- = 0$ and $\eta_+ = 4$,

$$\rho_0^c(\eta) = \frac{1}{2\pi} \sqrt{\frac{4-\eta}{\eta}}, \tag{7.127}$$

then we get a power-law $\xi^{-1/2}$ as $\xi \to \infty$, since $\rho_0^c(\eta) = O(\eta^{-1/2})$ as $\eta \to 0$. The corresponding "distribution" function is then not integrable, that indicates a singularity in the case $q = 1$, consistent with the transition phenomenon discussed before. More complicated integral transforms can be studied similarly.

The power-law distribution is an important research topic in the complexity subjects such as econophysics, for example, see [4, 18, 19, 22, 28, 29, 33]. The associated researches include random matrices and phase transitions, that can be found, for example, in [2, 9, 11, 17, 21, 27, 29, 34, 40–42]. Another related field is about the overreactions and herding behaviors in the behavioral sciences that have also drawn lot attentions in recent years, and many literatures in these subjects have been published, for example, see [6, 13, 16, 31, 32, 36]. Since this book is organized based on the integrable systems, we are not going to discuss the details of these researches. These references are listed here in order to remind the interested readers to pay attentions to the related new researches. In the following, we are going to show more properties about the density models and the power-laws.

To discuss the power-law by using the Fourier transform, let us review a basic property discussed in the method of stationary phase [26]. Consider the integral $\int_0^\delta e^{i\xi\eta} \eta^{\nu-1} d\eta$, where $\delta > 0$ is a fixed number and $0 < \nu < 1$. As shown in [26], the integration around the contour $\{\eta | \varepsilon \leq \mathrm{Re}\,\eta \leq R, \mathrm{Im}\,\eta = 0\} \cup \{\eta | 0 \leq \arg\eta \leq \pi/2, |\eta| = R\} \cup \{\eta | \varepsilon \leq \mathrm{Im}\,\eta \leq R, \mathrm{Re}\,\eta = 0\} \cup \{\eta | 0 \leq \arg \leq \pi/2, |\eta| = \varepsilon\}$ yields

$$\int_0^\infty e^{i\xi\eta}\eta^{\nu-1}d\eta = \frac{e^{\nu\pi i/2}\Gamma(\nu)}{\xi^\nu}. \tag{7.128}$$

Then

$$\int_0^\delta e^{i\xi\eta}\eta^{\nu-1}d\eta = \left\{\int_0^\infty - \int_\delta^\infty\right\}e^{i\xi\eta}\eta^{\nu-1}d\eta = \frac{e^{\nu\pi i/2}\Gamma(\nu)}{\xi^\nu} + o(\xi^{-\nu}),$$

as $\xi \to \infty$. Applying this asymptotic expansion to the densities discussed above, we can get the power-law $\xi^{-3/2}$ or $\xi^{-5/2}$ as $\xi \to \infty$.

Now, if we consider the moments $\int_{\eta_-}^{\eta_+}\eta^j\rho(\eta)d\eta$ discussed in the free energy function, it can be seen that they are in fact the Mellin transform $\int_0^\infty \eta^{\xi-1}\rho(\eta)d\eta$ when ξ is an integer, where ρ has a finite support. We are going to show the following relations for the moment quantities,

$$\int_0^\infty \frac{x^j}{n\pi}\,\mathrm{Re}\,\sqrt{\det A_n(x)}\,dx = \int_0^\infty x^{j+\alpha}e^{-x}K_n(x,x)\,dx, \tag{7.129}$$

for $j = 0, 1, 2$, where $K_n(x, y)$ is the Fredholm kernel for the Laguerre polynomials [35],

$$K_n(x, y) = \frac{1}{n}\sum_{k=0}^{n-1}\frac{1}{h_k}L_k^{(\alpha)}(x)L_k^{(\alpha)}(y) = \frac{1}{h_{n-1}}\frac{L_n^{(\alpha)}(x)L_{n-1}^{(\alpha)}(y) - L_n^{(\alpha)}(y)L_{n-1}^{(\alpha)}(x)}{x-y},$$

$$\tag{7.130}$$

with $h_k = \Gamma(\alpha+1)\binom{k+\alpha}{k}$. For $j = 0$, the right hand side of (7.129) is equal to 1 by the definition. By Theorem 7.1, the left hand side is also equal to 1 when $j = 0$.

To prove the relation for $j = 1, 2$, we first note that

$$K_n(x, x) = \frac{1}{h_{n-1}}\left(L_n^{(\alpha)}(x)\frac{d}{dx}L_{n-1}^{(\alpha)}(x) - L_{n-1}^{(\alpha)}(x)\frac{d}{dx}L_n^{(\alpha)}(x)\right).$$

By the recursion formula (7.2) and the differential equation (7.3), we can get

$$x\left(L_n^{(\alpha)}(x)\frac{d}{dx}L_{n-1}^{(\alpha)}(x) - L_{n-1}^{(\alpha)}(x)\frac{d}{dx}L_n^{(\alpha)}(x)\right)$$

$$= -L_n^{(\alpha)}(x)\left(L_{n-1}^{(\alpha)}(x) + (n-1+\alpha)L_{n-2}^{(\alpha)}(x)\right) + (n+\alpha)L_{n-1}^{(\alpha)}(x)L_{n-1}^{(\alpha)}(x),$$

and

$$x^2\left(L_n^{(\alpha)}(x)\frac{d}{dx}L_{n-1}^{(\alpha)}(x) - L_{n-1}^{(\alpha)}(x)\frac{d}{dx}L_n^{(\alpha)}(x)\right)$$

$$= \left[(n+1)L_{n+1}^{(\alpha)}(x) - \left(2n+1+\alpha+n(n+\alpha)\right)L_n^{(\alpha)}(x)\right]L_{n-1}^{(\alpha)}(x)$$

$$+ \left[(n+\alpha)(2n+\alpha)L_{n-1}^{(\alpha)}(x) - (n+\alpha)(n-1+\alpha)L_{n-2}^{(\alpha)}(x)\right]L_{n-1}^{(\alpha)}(x)$$

$$- (n-1+\alpha)xL_n^{(\alpha)}(x)L_{n-2}^{(\alpha)}(x).$$

Then

$$\int_0^\infty x^{1+\alpha} e^{-x} K_n(x,x)dx = n+\alpha,$$ (7.131)

and

$$\int_0^\infty x^{2+\alpha} e^{-x} K_n(x,x)dx = (n+\alpha)(2n+\alpha).$$ (7.132)

For the left hand side of (7.129), we have the asymptotics

$$\sqrt{-\det A_n(x)} = \frac{1}{2} - \left(n+\frac{\alpha}{2}\right)x^{-1} - n(n+\alpha)x^{-2} - n(n+\alpha)(2n+\alpha)x^{-3}$$
$$+ O\left(x^{-4}\right),$$

as $x \to \infty$. Consequently, there are the following results

$$\int_0^\infty \frac{x}{n\pi} \operatorname{Re} \sqrt{\det A_n(x)}dx = n+\alpha,$$ (7.133)

and

$$\int_0^\infty \frac{x^2}{n\pi} \operatorname{Re} \sqrt{\det A_n(x)}dx = (n+\alpha)(2n+\alpha).$$ (7.134)

Therefore we have proved (7.129) for $j = 1$ and 2. It is left to interested readers to verify whether (7.129) is true when $j > 2$.

The integral transforms of the density models have been discussed to give different power-laws connected to the different phases in the density transition models. In some sense, the integral transforms change the problem to be viewed from a different point so that the mathematical properties in the different aspects can be compared and referred, for example, when the uncertainty is involved in the consideration. We have seen that the integrable systems can be applied to solve many transition problems. And the importance of the integrable systems is not limited to these problems. One can associate other theories with the integrable systems to develop further methods and solve more application problems.

References

1. Ali, M., Rau, A.R.P., Alber, G.: Quantum discord for two-qubit X-states. Phys. Rev. A **81**, 042105 (2010), see also Erratum, Phys. Rev. A **82**, 069902(E) (2010)
2. Andriani, P., McKelvey, B.: Beyond Gaussian averages: redirecting organization science toward extreme events and power laws. J. Int. Bus. Stud. **38**, 1212–1230 (2007)
3. Binney, J.: The Theory of Critical Phenomena: an Introduction to the Renormalization Group. Oxford University Press, London (1992)
4. Biroli, G., Bouchaud, J.P., Potters, M.: On the top eigenvalue of heavy-tailed random matrices. Europhys. Lett. **78**, 10001 (2007)

5. Bouchaud, J.-P., Potters, M.: Theory of Financial Risks. Cambridge University Press, Cambridge (2001)
6. Caginalp, G., Porter, D., Smith, V.: Momentum and overreaction in experimental asset markets. Int. J. Ind. Organ. **18**, 187–204 (2000)
7. Dillenschneider, R.: Quantum discord and quantum phase transition in spin chains. Phys. Rev. B **78**, 224413 (2008)
8. Emerson, J., Weinstein, Y.S., Saraceno, M., Lloyd, S., Cory, D.G.: Pseudo-random unitary operators for quantum information processing. Science **302**, 2098 (2003)
9. Frahm, G., Jaekel, U.: Random matrix theory and robust covariance matrix estimation for financial data. arXiv:physics/0503007 (2005)
10. Gu, S.J., Tian, G.S., Lin, H.Q.: Local entanglement and quantum phase transition in spin models. New J. Phys. **8**, 61 (2006)
11. Harré, M., Bossomaier, T.: Phase-transition-like behaviour of information measures in financial markets. Europhys. Lett. **89**, 18009 (2009)
12. Harrow, A.W., Low, R.A.: Random quantum circuits are approximate 2-designs. Commun. Math. Phys. **291**, 257–302 (2009)
13. Hong, H., Stein, J.: A unified theory of underreaction, momentum, and overreaction in asset markets. J. Finance **54**, 2143–2184 (1999)
14. Its, A.R., Jin, B.Q., Korepin, V.E.: Entanglement in XY spin chain. J. Phys. A, Math. Gen. **38**, 2975 (2005)
15. Johansen, A., Sornette, D., Ledoit, O.: Crashes as critical points. Int. J. Theor. Appl. Finance **3**, 219–255 (2000)
16. Kahneman, D., Tversky, A.: Prospect theory: an analysis of decision under risk. Econometrica **XLVII**, 263–291 (1979)
17. Kasprzak, A., Kutner, R., Perelló, J., Masoliver, J.: Higher-order phase transitions on financial markets. Eur. Phys. J. B **76**, 513–527 (2010)
18. Kaulakys, B., Meskauskas, T.: Modeling 1/f noise. Phys. Rev. E **58**, 7013–7019 (1998)
19. Kaulakys, B., Ruseckas, J., Gontis, V., Alaburda, M.: Nonlinear stochastic models of 1/f noise and power-law distributions. Physica A **365**, 217–221 (2006)
20. Kitaev, A., Preskill, J.: Topological entanglement entropy. Phys. Rev. Lett. **96**, 110404 (2006)
21. Majumdar, S.N., Vergassola, M.: Large deviations of the maximum eigenvalue for Wishart and Gaussian random matrices. Phys. Rev. Lett. **102**, 060601 (2009)
22. Mantegna, R.N., Stanley, H.E.: An Introduction to Econophysics: Correlation and Complexity in Finance. Cambridge University Press, Cambridge (2000)
23. Marcenko, V.A., Pastur, L.A.: Distribution of eigenvalues for some sets of random matrices. Mat. Sb. **72(114)**(4), 507–536 (1967)
24. McKay, B.D.: The expected eigenvalue distribution of a large regular graph. Linear Algebra Appl. **40**, 203–216 (1981)
25. McLeod, J.B., Wang, C.B.: Eigenvalue density in Hermitian matrix models by the Lax pair method. J. Phys. A, Math. Theor. **42**, 205205 (2009)
26. Olver, F.W.J.: Asymptotics and Special Functions. Academic Press, New York (1974)
27. Ormerod, P., Mounfield, C.: Random matrix theory and the failure of macroeconomic forecasts. Physica A **280**, 497–504 (2000)
28. Plerou, V., Gopikrishnan, P., Rosenow, B., Amaral, L.A.N., Guhr, T., Stanley, H.E.: Random matrix approach to cross correlations in financial data. Phys. Rev. E **65**, 066126 (2002)
29. Plerou, V., Gopikrishnan, P., Stanley, H.E.: Two-phase behaviour of financial market. Nature **421**, 130 (2003)
30. Sengupta, A.M., Mitra, P.P.: Distributions of singular values for some random matrices. Phys. Rev. E **60**, 3389–3392 (1991)
31. Shefrin, H.: Beyond Greed and Fear: Understanding Behavioral Finance and the Psychology of Investing. Oxford University Press, London (2002)
32. Shefrin, H., Statman, M.: Behavioral portfolio theory. J. Financ. Quant. Anal. **35**, 127–151 (2000)

33. Simon, S.H., Moustakas, A.L.: Eigenvalue density of correlated complex random Wishart matrices. Phys. Rev. E **69**, 065101(R) (2004)
34. Sornette, D.: Critical Phenomena in Natural Sciences: Chaos, Fractals, Self Organization and Disorder: Concepts and Tools, 2nd edn. Springer, Berlin (2004)
35. Szegö, G.: Orthogonal Polynomials, 4th edn. American Mathematical Society Colloquium Publications, vol. 23. AMS, New York (1975)
36. Thaler, R.H.: Towards a positive theory of consumer choice. J. Econ Behav. Organ. **1**, 39–60 (1980)
37. Van Assche, W.: Discrete Painlevé equations for recurrence coefficients of orthogonal polynomials. In: Proceedings of the International Conference on Difference Equations, Special Functions and Orthogonal Polynomials, pp. 687–725. World Scientific, Hackensack (2007)
38. Vidal, G., Latorre, J.I., Rico, E., Kitaev, A.: Entanglement in quantum critical phenomena. Phys. Rev. Lett. **90**, 227902 (2003)
39. Vivo, P., Majumdar, S.N., Bohigas, O.: Large deviations of the maximum eigenvalue in Wishart random matrices. J. Phys. A, Math. Theor. **40**, 4317 (2007)
40. Wishart, J.: The generalized product moment distribution in samples from a normal multivariate population. Biometrika **20A**, 32–52 (1928)
41. Yalamova, R.: Correlations in Financial Time Series During Extreme Events—Spectral Clustering and Partition Decoupling Method. Proceedings of the World Congress on Engineering 2009, vol. II. WCE, London (2009)
42. Yalamova, R., McKelvey, B.: Explaining what leads up to stock market crashes: a phase transition model and scalability dynamics. In: 2nd International Financial Research Forum Paris (2009)

Appendix A
Some Integral Formulas

In Sect. 3.1, it is discussed that the free energy in the Hermitian matrix model can be calculated based on the moments $\int_{\eta_-}^{\eta_+} \eta^k \rho(\eta) d\eta$ and $\int_{\eta_-}^{\eta_+} \ln|\eta - a| \rho(\eta) d\eta$. These integrals can be expressed in terms of

$$R_{l,k} = \frac{i}{\pi} \int_{-\pi}^{\pi} (a + 2b\cos\theta)^k e^{-il\theta} \sin\theta d\theta, \qquad (A.1)$$

and

$$\Theta_l = \text{Re}\, \frac{i}{\pi} \int_0^{\pi} \theta e^{i\theta} \big[\big(e^{i\theta} + \sqrt{e^{2i\theta} - 1}\big)^l - \big(e^{i\theta} - \sqrt{e^{2i\theta} - 1}\big)^l \big] d\theta, \qquad (A.2)$$

where l is integer. The term $(a + 2b\cos\theta)^k$ in (A.1) is from the potential, and the term $e^{-il\theta}$ is from the eigenvalue density. It is seen that the integrals in (A.1) are the Fourier transforms, that reflect an element of correlation function since the correlation functions are usually studied by using the Fourier transforms and are related to the partition function which is reduced to the free energy function now. The integrals in (A.2) are modified transforms indicating that the discussion is not limited to the traditional Fourier transform.

It has been given in Sect. 3.1 that we need to use

$$Y_l = \frac{1}{2} \sum_{k=0}^{4} g_k R_{l,k} + \Theta_l, \qquad (A.3)$$

for $l = 1, 2$ and 3 to calculate the free energy when $m = 2$. If we denote $Y = (Y_1, Y_2, Y_3)^T$, $g = (g_0, g_1, g_2, g_3, g_4)^T$, $\Theta = (\Theta_1, \Theta_2, \Theta_3)^T$, and $R = (R_{l,k})_{3 \times 5}$, the above formula becomes

$$Y = \frac{1}{2} Rg + \Theta. \qquad (A.4)$$

C.B. Wang, *Application of Integrable Systems to Phase Transitions*,
DOI 10.1007/978-3-642-38565-0, © Springer-Verlag Berlin Heidelberg 2013

The $R_{l,k}$'s have the following results,

$$R_{1,0} = 1, \qquad R_{1,1} = a, \qquad R_{1,2} = a^2 + b^2, \qquad R_{1,3} = a(a^2 + 3b^2),$$

$$R_{1,4} = a^4 + 2b^4 + 6a^2b^2, \qquad R_{2,0} = 0, \qquad R_{2,1} = b, \qquad R_{2,2} = 2ab,$$

$$R_{2,3} = b(3a^2 + 2b^2), \qquad R_{2,4} = 4ab(a^2 + 2b^2), \qquad R_{3,0} = 0, \qquad R_{3,1} = 0,$$

$$R_{3,2} = b^2, \qquad R_{3,3} = 3ab^2, \qquad R_{3,4} = 3b^2(2a^2 + b^2).$$

Based on the discussions in Sect. 3.1, we have

$$\Theta_1 = -\frac{2}{\pi} \text{Re} \int_0^{\pi} \theta e^{i\theta} \left(1 - e^{2i\theta}\right)^{\frac{1}{2}} d\theta = \frac{4}{\pi} d_0,$$

$$\Theta_2 = -\frac{4}{\pi} \text{Re} \int_0^{\pi} \theta e^{2i\theta} \left(1 - e^{2i\theta}\right)^{\frac{1}{2}} d\theta = 0,$$

$$\Theta_3 = \frac{8}{\pi} \text{Re} \int_0^{\pi} \theta e^{i\theta} \left(1 - e^{2i\theta}\right)^{\frac{3}{2}} d\theta - \frac{6}{\pi} \text{Re} \int_0^{\pi} \theta e^{i\theta} \left(1 - e^{2i\theta}\right)^{\frac{1}{2}} d\theta$$

$$= \frac{4}{\pi}(3d_0 - 4d_1),$$

where

$$l_k = \int_0^1 \left(1 - x^2\right)^{k+\frac{1}{2}} dx, \tag{A.5}$$

$$d_k = \int_0^1 \int_0^1 \left(1 - x^2 y^2\right)^{k+\frac{1}{2}} dx dy \tag{A.6}$$

for $k = 0, 1, \ldots$. Then, we finally get $\Theta_1 = \frac{1}{2} + \ln 2$, $\Theta_2 = 0$, $\Theta_3 = -\frac{3}{4}$, based on the following values

$$l_0 = \frac{\pi}{4}, \qquad l_1 = \frac{3\pi}{16}, \qquad l_2 = \frac{15\pi}{96}, \tag{A.7}$$

$$d_0 = \frac{\pi}{8} + \frac{\pi}{4} \ln 2, \qquad d_1 = \frac{9\pi}{64} + \frac{3\pi}{16} \ln 2, \qquad d_2 = \frac{55\pi}{384} + \frac{15\pi}{96} \ln 2. \tag{A.8}$$

In addition, we have the following basic formulas,

$$\int_0^{2\pi} \theta \sqrt{1 - e^{i\theta}} \, d\theta = 2\pi^2 + \frac{4\pi}{i} (\ln 2 - 1), \tag{A.9}$$

$$\int_0^{2\pi} \theta \sqrt{1 - e^{2i\theta}} \, d\theta = 2\pi^2 + \frac{2\pi}{i} (\ln 2 - 1), \tag{A.10}$$

$$\int_0^{2\pi} \theta e^{i\theta} \sqrt{1 - e^{2i\theta}} \, d\theta = \frac{\pi^2}{2i}, \tag{A.11}$$

$$\int_0^{2\pi} \theta e^{2i\theta} \sqrt{1 - e^{2i\theta}} \, d\theta = \frac{2\pi}{3i}, \tag{A.12}$$

$$\int_0^{\pi} \theta e^{i\theta} \left(1 - e^{2i\theta}\right)^{-1/2} d\theta = -\pi \ln 2 + \frac{\pi^2}{2} i. \tag{A.13}$$

To show (A.13), we need $d_{-1} = \int_0^1 \int_0^1 (1 - x^2 y^2)^{-1/2} dx dy = \frac{\pi}{2} \ln 2$. These formulas and the similar formulas given in Sect. 3.1 are often needed in the calculations.

Appendix B
Properties of the Elliptic Integrals

B.1 Asymptotics of the Elliptic Integrals

Now, we want to show that for $0 < x_1 < x_2$, there is

$$\int_{x_1}^{x_2} \sqrt{(x_2^2 - x^2)(x^2 - x_1^2)}\, dx = \frac{x_2^3}{3} - \frac{x_1^2 x_2}{2}\left[\ln\left(\frac{4x_2}{x_1}\right) - \frac{1}{2}\right] + O\left(x_1^4 \ln x_1\right), \quad \text{(B.1)}$$

as $x_1 \to 0$.

The expansion for the elliptic integral is achieved by using the contour integration such that the different orders of the small terms can be separated. For this problem, we consider the function $f(z) = \ln z \sqrt{(x_2^2 - z^2)(z^2 - x_1^2)}$, where the branch cut for $\ln z$ is the upper edge of the negative real axis. The function $f(z)$ is analytic in the domain enclosed by the contour γ, where

$$\gamma = [x_1, x_2]^+ \cup \left\{x_2 e^{i\theta} \,|\, 0 \le \theta \le \pi\right\} \cup [-x_2, -x_1]^+ \cup \left\{x_1 e^{i\theta} \,|\, \pi \ge \theta \ge 0\right\},$$

that implies $\int_\gamma f(z)dz = 0$, or

$$\int_{-x_2}^{-x_1} f(z)dz + x_1 \int_\pi^0 f\left(x_1 e^{i\theta}\right) i e^{i\theta}\, d\theta + \int_{x_1}^{x_2} f(z)dz + x_2 \int_0^\pi f\left(x_2 e^{i\theta}\right) i e^{i\theta}\, d\theta$$
$$= 0.$$

The contour integrals can separate the different orders of the small terms so that the analysis become easier. By taking the imaginary parts on both sides and noting that $\sqrt{(x_2^2 - z^2)(z^2 - x_1^2)}$ is negative on $[-x_2, -x_1]$ as discussed in Sect. 3.3, we obtain

$$\int_{x_1}^{x_2} \sqrt{(x_2^2 - x^2)(x^2 - x_1^2)}\, dx = \frac{x_2^2}{\pi} \operatorname{Re} I_2 - \frac{x_1^2}{\pi} \operatorname{Re} I_1, \quad \text{(B.2)}$$

C.B. Wang, *Application of Integrable Systems to Phase Transitions*,
DOI 10.1007/978-3-642-38565-0, © Springer-Verlag Berlin Heidelberg 2013

where

$$I_1 = \int_0^\pi (\ln x_1 + i\theta)\sqrt{(x_2^2 - x_1^2 e^{2i\theta})(e^{2i\theta} - 1)}\, e^{i\theta}\, d\theta,$$

$$I_2 = \int_0^\pi (\ln x_2 + i\theta)\sqrt{(1 - e^{2i\theta})(x_2^2 e^{2i\theta} - x_1^2)}\, e^{i\theta}\, d\theta.$$

By using the integral formulas given in Appendix A, we can get

$$\frac{1}{\pi}\,\mathrm{Re}\,I_1 = -\frac{x_2}{2}\left[\ln(x_1/2) - \frac{1}{2}\right] + O(x_1^2 \ln x_1),$$

$$\frac{1}{\pi}\,\mathrm{Re}\,I_2 = \frac{x_2}{2} - \frac{x_1^2}{2x_2}(\ln(2x_2) - 1) + O(x_1^4),$$

and then the asymptotic expansion above can be obtained.

The asymptotics above is applied to study the third-order transition in Sect. 3.4 for the Hermitian matrix model. The following asymptotics is applied to study the discontinuity in the Laguerre model for the generalized Marcenko-Pastur distribution discussed in Sect. 7.2.

For $0 < x_1 < x_2 < u$, there is

$$\int_{x_1}^{x_2} \frac{x}{x+u}\sqrt{(x_2^2 - x^2)(x^2 - x_1^2)}\, dx = \frac{u}{4}(2u^2 - x_2^2)\pi + \frac{x_2^3}{3} - x_2 u^2$$

$$+ iu^2\sqrt{u^2 - x_2^2}\ln\left(\frac{u + x_2 - i\sqrt{u^2 - x_2^2}}{u + x_2 + i\sqrt{u^2 - x_2^2}} \cdot \frac{x_2 + i\sqrt{u^2 - x_2^2}}{x_2 - i\sqrt{u^2 - x_2^2}}\right) + O(x_1^2), \quad \text{(B.3)}$$

as $x_1 \to 0$. To show this asymptotics, consider $f(z) = \frac{z}{z+u}\ln z\sqrt{(x_2^2 - z^2)(z^2 - x_1^2)}$, where the branch cut for $\ln z$ is the lower edge of the positive real axis. Then there is $\int_{\gamma_1 \cup \gamma_2} f(z)dz = 0$, where

$$\gamma_1 = [x_1, x_2]^+ \cup \{x_2 e^{i\theta}\,|\,0 \le \theta \le \pi\} \cup [-x_2, -x_1]^+ \cup \{x_1 e^{i\theta}\,|\,\pi \ge \theta \ge 0\},$$

and

$$\gamma_2 = (-[-x_2, -x_1]^-) \cup \{x_2 e^{i\theta}\,|\,\pi \le \theta \le 2\pi\} \cup (-[x_1, x_2]^-) \cup \{x_1 e^{i\theta}\,|\,2\pi \ge \theta \ge \pi\}.$$

By taking the imaginary parts on both sides of $\int_{\gamma_1 \cup \gamma_2} f(z)dz = 0$, we can get

$$\int_{x_1}^{x_2} \frac{x}{x+u}\sqrt{(x_2^2 - x^2)(x^2 - x_1^2)}\, dx = \frac{x_2^3}{2\pi}\,\mathrm{Re}\,\hat{I}_2 - \frac{x_1^3}{2\pi}\,\mathrm{Re}\,\hat{I}_1, \quad \text{(B.4)}$$

where

$$\hat{I}_1 = \int_0^{2\pi} \frac{\ln x_1 + i\theta}{u + x_1 e^{i\theta}}\sqrt{(x_2^2 - x_1^2 e^{2i\theta})(e^{2i\theta} - 1)}\, e^{2i\theta}\, d\theta,$$

$$\hat{I}_2 = \int_0^\pi \frac{\ln x_2 + i\theta}{u + x_2 e^{i\theta}} \sqrt{(1 - e^{2i\theta})(x_2^2 e^{2i\theta} - x_1^2)} e^{2i\theta} d\theta.$$

Based on the integral formulas given in Appendix A and

$$\int_0^{2\pi} \theta \frac{\sqrt{1 - e^{2i\theta}}}{1 + r^{-1} e^{i\theta}} d\theta$$

$$= \frac{2\pi}{i}\left(\ln 2 - \frac{\pi}{2} r - i\sqrt{r^2 - 1} \ln \frac{r + 1 - i\sqrt{r^2 - 1}}{r + 1 + i\sqrt{r^2 - 1}} \cdot \frac{1 + i\sqrt{r^2 - 1}}{1 - i\sqrt{r^2 - 1}}\right), \quad \text{(B.5)}$$

where $r = u/x_2$ and $\text{Re}\,\hat{I}_1 = O(\ln x_1)$ as $x_1 \to 0$, we can get (B.3).

B.2   Elliptic Integrals Associated with Legendre's Relation

Consider

$$\rho(x) = \frac{1}{2\pi}(3g_3 + 4g_4(x + 3u))\sqrt{(x_2^2 - x^2)(x^2 - x_1^2)}, \quad \text{(B.6)}$$

for $x \in \Omega_\rho = [-x_2, -x_1] \cup [x_1, x_2]$, and

$$\hat{\omega}(x) = \frac{1}{2}(3g_3 + 4g_4(x + 3u))\sqrt{(x_1^2 - x^2)(x_2^2 - x^2)}, \quad \text{(B.7)}$$

for $x \in \Omega_{\hat{\omega}} = [-x_1, x_1]$, where $0 \le x_1 < x_2$ defined by

$$x_1^2 = \frac{1}{4}(a_1 - a_2)^2 + (b_1 - b_2)^2, \qquad x_2^2 = \frac{1}{4}(a_1 - a_2)^2 + (b_1 + b_2)^2, \quad \text{(B.8)}$$

$u = (a_1 + a_2)/2$, and the parameters satisfy the conditions

$$3g_3 + 4g_4(x_1 + 3u) \ge 0, \qquad 3g_3 + 4g_4(-x_1 + 3u) \le 0, \quad \text{(B.9)}$$

where $4g_4 b_1^2 b_2^2 = 1$ as discussed in Chap. 3, such that $\rho(x)$ is non-negative on Ω_ρ.
  We want to discuss the minimum value of the quantity

$$l_0 = c_1 \int_{-x_1}^{x_1} \hat{\omega}(x)dx + c_2 \int_{x_1}^{x_2} \rho(x)dx, \quad \text{(B.10)}$$

for the given constants c_1 and c_2 when the parameters change in the allowed region(s). The above formula can be changed to

$$l_0 - \frac{c_2}{2} = \frac{3}{2}(g_3 + 4g_4 u)\left(c_1 I_1 + \frac{c_2}{\pi} I_2\right), \quad \text{(B.11)}$$

where

$$I_1 = \int_{-x_1}^{x_1} \sqrt{(x_1^2 - x^2)(x_2^2 - x^2)}\,dx \geq 0,$$

$$I_2 = \int_{x_1}^{x_2} \sqrt{(x_2^2 - x^2)(x^2 - x_1^2)}\,dx \geq 0,$$

(B.12)

since $[-x_1, x_1]$ is between the two cuts $[-x_2, -x_1]$ and $[x_1, x_2]$. Here, we have used

$$\frac{2g_4}{\pi} \int_{x_1}^{x_2} x\sqrt{(x_2^2 - x^2)(x^2 - x_1^2)}\,dx = \frac{g_4}{8}(x_2^2 - x_1^2)^2 = \frac{1}{2}. \qquad (B.13)$$

Both I_1 and I_2 are functions of X and Y, where

$$X = x_1^2 + x_2^2, \qquad Y = x_1^2 x_2^2. \qquad (B.14)$$

We consider l_0 in the domain

$$D = \left\{ (X, Y) \,|\, X > 0, Y > 0, X^2 - 4Y > 0 \right\}. \qquad (B.15)$$

If the parameters satisfy

$$0 \leq g_3 + 4g_4 u \leq \frac{4}{3} g_4 x_1, \qquad (B.16)$$

then the conditions given by (B.9) are satisfied. Further, if $c_1 > 0$ and $c_2 > 0$, then $l_0 - c_2/2$ is non-negative. The minimum value of $l_0(X, Y)$ in $\bar{D}$, the closure of D, is $c_2/2$, where the minimum is reached when $g_3 + 4g_4 u = 0$ which includes different cases. The critical point $a_1 = a_2$ and $b_1 = b_2$ ($Y = x_1 = 0$) is a special case in the minimum state. In the following, let us discuss the behaviors of the function $l_0(X, Y)$ around the critical point.

First, we discuss the derivatives of $l_0(X, Y)$ in the X and Y directions. In the following, we use the second order derivatives to analyze the monotonicity and find the minimum value of the linear combination of the elliptic integrals. Let us consider

$$I = \int \sqrt{x^4 - Xx^2 + Y}\,dx, \qquad J = \int x^2\sqrt{x^4 - Xx^2 + Y}\,dx, \qquad (B.17)$$

in order to analyze the elliptic integral I_1 or I_2. By integration by parts as discussed in Appendix D, we can get

$$(2X^2 - 4Y)I_X + 2XY I_Y = XI + 5J, \qquad (B.18)$$

$$2X I_X + 4Y I_Y = 3I, \qquad (B.19)$$

which imply that these integrals satisfy the hypergeometric (or this type) differential equations

$$(X^2 - 4Y)I_{XX} = \frac{3}{4}I, \qquad (B.20)$$

$$Y(X^2 - 4Y)I_{YY} = \frac{3}{4}I. \qquad (B.21)$$

Based on these properties, we see that the I_1 and I_2 as positive functions of X and Y in the domain D are convex in each of the X and Y directions. It is easy to see that I_1 is decreasing in X direction and increasing in Y direction, and I_2 is decreasing in Y direction and increasing in X direction by calculating their firs-order derivatives. Then, the monotonicity of a linear combination of I_1 and I_2 is a confusing problem. We want to see whether the function

$$l_1 = l_1(X, Y) = c_1 I_1 + \frac{c_2}{\pi} I_2 \tag{B.22}$$

has extreme point in the domain D.

It can be calculated that

$$\begin{pmatrix} \partial l_1/\partial X \\ \partial l_1/\partial Y \end{pmatrix} = \frac{1}{2} \begin{pmatrix} \xi_2 Y^{1/4} & 0 \\ 0 & \zeta_2^{-1} X^{-1/2} \end{pmatrix} \begin{pmatrix} 2(E(k) - K(k)) & \pi^{-1} E(k') \\ 2K(k) & -\pi^{-1} K(k') \end{pmatrix} \begin{pmatrix} c_1 \\ c_2 \end{pmatrix}, \tag{B.23}$$

where the matrix on the left side of $\begin{pmatrix} c_1 \\ c_2 \end{pmatrix}$ is denoted as B,

$$K(k) = \int_0^1 \frac{1}{\sqrt{(1 - t^2)(1 - k^2 t^2)}} dt, \qquad E(k) = \int_0^1 \sqrt{\frac{1 - k^2 t^2}{1 - t^2}} dt \tag{B.24}$$

are the complete elliptic integrals of the first and second kinds,

$$k = \frac{\xi_1}{\xi_2} = \frac{\zeta_1}{\zeta_2}, \qquad k' = \sqrt{1 - k^2}, \tag{B.25}$$

and

$$\xi_{1,2} = \left(\frac{\tau \mp \sqrt{\tau^2 - 4}}{2} \right)^{1/2}, \qquad \zeta_{1,2} = \left(\frac{1 \mp \sqrt{1 - 4\tau^{-2}}}{2} \right)^{1/2}, \tag{B.26}$$

with $\tau = X/\sqrt{Y} > 2$. The coefficient matrix B above has non vanishing determinant

$$\det B = -\frac{\xi_2}{2\pi \zeta_2 \sqrt{\tau}} \left(K(k) E(k') + E(k) K(k') - K(k) K(k') \right) = -\frac{1}{4}, \tag{B.27}$$

by using the Legendre's relation, where $\xi_2/\zeta_2 = \sqrt{\tau}$. That implies $\partial l_1/\partial X$ and $\partial l_1/\partial Y$ can not vanish simultaneously in the domain D, and then l_1 does not have extreme point in D. Since l_1 is positive in D and $l_1(0, 0) = 0$, we see that the minimum of l_1 in $\bar{D}$ is 0.

Also, the integral

$$y(\tau) = \int_0^{\xi_1} \sqrt{\xi^4 - \tau \xi^2 + 1} d\xi, \quad \text{or} \quad y(\tau) = \int_{\xi_1}^{\xi_2} \sqrt{\xi^4 - \tau \xi^2 + 1} d\xi \tag{B.28}$$

satisfies

$$2\tau y' - 4y_1' = -y, \tag{B.29}$$

$$(4 + 2\tau^2)\tau y' - 2\tau y_1' = -5y_1, \tag{B.30}$$

where $y_1 = \int \xi^2 \sqrt{\xi^4 - \tau\xi^2 + 1}\,d\xi$ and $' = d/d\tau$. Then the function $y(\tau)$ satisfies a second order differential equation

$$(4 + \tau^2)y'' + 4\tau y' + \frac{3}{4}y = 0. \tag{B.31}$$

By changing the variable, it can be found that this is a hypergeometric differential equation. The asymptotics of the hypergeometric function shows the behaviors of y.

In addition, if one is interested in the monotonicity of the integrals

$$f_{\hat{\omega}}(r) = \int_{-x_1}^{x_1} \hat{\omega}(x)\,dx, \qquad f_{\rho}(r) = \int_{x_1}^{x_2} \rho(x)\,dx, \tag{B.32}$$

in the direction

$$r = |a_1 - a_2|, \tag{B.33}$$

there are the following properties. First, if $g_3 + 4g_4 u = 0$, then $f_{\hat{\omega}}(r) = 0$ and $f_{\rho}(r) = 1/2$. In the following we discuss the case when $g_3 + 4g_4 u > 0$ such that $f_{\hat{\omega}}$ is not negative.

Since $x_1^2 = \frac{1}{4}r^2 + (b_1 - b_2)^2$ and $x_2^2 = \frac{1}{4}r^2 + (b_1 + b_2)^2$, there is

$$\frac{d}{dr}\sqrt{(x_1^2 - x^2)(x_2^2 - x^2)} = \frac{r}{4}(x_1^2 + x_2^2 - 2x^2)[(x_1^2 - x^2)(x_2^2 - x^2)]^{-1/2} > 0, \tag{B.34}$$

for $x \in \Omega_{\hat{\omega}}$. Based on that, it can be seen

$$f_{\hat{\omega}}'(r) = \frac{r}{8}\int_{-x_1}^{x_1} (3g_3 + 4g_4(x + 3u))(x_1^2 + x_2^2 - 2x^2)$$

$$\times [(x_1^2 - x^2)(x_2^2 - x^2)]^{-1/2}dx > 0, \tag{B.35}$$

for $r > 0$, where $' = d/dr$. Therefore, $\min_{r \geq 0} f_{\hat{\omega}}(r) = f_{\hat{\omega}}(0)$.

For the f_{ρ} with $x \in \Omega_{\rho}$, the corresponding derivative like (B.34) does not have straight sign since $x_1 \leq x \leq x_2$. To analyze the sign of f_{ρ}', let us choose a point x_* between x_1 and x_2 defined by

$$x_*^2 = \frac{1}{2}(x_1^2 + x_2^2). \tag{B.36}$$

Then it is not hard to get

$$f_1 \equiv \int_{x_1}^{x_2} \sqrt{\frac{x^2 - x_1^2}{x_2^2 - x^2}}\,dx = \int_{x_*}^{x_2} \sqrt{\frac{x_2^2 - t^2}{t^2 - x_1^2}} \frac{t\,dt}{\sqrt{x_1^2 + x_2^2 - t^2}} + \int_{x_*}^{x_2} \sqrt{\frac{x^2 - x_1^2}{x_2^2 - x^2}}\,dx$$

and

$$f_2 \equiv \int_{x_1}^{x_2} \sqrt{\frac{x_2^2 - x^2}{x^2 - x_1^2}} dx = \int_{x_*}^{x_2} \sqrt{\frac{t^2 - x_1^2}{x_2^2 - t^2}} \frac{t \, dt}{\sqrt{x_1^2 + x_2^2 - t^2}} + \int_{x_*}^{x_2} \sqrt{\frac{x_2^2 - x^2}{x^2 - x_1^2}} dx,$$

by using the transformation $x^2 - x_1^2 = x_2^2 - t^2$ for $x_1 \leq x \leq x_*$. Then we have

$$f_1 - f_2 = \int_{x_*}^{x_2} \left(\sqrt{\frac{x_2^2 - x^2}{x^2 - x_1^2}} - \sqrt{\frac{x^2 - x_1^2}{x_2^2 - x^2}} \right) \left(\frac{x}{\sqrt{x_1^2 + x_2^2 - x^2}} - 1 \right) dx < 0.$$

Based on this property, we can get

$$f_\rho'(r) = \frac{r}{8} \int_{x_1}^{x_2} \left(3g_3 + 4g_4(x + 3u) \right) \left(2x^2 - x_1^2 - x_2^2 \right) \left[\left(x_2^2 - x^2 \right) \left(x^2 - x_1^2 \right) \right]^{-1/2} dx < 0,$$

$$(B.37)$$

for $r < 0$. Hence we have that $\max_{r \geq 0} f_\rho(r) = f_\rho(0)$.

For the density models in the first-order transition model discussed in Sect. 4.1.2, there are similar results for the problem like (B.10). The elliptic integrals in the large-N transition models have different properties comparing with the elliptic integrals in the bifurcation transition models. In Appendix D, we will discuss that there is a constant Wronskian for the elliptic integrals associated with the large-N transition models. For the bifurcation transition model, we have discussed above that there is a Legendre's relation for the elliptic integrals.

Appendix C
Lax Pairs Based on the Potentials

The Lax Pairs for the string equation and Toda lattice in the Hermitian matrix model have been discussed in Sects. 2.2 and 4.3.1, and the Lax pairs for the string equation and Toda lattice in the unitary matrix model have been discussed in Sect. 5.3. In this appendix, Lax pairs for continuum Painlevé II, III, IV and V equations will be discussed for completeness. The consistency condition for the Lax pair is the continuum Painlevé equation or its equivalent version. The results are obtained based on the corresponding orthogonal polynomials, technically differing from Lax pairs discussed in other literatures since we start from the potential of the model.

C.1 Cubic Potential

For the potential $V(z) = tz + \frac{2}{3}z^3$ considered in Sect. 4.1, there are two continuum variables z and t. We are going to discuss the differential equations in these directions. The coefficients u_n and v_n in the recursion formula of the orthogonal polynomials satisfy the evolution equations

$$u'_n = v_n - v_{n+1}, \tag{C.1}$$

$$v'_n = (u_{n-1} - u_n)v_n, \tag{C.2}$$

where $' = d/dt$, that have been discussed in Sect. 3.2. Denote $\Phi_n = e^{-\frac{1}{2}V(z)} \times (p_n, p_{n-1})^T$. By using the recursion formula, there is

$$p_{n,t} = v_n p_{n-1}, \tag{C.3}$$

where, t means the derivative with respect to t, which can be changed to

$$\Phi_{n,t} = M_n \Phi_n, \tag{C.4}$$

where

$$M_n = \begin{pmatrix} -\frac{1}{2}z & v_n \\ -1 & \frac{1}{2}z - u_{n-1} \end{pmatrix}. \tag{C.5}$$

C.B. Wang, *Application of Integrable Systems to Phase Transitions*,
DOI 10.1007/978-3-642-38565-0, © Springer-Verlag Berlin Heidelberg 2013

We have obtained that in the z direction, there is the following equation,

$$\Phi_{n,z} = A_n \Phi_n, \tag{C.6}$$

where

$$A_n = \begin{pmatrix} -z^2 - \frac{t}{2} - 2v_n & 2v_n(z + u_n) \\ -2(z + u_{n-1}) & z^2 + \frac{t}{2} + 2v_n \end{pmatrix}. \tag{C.7}$$

By the associated string equations,

$$2(u_n + u_{n-1})v_n = n, \tag{C.8}$$

$$t + 2\big(u_n^2 + v_n + v_{n+1}\big) = 0, \tag{C.9}$$

and the evolution equations given above, the u_n's and v_n's can be written in terms of one variable. In terms of u_n, the consistency condition for the Lax pair (C.4) and (C.6) is the following continuum Painlevé II equation,

$$u_n'' = tu_n + 2u_n^3 + n + \frac{1}{2}. \tag{C.10}$$

In terms of v_n, the consistency condition for the Lax pair above is the following equation,

$$v_n'' = \frac{(v_n')^2}{2v_n} - 4v_n^2 - tv_n - \frac{n^2}{8v_n}, \tag{C.11}$$

which is a different version of the continuum Painlevé II equation. By using the double scaling given in Sect. 4.1.1, this equation can be reduced to

$$v_{\xi\xi} = \frac{v_\xi^2}{2v} - 4v^2 + gv - \frac{1}{8v}, \tag{C.12}$$

which is close to (C.11) for $n = 1$, but in (C.12) g and ξ are independent. It can be calculated that

$$v_{\xi\xi} - \frac{v_\xi^2}{2v} + 4v^2 - gv + \frac{1}{8v} = \frac{v}{2v_\xi} \frac{d}{d\xi}\left(\frac{v_\xi^2}{v} + \left(2v - \frac{g}{2}\right)^2 - \frac{1}{4v}\right). \tag{C.13}$$

This relation indicates that the second-order equation can be reduced to the elliptic differential equation satisfied by the Weierstrass elliptic $\wp$-function discussed in Sect. 4.1.2. It is obtained in Sect. 4.1.1 that

$$-\det A_n(z) = n^{4/3}\big(\eta^4 - g\eta^2 - 2\eta - X + O\big(n^{-1}\big)\big), \tag{C.14}$$

as n is large, where $z = n^{1/3}\eta$, g is $-t/n^{2/3}$, and X is given in Sect. 4.1,

$$X = \frac{1}{4v} - \left(2v - \frac{g}{2}\right)^2 - \frac{v_\xi^2}{v}. \tag{C.15}$$

We will discuss the differential equations satisfied by the integrals in this model in Sect. D.1. The right hand side of (C.13) is corresponding to the $X = 0$ case discussed in Sects. 4.1.2 and D.1.

C.2 Quartic Potential

For the potential $V(z) = tz^2 + \frac{1}{4}z^4$ discussed by Fokas, Its and Kitaev in 1991, by the orthogonality of the polynomials $p_n(z)$ discussed in Sect. 4.2, there is

$$p_{n,t} = -v_n p_n + v_n z p_{n-1}, \tag{C.16}$$

and the v_n's satisfy the following relations,

$$(2t + v_n + v_{n-1} + v_{n+1})v_n = n, \tag{C.17}$$

$$v_n' = v_n(v_{n-1} - v_{n+1}), \tag{C.18}$$

where $' = d/dt$. Denote $\Phi_n = e^{-\frac{1}{2}V(z)}(p_n, p_{n-1})^T$. We then have the following equation,

$$\Phi_{n,t} = M_n \Phi_n, \tag{C.19}$$

where

$$M_n = \begin{pmatrix} -\frac{1}{2}z^2 - v_n & v_n z \\ -z & \frac{1}{2}z^2 - v_{n-1} \end{pmatrix}. \tag{C.20}$$

Also, we have obtained in Sect. 4.3.1 that in the z direction there is

$$\Phi_{n,z} = A_n \Phi_n, \tag{C.21}$$

where

$$A_n = \begin{pmatrix} -\frac{1}{2}z^3 - tz - v_n z & v_n(z^2 + 2t + v_n + v_{n+1}) \\ -z^2 - 2t - v_n - v_{n-1} & \frac{1}{2}z^3 + tz + v_n z \end{pmatrix}. \tag{C.22}$$

The consistency condition for the Lax pair (C.19) and (C.21) is the following continuum Painlevé IV equation,

$$v_n'' = \frac{(v_n')^2}{2v_n} + \frac{3}{2}v_n^3 + 4tv_n^2 + (2t^2 + n)v_n - \frac{n^2}{2v_n}. \tag{C.23}$$

The v_{n-1} and v_{n+1} in A_n and M_n can be written in terms of v_n according to the relations given above. Interested readers can refer the works of Peter A. Clarkson for more discussions on the Painlevé IV equation.

C.3 Potential in the Unitary Model

Now, let us discuss the differential equation in the s direction for the potential
$V(z) = s(z + z^{-1})$ explained in Sect. 6.1. Based on the discussion in Sect. 6.1 on
the unit circle $|z| = 1$, the equation in the s direction can be written in the following
matrix form,

$$
\begin{pmatrix} p_{n,s} \\ \tilde{p}_{n,s} \end{pmatrix} = \begin{pmatrix} -z^{-1} - x_n x_{n+1} & x_{n+1} + x_n/z \\ x_{n+1} + zx_n & -z - x_n x_{n+1} \end{pmatrix} \begin{pmatrix} p_n \\ \tilde{p}_n \end{pmatrix},
\tag{C.24}
$$

where $, s$ means the derivative with respect to s and $zp_n = p_{n+1}(z) - x_{n+1}\tilde{p}_n(z)$,
where $\tilde{p}_n(z) = z^n \overline{p_n(z)}$ (Sect. 5.2, with $\gamma_0 = -1/2$). Let

$$
\Psi_n = e^{\frac{s}{2}(z+z^{-1})} \begin{pmatrix} z^{-n/2+1/4} & 0 \\ 0 & z^{-n/2-1/4} \end{pmatrix} \begin{pmatrix} p_n \\ \tilde{p}_n \end{pmatrix} = e^{\frac{s}{2}(z+z^{-1})} \begin{pmatrix} \chi_n \\ \hat{\chi}_n \end{pmatrix},
\tag{C.25}
$$

for the $(\chi_n, \hat{\chi}_n)^T$ in the discrete AKNS-ZS system discussed in Sect. 5.2. then we
have

$$
\Psi_{n,s} = M_n \Psi_n,
\tag{C.26}
$$

where

$$
M_n = \begin{pmatrix} \frac{1}{2}(z - z^{-1}) - x_n x_{n+1} & x_{n+1}z^{1/2} + x_n z^{-1/2} \\ x_n z^{1/2} + x_{n+1}z^{-1/2} & -\frac{1}{2}(z - z^{-1}) - x_n x_{n+1} \end{pmatrix}.
\tag{C.27}
$$

In the z direction, we have obtained in Sect. 5.3.1 that

$$
\Phi_{n,z} = A_n(z)\Phi_n,
\tag{C.28}
$$

where $\Phi_n(z) = e^{\frac{s}{2}(z+z^{-1})}(z^{-n/2} p_n(z), z^{n/2}\overline{p_n(z)})^T$ and

$$
A_n = \begin{pmatrix} \frac{s}{2} + \frac{s}{2z^2} + \frac{n-2sx_nx_{n+1}}{2z} & s(x_{n+1} - \frac{x_n}{z})z^{-1} \\ s(x_n - \frac{x_{n+1}}{z}) & -\frac{s}{2} - \frac{s}{2z^2} - \frac{n-2sx_nx_{n+1}}{2z} \end{pmatrix}.
\tag{C.29}
$$

Since $\tilde{p}_n(z) = z^n \overline{p_n(z)}$, it implies Ψ_n and Φ_n have the following relation

$$
\Psi_n = e^{\frac{s}{2}(z+z^{-1})} \begin{pmatrix} z^{-n/2+1/4} & 0 \\ 0 & z^{-n/2-1/4} \end{pmatrix} \begin{pmatrix} 1 & 0 \\ 0 & z^n \end{pmatrix} \begin{pmatrix} p_n \\ \tilde{p}_n \end{pmatrix} = \begin{pmatrix} z^{1/4} & 0 \\ 0 & z^{-1/4} \end{pmatrix} \Phi_n.
\tag{C.30}
$$

Direct calculations show that the consistency condition $\Phi_{n,zs} = \Phi_{n,sz}$ holds if the
following two equations hold,

$$
x'_n - 2x_{n+1} - x_n\left(\frac{n}{s} - 2x_n x_{n+1}\right) = 0,
\tag{C.31}
$$

$$
x'_{n+1} + 2x_n + x_{n+1}\left(\frac{n+1}{s} - 2x_n x_{n+1}\right) = 0,
\tag{C.32}
$$

where $' = d/ds$. These two equations are equivalent to the Toda lattice combined with the string equation. We are going to discuss in the following that these two equations are a different version of the continuum Painlevé III or V equation.

Let us first consider the continuum Painlevé III equation,

$$Y' = -4Y^2 Z + 2Y^2 + \frac{2\theta_\infty - 1}{s} Y - 2, \tag{C.33}$$

$$Z' = 4Y Z^2 - 4Y Z - \frac{2\theta_\infty}{s} Z + \frac{\theta_0 + \theta_\infty}{s}, \tag{C.34}$$

$$(\log W)' = -\frac{\theta_0 + \theta_\infty}{2s Z} + \frac{\theta_0 - \theta_\infty}{2s(1 - Z)} + \frac{\theta_\infty}{s}, \tag{C.35}$$

where $Y = Y(s)$, $Z = Z(s)$, $' = d/ds$, and θ_0, θ_∞ are constants. The variables y, z, w and t in the paper of Jimbo and Miwa (1981) are changed to $-Y, sZ$, $W(1 - Z)^{1/2} Z^{-1/2}$ and s respectively here. It can be verified that Y satisfies the standard form of the continuum Painlevé III equation

$$Y'' = \frac{1}{Y} Y'^2 - \frac{1}{s} Y' + \frac{1}{s} \left(\alpha_{III} Y^2 + \beta_{III} \right) + \gamma_{III} Y^3 + \frac{\delta_{III}}{Y}, \tag{C.36}$$

where

$$\alpha_{III} = -4\theta_0, \qquad \beta_{III} = -4(1 - \theta_\infty), \qquad \gamma_{III} = 4, \qquad \delta_{III} = -4,$$

and Z satisfies

$$Z'' = \frac{1}{2} \left(\frac{1}{Z} + \frac{1}{Z - 1} \right) Z'^2 - \frac{1}{s} Z' - \frac{1}{2s^2} \frac{2Z - 1}{Z(Z - 1)} (2\theta_\infty Z - \theta_0 - \theta_\infty)^2$$

$$+ \frac{2\theta_\infty}{s^2} (2\theta Z_\infty - \theta_0 - \theta_\infty) + 8Z(1 - Z). \tag{C.37}$$

Equation (C.37) can be changed to the continuum Painlevé V equation by the transformation $Z_1 = Z/(Z - 1)$ with $t = s^2$,

$$Z_1'' = \left(\frac{1}{2Z_1} + \frac{1}{Z_1 - 1} \right) Z_1'^2 - \frac{1}{t} Z_1' + \frac{(Z_1 - 1)^2}{t^2} \left(\alpha_V Z_1 + \frac{\beta_V}{Z_1} \right) + \frac{\gamma_V Z_1}{t}$$

$$+ \frac{\delta_V Z_1 (Z_1 + 1)}{Z_1 - 1}, \tag{C.38}$$

where

$$\alpha_V = \frac{(\theta_0 - \theta_\infty)^2}{8}, \qquad \beta_V = -\frac{(\theta_0 + \theta_\infty)^2}{8}, \qquad \gamma_V = 2, \qquad \delta_V = 0,$$

and $Z_1' = dZ_1/dt$.

The string equation and the Toda lattice given in the following

$$\frac{n}{s}x_n = -\left(1 - x_n^2\right)(x_{n+1} + x_{n-1}),\tag{C.39}$$

$$x_n' = \left(1 - x_n^2\right)(x_{n+1} - x_{n-1}),\tag{C.40}$$

involve the functions x_{n-1}, x_n and x_{n+1}. We want to eliminate x_{n-1} and x_{n+1} to get an equation for x_n with fixed n. Let

$$y_n = x_{n-1}x_n.\tag{C.41}$$

Then

$$y_n = -\frac{(x_n^2)' + 2nx_n^2/s}{4(1 - x_n^2)},\qquad y_{n+1} = \frac{(x_n^2)' - 2nx_n^2/s}{4(1 - x_n^2)}.\tag{C.42}$$

Applying the Toda lattice to $y_n' = x_{n-1}x_n' + x_nx_{n-1}'$, we have

$$y_n' = \frac{1 - x_n^2}{x_n^2}y_ny_{n+1} + x_n^2 - \frac{y_n^2}{x_n^2} - \left(1 - x_{n-1}^2\right)x_{n-2}x_n.\tag{C.43}$$

By using the string equation on the last term above, the equation becomes

$$y_n' = \frac{1 - x_n^2}{x_n^2}y_ny_{n+1} + 2x_n^2 - \frac{1 + x_n^2}{x_n^2}y_n^2 + \frac{n - 1}{s}y_n.\tag{C.44}$$

Substituting the formulas for y_n and y_{n+1} given above into (C.44), we get

$$Z_2'' = \frac{1}{2}\left(\frac{1}{Z_2} + \frac{1}{Z_2 - 1}\right)Z_2'^2 - \frac{1}{s}Z_2' + \frac{2n^2}{s^2}\frac{Z_2}{1 - Z_2} - 8Z_2(1 - Z_2),\tag{C.45}$$

where $Z_2 = x_n^2$. This is the equation (C.37) for $Z = 1 - Z_2$ and $\theta_0 = \theta_\infty = n$.

It can be shown by using the string equation, Toda lattice and $h_n' = 2h_n(u_n - v_n) = -2h_nx_nx_{n+1}$ (McLeod and Wang 2004) that

$$Y = -\frac{x_{n-1}}{x_n} = \frac{1}{u_{n-1}},\tag{C.46}$$

$$Z = 1 - x_n^2 = \frac{v_n}{u_n},\tag{C.47}$$

$$W = \frac{1}{(h_nh_{n-1})^{1/2}} = \kappa_n\kappa_{n-1}\tag{C.48}$$

satisfy (C.33), (C.34) and (C.35) with $\theta_0 = \theta_\infty = n$, where u_n and v_n are the coefficients in the recursion formula $z(p_n + v_np_{n-1}) = p_{n+1} + u_np_n$ of the orthogonal polynomials, and $\kappa_n = 1/\sqrt{h_n}$. Thus $1/u_{n-1}$ solves the continuum Painlevé III equation (C.36), and $v_n/(v_n - u_n)$ solves the continuum Painlevé V equation (C.38).

The above discussions indicate that the continuum Painlevé III and V equations are the results of the Toda lattice and the string equation. The Toda lattice and string equation can help us to analyze the singularity of the derivatives of the free energy functions, that implies the continuum Painlevé equations are related to the phase transition problems. It is discussed in Sect. 6.1 that

$$-\frac{z^2}{s^2} \det A_n = \frac{1}{4}\left(z + z^{-1} + T\right)^2 - \frac{X}{4}, \tag{C.49}$$

as n is large, where $T = n/s$ and X is given in Sect. 6.1. We will discuss the differential equations satisfied by the integrals in this model in Sect. D.2.

Appendix D
Hypergeometric-Type Differential Equations

In this appendix, we discuss elliptic integrals reduced from $\sqrt{\det A_n}$ satisfying hypergeometric-type differential equations with an independent variable X introduced in the large-N asymptotics. The singularities of the hypergeometric-type differential equations can be applied to find different phases and corresponding critical points in the phase transitions. The related models can be found in Sects. 4.1, 4.2, 6.2 and 6.3.

D.1 Singular Points in the Hermitian Model

Based on the phase model obtained in Sect. C.1 in the Lax pair system for the continuum Painlevé II, let us now consider the following moment quantities in terms of the elliptic integrals,

$$M_j = \int_\Omega \eta^j \sqrt{\mu(\eta)} d\eta, \quad j = 0, 1, 2, 3, \tag{D.1}$$

where Ω is a cut for the $\mu(\eta)$ defined by

$$\mu(\eta) = \eta^4 - g\eta^2 - 2\eta - X. \tag{D.2}$$

Then, it can be obtained by using integration by parts that

$$C \frac{d}{dX} \begin{pmatrix} M_0 \\ M_1 \\ M_2 \\ M_3 \end{pmatrix} = \begin{pmatrix} 3 & 0 & 0 & 0 \\ 0 & 3 & 0 & 0 \\ g & 0 & 3 & 0 \\ 3 & g & 0 & 3 \end{pmatrix} \begin{pmatrix} M_0 \\ M_1 \\ M_2 \\ M_3 \end{pmatrix}, \tag{D.3}$$

where

$$C = \begin{pmatrix} 4X & 6 & 2g & 0 \\ 0 & 4X & 6 & 2g \\ 2gX & 4g & 4X + 2g^2 & 6 \\ 6X & 12 + 2gX & 10g & 4X + 2g^2 \end{pmatrix}. \tag{D.4}$$

C.B. Wang, *Application of Integrable Systems to Phase Transitions*,
DOI 10.1007/978-3-642-38565-0, © Springer-Verlag Berlin Heidelberg 2013

We want to use the singular points of the determinant of the coefficient matrix C to find the different phases in the transition model. There would be a complicated background about the relation between the roots of $\det C$ and the phases in the transition models. The Gross-Witten model in the unitary matrix model just meet with the singularities obtained from the corresponding hypergeometric-type differential equation. In the Hermitian matrix models, the singular points obtained from the hypergeometric-type differential equations given below have been experienced working very well for the phase transition problems. There would be more interesting properties for the hypergeometric-type differential equations in this method, that are left for further investigations.

We want to factorize $\det C$ to find the roots, that is in fact not obvious because it is about the factorization of a polynomial in X of degree 4. However, if we change the parameter g to a new parameter a by the following transformation,

$$g = 2a^2 + a^{-1}, \tag{D.5}$$

obtained based on the relations of u_n, v_n and t discussed in Sect. C.1, then the problem becomes easier. The result is

$$\det C = 4^4 X \left(X - a + a^4 \right) \left(X - A(a) \right) \left(X - B(a) \right), \tag{D.6}$$

where

$$2A(a) = -\left(a^4 + 3a + \frac{1}{2}a^{-2} \right) + \left(a^2 + 2a^{-1} \right)\sqrt{a^4 + 2a}, \tag{D.7}$$

$$2B(a) = -\left(a^4 + 3a + \frac{1}{2}a^{-2} \right) - \left(a^2 + 2a^{-1} \right)\sqrt{a^4 + 2a}. \tag{D.8}$$

It is interesting to investigate what type of differential equation each M_j would satisfy. In the following, we discuss an easy case of the elliptic integrals, and show that the corresponding elliptic integrals satisfy the hypergeometric differential equation, that indicates (D.3) is a hypergeometric-type differential equation. Denote $\Omega_1 = [-\zeta_1 i, \zeta_1 i]$ and $\Omega_2 = [\zeta_1 i, \zeta_2 i]$, where $\zeta_1^2 = (1 - \sqrt{1 - X_1})/2$ and $\zeta_2^2 = (1 + \sqrt{1 - X_1})/2$ with $X_1 \in [0, 1]$. Consider

$$I = \int_\Omega \sqrt{\mu_1} d\zeta, \qquad J = \int_\Omega \zeta^2 \sqrt{\mu_1} d\zeta, \tag{D.9}$$

where Ω is Ω_1 or Ω_2, and

$$\mu_1 = \zeta^4 - \zeta^2 + \frac{X_1}{4}, \tag{D.10}$$

which is obtained from another form of the Lax pair of the continuum Painlevé II equation (Its and Novokshenov 1986). It can be verified that

$$\begin{pmatrix} 4X_1 & -8 \\ 2X_1 & 4X_1 - 8 \end{pmatrix} \begin{pmatrix} I' \\ J' \end{pmatrix} = \begin{pmatrix} 3 & 0 \\ 1 & 5 \end{pmatrix} \begin{pmatrix} I \\ J \end{pmatrix}, \tag{D.11}$$

where $' = d/dX_1$. Then, we have

$$X_1(1 - X_1)\begin{pmatrix} I'' \\ J'' \end{pmatrix} = \frac{1}{16}\begin{pmatrix} 3 & 0 \\ 2 & -5 \end{pmatrix}\begin{pmatrix} I \\ J \end{pmatrix}. \tag{D.12}$$

Therefore, I satisfies the hypergeometric differential equation

$$X_1(1 - X_1)I'' = \frac{3}{16}I. \tag{D.13}$$

If one denotes I_1 and I_2 for I on Ω_1 and Ω_2 respectively, then the Wronskian $W(I_1, I_2) = I_1 I_2' - I_2 I_1'$ is a constant, that can be calculated at $X_1 = 0$ or 1.

D.2 Singular Points in the Unitary Model

For the phase model discussed in Sect. C.3 associated with the continuum Painlevé III, let us consider the following moment quantities,

$$M_j^{\pm} = \int_{\Omega} z^{\pm j}\sqrt{\mu(z)}dz, \quad j = 1, 2, \tag{D.14}$$

where

$$\mu(\eta) = (z + z^{-1} + T)^2 - X, \tag{D.15}$$

and Ω is a cut for the $\mu(\eta)$. Then it can be obtained by using integration by parts that

$$C\frac{d}{dX}\begin{pmatrix} M_1^- \\ M_1^+ \\ M_2^- \\ M_2^+ \end{pmatrix} = \begin{pmatrix} -2 & 0 & 0 & 0 \\ 0 & -6 & 0 & 0 \\ 0 & 0 & -4 & 0 \\ 0 & -2T & 0 & 8 \end{pmatrix}\begin{pmatrix} M_1^- \\ M_1^+ \\ M_2^- \\ M_2^+ \end{pmatrix}, \tag{D.16}$$

where

$$C = \begin{pmatrix} T^2 + 2 - X & 2 & 4T & 0 \\ 2 & T^2 + 2 - X & 3T & T \\ 3T & T & T^2 + 4 - X & 0 \\ T & T(T^2 - 1 - X) & 2T^2 - 2 & T^2 - 2 + X \end{pmatrix}. \tag{D.17}$$

We have the following factorization for the determinant of the coefficient matrix for the hypergeometric-type differential equation

$$\det C = -X(X - T^2 - 2)(X - (T - 2)^2)(X - (T + 2)^2). \tag{D.18}$$

These singular values of X are applied in Chap. 6 to discuss the phase transition or discontinuity problems in the unitary matrix models. The cases $X = 0$ and $X =$

$(T-2)^2$ are corresponding to the strong and weak coupling cases respectively in the Gross-Witten third-order phase transition model.

Let us replace X by $4X_1$ and consider a special case $T = 0$ according to the Lax pair for the continuum Painlevé III discussed in Sect. C.3. In this special case, we will get a second-order hypergeometric differential equation. Denote $\Omega_1 = \{e^\theta | \theta_1 \le \theta \le \theta_2\}$ and $\Omega_2 = \{e^\theta | -\theta_1 \le \theta \le \theta_1\}$, where $\cos\theta_1 = \sqrt{X_1}$ and $\cos\theta_2 = -\sqrt{X_1}$ with $X_1 \in [0, 1]$, $0 \le \theta_1 \le \theta_2 \le \pi$ and $\theta_1 + \theta_2 = \pi$. Consider

$$I = \int_\Omega \eta^{-2}\sqrt{\mu_1}d\eta, \qquad J = \int_\Omega \sqrt{\mu_1}d\eta, \qquad (D.19)$$

where Ω is Ω_1 or Ω_2 and

$$\mu_1(\eta) = \left(\eta^2 + 1\right)^2 - 4X_1\eta^2. \qquad (D.20)$$

If we want to associate these integrals with the integrals of ρ and ω discussed in Sect. B.2, it should be noticed that we take Ω_2 and $-\Omega_2$ as the arc cuts on the unit circle, and Ω_1 and $-\Omega_1$ as the complement in the unit circle. Also, these integrals are based on the general phase models in Sect. 6.1, but the X_1 variable is not fixed to the singular value(s), 0 or 1. It can be verified that the I and J defined above satisfy

$$\begin{pmatrix} 2X_1 - 1 & -1 \\ 1 & 1 - 2X_1 \end{pmatrix}\begin{pmatrix} I' \\ J' \end{pmatrix} = \begin{pmatrix} 1 & 0 \\ 0 & -3 \end{pmatrix}\begin{pmatrix} I \\ J \end{pmatrix}, \qquad (D.21)$$

where $' = d/dX_1$. We can further get

$$X_1(1 - X_1)\begin{pmatrix} I'' \\ J'' \end{pmatrix} = \frac{1}{4}\begin{pmatrix} 1 & 0 \\ 0 & -3 \end{pmatrix}\begin{pmatrix} I \\ J \end{pmatrix}. \qquad (D.22)$$

Therefore, I_j $(j = 1, 2)$ are two linearly independent solutions for the following hypergeometric differential equation,

$$X_1(1 - X_1)I'' = \frac{1}{4}I. \qquad (D.23)$$

Because I_1 and I_2 both satisfy the same differential equation above which does not have the first-order derivative term, the Wronskian of I_1 and I_2 is a constant,

$$W(I_1, I_2) = I_1 I_2' - I_1' I_2 = \text{const}. \qquad (D.24)$$

Specifically the constant is equal to $I_1(1)I_2'(1) - I_1'(1)I_2(1)$. To calculate its value, recall that

$$\eta_1^2, \eta_2^2 = 2X_1 - 1 \pm 2\sqrt{X_1^2 - X_1}, \qquad (D.25)$$

or $\eta_1 = e^{i\theta_1}$ and $\eta_2 = -e^{i\theta_2}$. As $X_1 \to 1$, there are $\eta_1 \to 1$ and $\eta_2 \to 1$, or $\theta_1 \to 0$ and $\theta_2 \to \pi$. And then

$$I_1(1) = 2\int_0^\pi \sqrt{1 - \cos^2\theta}\,d\theta = 4, \qquad \eta = e^{i\theta},$$

$$I_2'(1) = 2 \lim_{X_1 \to 1} \int_{\eta_1}^{\eta_2} \frac{d\eta}{\sqrt{(\eta^2 + 1)^2 - 4X_1\eta^2}} = 2 \lim_{X_1 \to 1} \int_{\eta_1}^{\eta_2} \frac{d\eta}{\sqrt{(\eta^2 - \eta_1^2)(\eta^2 - \eta_2^2)}}.$$

It is the standard argument that if we make a transformation $\eta^2 = \eta_2^2 + (\eta_1^2 - \eta_2^2)\sigma^2$, then

$$\int_{\eta_1}^{\eta_2} \frac{d\eta}{\sqrt{(\eta^2 - \eta_1^2)(\eta^2 - \eta_2^2)}} = \frac{1}{i\eta_2} \int_0^1 \frac{d\sigma}{\sqrt{(1 - \sigma^2)(1 - k^2\sigma^2)}},$$

with

$$k^2 = 1 - k'^2, \quad k'^2 = \eta_1^2/\eta_2^2. \tag{D.26}$$

As $X_1 \to 1$, there are $\eta_1 \to 1$, $\eta_2 \to 1$, $k' \to 1$ and $k \to 0$, that imply

$$I_2'(1) = 2 \lim_{X_1 \to 1} \frac{1}{i\eta_2} \int_0^1 \frac{d\sigma}{\sqrt{(1 - \sigma^2)(1 - k^2\sigma^2)}} = -\pi i, \tag{D.27}$$

and then $I_1(1)I_2'(1) = -4\pi i$.

For the second term $I_1'(1)I_2(1)$, we have that as $X_1 \to 1$,

$$I_1'(X_1) = -2 \int_{\eta_1}^{-\eta_2} \frac{d\eta}{\sqrt{(\eta^2 - \eta_1^2)(\eta^2 - \eta_2^2)}} = O\left(\log(\eta_1 - \eta_2)\right),$$

$$I_2(X_1) = \int_{\eta_2}^{\eta_1} \eta^{-2}\sqrt{(\eta^2 + 1)^2 - 4X_1\eta^2}\, d\eta = O(\eta_1 - \eta_2).$$

So we finally get

$$I_1 I_2' - I_1' I_2 = -4\pi i. \tag{D.28}$$

The hypergeometric differential equations (D.12) for the Painlevé II equation and the corresponding Wronskian and conformal mappings were studied early by Professor J. Bryce McLeod from the University of Pittsburgh. Professor McLeod guided me (Chie Bing) to get the hypergeometric differential equations (D.22) for the Painlevé III equation discussed in my thesis at the University of Pittsburgh. The hypergeometric-type differential equations (fourth-order differential equations) discussed in this appendix are the consequence of the second-order hypergeometric differential equations as explained above.

The elliptic integrals and conformal mapping theory are important to study the mathematical problems in the phase transition models, and there are various unsolved problems in this area. When the hypergeometric-type differential equations are applied to other models, it should be noted that the coefficients in the $\mu(\eta)$ polynomial is generally not so simple. There may be several variables like the X variable discussed above. The general discussion is not easy.

Index

A

AKNS-ZS system (discrete), 111
Asymptotic expansion, 62, 172, 186, 196
Asymptotics
    as $\eta \to \infty$, 44, 60, 69, 70, 87, 179
    as $x \to \infty$, 184
    as $z \to 0$ or $z \to \infty$, 136
    as $z \to \infty$, 24, 25, 35, 36, 75
    large-N, 90, 91, 98, 143, 146

B

Bifurcation, 14, 15, 18, 52, 133

C

Cauchy theorem, 46, 137, 166, 180
Cayley-Hamilton theorem, 24, 29, 31, 39
Center parameter, 37, 55
Complement, 11, 108, 125, 127
Contour integral, 40, 46, 60, 62, 94, 109, 137, 150, 163, 166, 170, 180
Correlation function, 12, 158
Critical exponent, 158
Critical phenomenon
    Hermitian model, 16, 89, 90, 98
    unitary model, 12, 142, 156
Critical point
    1st-order, Hermitian, 79
    1st-order, unitary, 138
    2nd-order, Hermitian, 88
    3rd-order, Hermitian, 16, 57, 66, 71, 72, 89, 95
    3rd-order, unitary, 12, 129, 143, 148, 149, 151, 155, 156
    4th-order, Hermitian, 101, 105
    5th-order, Hermitian, 84

critical phenomenon, Hermitian, 89
critical phenomenon, unitary, 12, 142, 156
Cut, 40, 60, 83, 136, 139, 149, 153, 211, 213

D

Determinant
    for hypergeometric-type differential equation, 212, 213
    of A_n matrix, 24, 35, 36, 57, 75, 76, 86, 94, 103
    of J matrix, 29, 30
    of Lax matrix, 132, 141, 162, 177, 186
    of L matrix, 129
    related to Legendre's relation, 199
Discontinuity, 16, 58, 95, 103, 141
Divergence, 12, 18, 95, 133
Double scaling
    Hermitian model, 98, 99
    unitary model, 143
Duality, 148

E

Eigenvalue density
    Hermitian model, multi-interval, 22–24, 40, 67
    Hermitian model, one-interval, 24, 28, 83, 183
    Jacobi model, 177
    Laguerre model, 162
    planar diagram model, 7, 22
    unitary model, multi-arc, 107, 150
    unitary model, one-arc, 11, 127, 135, 141
Elliptic integral
    Hermitian model, 15, 52, 61, 198, 199, 211
    unitary model, 132, 212
Entanglement, 169
Equation of state, 13
Euler beta function, 48, 184

C.B. Wang, *Application of Integrable Systems to Phase Transitions*,
DOI 10.1007/978-3-642-38565-0, © Springer-Verlag Berlin Heidelberg 2013